Lecture Notes in Mathematics 1845

Editors:
J.-M. Morel, Cachan
F. Takens, Groningen
B. Teissier, Paris

T0240560

Springer
Berlin
Heidelberg
New York
Hong Kong
London
Milan
Paris
Tokyo

Min Ho Lee

Mixed Automorphic Forms, Torus Bundles, and Jacobi Forms

 Springer

Author

Min Ho LEE

Department of Mathematics
University of Northern Iowa
Cedar Falls
IA 50614, U.S.A.

e-mail: lee@math.uni.edu

Library of Congress Control Number: 2004104067

Mathematics Subject Classification (2000):
11F11, 11F12, 11F41, 11F46, 11F50, 11F55, 11F70, 14C30, 14D05, 14D07, 14G35

ISSN 0075-8434
ISBN 3-540-21922-6 Springer-Verlag Berlin Heidelberg New York

Springer-Verlag is a part of Springer Science + Business Media

http://www.springeronline.com

Typesetting: Camera-ready TeX output by the authors

SPIN: 11006329   41/3142/du - 543210 - Printed on acid-free paper

To Virginia, Jenny, and Katie

Preface

This book is concerned with various topics that center around equivariant holomorphic maps of Hermitian symmetric domains and is intended for specialists in number theory and algebraic geometry. In particular, it contains a comprehensive exposition of mixed automorphic forms that has never appeared in book form.

The period map $\omega : \mathcal{H} \to \mathcal{H}$ of an elliptic surface E over a Riemann surface X is a holomorphic map of the Poincaré upper half plane $\mathcal{H}$ into itself that is equivariant with respect to the monodromy representation $\chi : \Gamma \to SL(2, \mathbb{R})$ of the discrete subgroup $\Gamma \subset SL(2, \mathbb{R})$ determined by X. If ω is the identity map and χ is the inclusion map, then holomorphic 2-forms on E can be considered as an automorphic form for Γ of weight three. In general, however, such holomorphic forms are mixed automorphic forms of type $(2, 1)$ that are defined by using the product of the usual weight two automorphy factor and a weight one automorphy factor involving ω and χ. Given a positive integer m, the elliptic variety E^m can be constructed by essentially taking the fiber product of m copies of E over X, and holomorphic $(m + 1)$-forms on E^m may be regarded as mixed automorphic forms of type $(2, m)$. The generic fiber of E^m is the product of m elliptic curves and is therefore an abelian variety, or a complex torus. Thus the elliptic variety E^m is a complex torus bundle over the Riemann surface X.

An equivariant holomorphic map $\tau : \mathcal{D} \to \mathcal{D}'$ of more general Hermitian symmetric domains $\mathcal{D}$ and $\mathcal{D}'$ can be used to define mixed automorphic forms on $\mathcal{D}$. When $\mathcal{D}'$ is a Siegel upper half space, the map τ determines a complex torus bundle over a locally symmetric space $\Gamma \backslash \mathcal{D}$ for some discrete subgroup Γ of the semisimple Lie group G associated to $\mathcal{D}$. Such torus bundles are often families of polarized abelian varieties, and they are closely related to various topics in number theory and algebraic geometry. Holomorphic forms of the highest degree on such a torus bundle can be identified with mixed automorphic forms on $\mathcal{D}$ of certain type. Mixed automorphic forms can also be used to construct an embedding of the same torus bundle into a complex projective space. On the other hand, sections of a certain line bundle over this torus bundle can be regarded as Jacobi forms on the Hermitian symmetric domain $\mathcal{D}$.

The main goal of this book is to explore connections among complex torus bundles, mixed automorphic forms, and Jacobi forms of the type described above. Both number-theoretic and algebro-geometric aspects of such connections and related topics are discussed.

This work was supported in part by a 2002–2003 Professional Development Assignment award from the University of Northern Iowa.

Cedar Falls, Iowa, April 5, 2004   *Min Ho Lee*

Contents

Introduction

Let E be an elliptic surface in the sense of Kodaira [52]. Thus E is a compact smooth surface over $\mathbb{C}$, and it is the total space of an elliptic fibration $\pi :$ $E \to X$ over a compact Riemann surface X whose generic fiber is an elliptic curve. Let $\Gamma \subset SL(2, \mathbb{R})$ be a Fuchsian group of the first kind acting on the Poincaré upper half plane $\mathcal{H}$ by linear fractional transformations such that the base space for the fibration π is given by $X = \Gamma \backslash \mathcal{H}^*$, where $\mathcal{H}^*$ is the union of $\mathcal{H}$ and the set of cusps for Γ. Given $z \in X_0 = \Gamma \backslash \mathcal{H}$, let Φ be a holomorphic 1-form on the fiber $E_z = \pi^{-1}(z)$, and choose an ordered basis $\{\gamma_1(z), \gamma_2(z)\}$ for $H_1(E_z, \mathbb{Z})$ that depends on the parameter z in a continuous manner. Consider the periods ω_1 and ω_2 of E given by

$$\omega_1(z) = \int_{\gamma_1(z)} \Phi, \quad \omega_2(z) = \int_{\gamma_2(z)} \Phi.$$

Then the imaginary part of the quotient $\omega_1(z)/\omega_2(z)$ is nonzero for each z, and therefore we may assume that $\omega_1(z)/\omega_2(z) \in \mathcal{H}$. In fact, ω_1/ω_2 is a many-valued holomorphic function on X_0, and the period map $\omega : \mathcal{H} \to \mathcal{H}$ is obtained by lifting the map $\omega_1/\omega_2 : X_0 \to \mathcal{H}$ from X_0 to its universal covering space $\mathcal{H}$. If Γ is identified with the fundamental group of X_0, the natural connection on E_0 determines the monodromy representation $\chi : \Gamma \to SL(2, \mathbb{R})$ of Γ, and the period map is equivariant with respect to χ, that is, it satisfies

$$\omega(\gamma z) = \chi(\gamma)\omega(z)$$

for all $\gamma \in \Gamma$ and $z \in \mathcal{H}$. Given nonnegative integers k and ℓ, we consider a holomorphc function f on $\mathcal{H}$ satisfying

$$f(\gamma z) = (cz + d)^k (c_\chi \omega(z) + d_\chi)^\ell f(z) \tag{0.1}$$

for all $z \in \mathcal{H}$ and $\gamma = \left(\begin{smallmatrix} a & b \\ c & d \end{smallmatrix} \right) \in \Gamma$ with $\chi(\gamma) = \left(\begin{smallmatrix} a_\chi & b_\chi \\ c_\chi & d_\chi \end{smallmatrix} \right) \in SL(2, \mathbf{R})$. Such a function becomes a mixed automorphic or cusp form for Γ of type (k, ℓ) if in addition it satisfies an appropriate cusp condition. It was Hunt and Meyer [43] who observed that a holomorphic form of degree two on the elliptic surface E can be interpreted as a mixed cusp form for Γ of type $(2, 1)$ associated to ω and χ. If χ is the inclusion map of Γ into $SL(2, \mathbb{R})$ and if ω is the identity map on $\mathcal{H}$, then E is called an elliptic modular surface. The observation of

Hunt and Meyer [43] in fact generalizes the result of Shioda [115] who showed that a holomorphic 2-form on an elliptic modular surface is a cusp form of weight three. Given a positive integer m, the elliptic variety E^m associated to the elliptic fibration $\pi : E \to X$ can be obtained by essentially taking the fiber product of m copies of E over X (see Section 2.2 for details), and holomorphic $(m+1)$-forms on E^m provide examples of mixed automorphic forms of higher weights (cf. [18, 68]). Note that the generic fiber of E^m is an abelian variety, and therefore a complex torus, obtained by the product of elliptic curves. Thus the elliptic variety E^m can be regarded as a family of abelian varieties parametrized by the Riemann surface X or as a complex torus bundle over X.

Another source of examples of mixed automorphic forms comes from the theory of linear ordinary differential equations on a Riemann surface (see Section 1.4). Let $\Gamma \subset SL(2,\mathbb{R})$ be a Fuchsian group of the first kind as before. Then the corresponding compact Riemann surface $X = \Gamma \backslash \mathcal{H}^*$ can be regarded as a smooth algebraic curve over $\mathbb{C}$. We consider a second order linear differential equation $\Lambda_X^2 f = 0$ with

$$\Lambda_X^2 = \frac{d^2}{dx^2} + P_X(x)\frac{d}{dx} + Q_X(x), \qquad (0.2)$$

where $P_X(x)$ and $Q_X(x)$ are rational functions on X. Let ω_1 and ω_2 be linearly independent solutions of $\Lambda_X^2 f = 0$, and for each positive integer m let $S^m(\Lambda_X^2)$ be the linear ordinary differential operator of order $m+1$ such that the $m+1$ functions

$$\omega_1^m, \omega_1^{m-1}\omega_2, \ldots, \omega_1\omega_2^{m-1}, \omega_2^m$$

are linearly independent solutions of the corresponding homogeneous equation $S^m(\Lambda_X^2)f = 0$. By pulling back the operator in (0.2) via the natural projection map $\mathcal{H}^* \to X = \Gamma \backslash \mathcal{H}^*$ we obtain a differential operator

$$\Lambda^2 = \frac{d^2}{dz^2} + P(z)\frac{d}{dz} + Q(z) \qquad (0.3)$$

such that $P(z)$ and $Q(z)$ are meromorphic functions on $\mathcal{H}^*$. Let $\omega_1(z)$ and $\omega_2(z)$ for $z \in \mathcal{H}$ be the two linearly independent solutions of $\Lambda^2 f = 0$ corresponding to ω_1 and ω_2 above. Then the monodromy representation for the differential equation $\Lambda^2 f = 0$ is the group homomorphism $\chi : \Gamma \to GL(2,\mathbb{C})$ which can be defined as follows. Given elements $\gamma \in \Gamma$ and $z \in \mathcal{H}$, we assume that the elements $\omega_1(\gamma z)$ and $\omega_2(\gamma z)$ can be written in the form

$$\omega_1(\gamma z) = a_\chi \omega_1(z) + b_\chi \omega_2(z), \quad \omega_2(\gamma z) = c_\chi \omega_1(z) + d_\chi \omega_2(z).$$

Then the image of $\gamma \in \Gamma$ under the monodromy representation χ is given by

$$\chi(\gamma) = \begin{pmatrix} a_\chi & b_\chi \\ c_\chi & d_\chi \end{pmatrix} \in GL(2,\mathbb{C}). \qquad (0.4)$$

We assume that $\chi(\Gamma) \subset SL(2, \mathbb{R})$ and that

$$w(z) = \omega_1(z)/\omega_2(z) \in \mathcal{H}$$

for all $z \in \mathcal{H}$. Then the resulting map $w : \mathcal{H} \to \mathcal{H}$ satisfies

$$w(\gamma z) = \frac{a_\chi w(z) + b_\chi}{c_\chi w(z) + d_\chi} = \chi(\gamma) w(z)$$

for all $z \in \mathcal{H}$ and $\gamma \in \Gamma$. Thus the map w is equivariant with respect to χ, and we may consider the associated meromorphic mixed automorphic or cusp forms as meromorphic functions satisfying the transformation formula in (0.1) and a certain cusp condition. If $S^m(\Lambda^2)$ is the differential operator acting on the functions on $\mathcal{H}$ obtained by pulling back $S^m(\Lambda_X^2)$ via the projection map $\mathcal{H}^* \to X$, then the solutions of the equation $S^m(\Lambda^2)f = 0$ are of the form

$$\sum_{i=0}^{m} c_i \omega_1(z)^{m-i} \omega_2(z)^i$$

for some constants $c_0, \ldots, c_m$. Let ψ be a meromorphic function on $\mathcal{H}^*$ corresponding to an element ψ_X in $K(X)$, and let f^ψ be a solution of the non-homogeneous equation

$$S^m(\Lambda^2)f = \psi.$$

If k is a nonnegative integer k, then it can be shown the function

$$\Phi_k^\psi(z) = w'(z)^k \frac{d^{m+1}}{dw(z)^{m+1}} \left(\frac{f^\psi(z)}{\omega_2(z)^m} \right)$$

for $z \in \mathcal{H}$ is independent of the choice of the solution f^ψ and is a mixed automorphic form of type $(2k, m - 2k + 2)$ associated to Γ, w and the monodromy representation χ.

If f is a cusp form of weight w for a Fuchsian group $\Gamma \subset SL(2, \mathbb{R})$, the periods of f are given by the integrals

$$\int_0^{i\infty} f(z) z^k dz$$

with $0 \leq k \leq w - 2$, and it is well-known that such periods of cusp forms are closely related to the values at the integer points in the critical strip of the Hecke L-series. In [22] Eichler discovered certain relations among the periods of cusp forms, which were extended later by Shimura [112]; these relations are called Eichler-Shimura relations. More explicit connections between the Eichler-Shimura relations and the Fourier coefficients of cusp forms were found by Manin [91]. If f is a mixed cusp form of type $(2, m)$ associated to Γ and an equivariant pair (w, χ), then the periods of f can be defined by the integrals

$$\int_0^{i\infty} f(z)\omega(z)^k dz$$

with $0 \le k \le m$. The interpretation of mixed automorphic forms as holomorphic forms on an elliptic variety described earlier can be used to obtain a relation among such periods, which may be regarded as the Eichler-Shimura relation for mixed cusp forms (see Section 2.4).

Connections between the cohomology of a discrete subgroup Γ of $SL(2,\mathbb{R})$ and automorphic forms for Γ were made by Eichler [22] and Shimura [112] decades ago. Indeed, they established an isomorphism between the space of cusp forms of weight $m+2$ for Γ and the parabolic cohomology space of Γ with coefficients in the space of homogeneous polynomials of degree m in two variables over $\mathbb{R}$. To be more precise, let $\mathrm{Sym}^m(\mathbb{C}^2)$ denote the m-th symmetric power of $\mathbb{C}^2$, and let $H_P^1(\Gamma, \mathrm{Sym}^m(\mathbb{C}^2))$ be the associated parabolic cohomology of Γ, where the Γ-module structure of $\mathrm{Sym}^m(\mathbb{C}^2)$ is induced by the standard representation of $\Gamma \subset SL(2,\mathbb{R})$ on $\mathbb{C}^2$. Then the Eichler-Shimura isomorphism can be written in the form

$$H_P^1(\Gamma, \mathrm{Sym}^m(\mathbb{C}^2)) = S_{m+2}(\Gamma) \oplus \overline{S_{m+2}(\Gamma)},$$

where $S_{m+2}(\Gamma)$ is the space of cusp forms of weight $m+2$ for Γ (cf. [22, 112]). In particular, there is a canonical embedding of the space of cusp forms into the parabolic cohomology space. The Eichler-Shimura isomorphism can also be viewed as a Hodge structure on the parabolic cohomology (see e.g. [6]). If (ω, χ) is an equivariant pair considered earlier, we may consider another action of Γ on $\mathrm{Sym}^m(\mathbb{C}^2)$ which is determined by the composition of the homomorphism $\chi : \Gamma \to SL(2,\mathbb{R})$ with the standard representation of $SL(2,\mathbb{R})$ in $\mathrm{Sym}^m(\mathbb{C}^2)$. If we denote the resulting Γ-module by $\mathrm{Sym}_\chi^m(\mathbb{C}^2)$, the associated parabolic cohomology $H_P^1(\Gamma, \mathrm{Sym}_\chi^m(\mathbb{C}^2))$ is linked to mixed automorphic forms for Γ associated to the equivariant pair (ω, χ). Indeed, the space of certain mixed cusp forms can be embedded into such parabolic cohomology space, and a Hodge structure on $H_P^1(\Gamma, \mathrm{Sym}_\chi^m(\mathbb{C}^2))$ can be determined by an isomorphism of the form

$$H_P^1(\Gamma, \mathrm{Sym}_\chi^m(\mathbb{C}^2)) \cong S_{2,m}(\Gamma, \omega, \chi) \oplus W \oplus \overline{S_{2,m}(\Gamma, \omega, \chi)}, \qquad (0.5)$$

where W is a certain subspace of $H_P^1(\Gamma, \mathrm{Sym}_\chi^m(\mathbb{C}^2))$ and $S_{2,m}(\Gamma, \omega, \chi)$ is the space of mixed cusp forms of type $(2, m)$ associated to Γ, ω and χ (see Chapter 3). The space W in (0.5) is not trivial in general as can be seen in [20, Section 3], where mixed cusp forms of type $(0, 3)$ were studied in connection with elliptic surfaces. The isomorphism in (0.5) may be regarded as a generalized Eichler-Shimura isomorphism.

The correspondence between holomorphic forms of the highest degree on an elliptic variety and mixed automorphic forms of one variable described above can be extended to the case of several variables by introducing mixed Hilbert and mixed Siegel modular forms. For the Hilbert modular case we consider a totally real number field F of degree n over $\mathbb{Q}$, so that $SL(2, F)$ can be embedded in $SL(2, \mathbb{R})^n$. Given a subgroup Γ of $SL(2, F)$ whose embedded image in $SL(2, \mathbb{R})^n$ is a discrete subgroup, we can consider the associated Hilbert modular variety $\Gamma \backslash \mathcal{H}^n$ obtained by the quotient of the n-fold product $\mathcal{H}^n$ of the Poincaré upper half plane $\mathcal{H}$ by the action of Γ given by linear fractional transformations. If $\omega : \mathcal{H}^n \to \mathcal{H}^n$ is a holomorphic map equivariant with respect to a homomorphism $\chi : \Gamma \to SL(2, F)$, then the equivariant pair (ω, χ) can be used to define mixed Hilbert modular forms, which can be regarded as mixed automorphic forms of n variables. On the other hand, the same equivariant pair also determines a family of abelian varieties parametrized by $\Gamma \backslash \mathcal{H}^n$. Then holomorphic forms of the highest degree on such a family correspond to mixed Hilbert modular forms of certain type (see Section 4.2). Another type of mixed automorphic forms of several variables can be obtained by generalizing Siegel modular forms (see Section 4.3). Let $\mathcal{H}_m$ be the Siegel upper half space of degree m on which the symplectic group $Sp(m, \mathbb{R})$ acts as usual, and let Γ_0 be a discrete subgroup of $Sp(m, \mathbb{R})$. If $\tau : \mathcal{H}_m \to \mathcal{H}_{m'}$ is a holomorphic map of $\mathcal{H}_m$ into another Siegel upper half space $\mathcal{H}_{m'}$ that is equivariant with respect to a homomorphism $\rho : \Gamma_0 \to Sp(m', \mathbb{R})$, then the equivariant pair (τ, ρ) can be used to define mixed Siegel modular forms. The same pair can also be used to construct a family of abelian varieties parametrized by the Siegel modular variety $\Gamma \backslash \mathcal{H}_m$ such that holomorphic forms of the highest degree on the family correspond to mixed Siegel modular forms (see Section 4.3).

A further generalization of mixed automorphic forms can be considered by using holomorphic functions on more general Hermitian symmetric domains which include the Poincaré upper half plane or Siegel upper half spaces. Let G and G' be semisimple Lie groups of Hermitian type, so that the associated Riemannian symmetric spaces $\mathcal{D}$ and $\mathcal{D}'$, respectively, are Hermitian symmetric domains. We consider a holomorphic map $\tau : \mathcal{D} \to \mathcal{D}'$, and assume that it is equivariant with respect to a homomorphism $\rho : G \to G'$. Let Γ be a discrete subgroup of G. Note that, unlike in the earlier cases, we are assuming that τ is equivariant with respect to a homomorphism ρ defined on the group G itself rather than on the subgroup Γ. This provides us with more flexibility in studying associated mixed automorphic forms. Various aspects of such equivariant holomorphic maps were studied extensively by Satake in [108]. Given complex vector spaces V and V' and automorphy factors $J : G \times \mathcal{D} \to GL(V)$ and $J' : G' \times \mathcal{D}' \to GL(V')$, a mixed automorphic form on $\mathcal{D}$ for Γ is a holomorphic function $f : \mathcal{D} \to V \otimes V'$ satisfying

$$f(\gamma z) = J(\gamma, z) \otimes J'(\rho(\gamma), \tau(z)) f(z)$$

for all $z \in \mathcal{D}$ and $\gamma \in \Gamma$ (see Section 5.1). Another advantage of considering an equivariant pair (τ, ρ) with ρ defined on G instead of Γ is that it allows us to introduce a representation-theoretic description of mixed automorphic forms. Such interpretation includes not only the holomorphic mixed automorphic forms described above but also nonholomorphic ones. Given a semisimple Lie group G, a maximal compact subgroup K, and a discrete subgroup Γ of finite covolume, automorphic forms on G can be described as follows. Let $Z(\mathfrak{g})$ be the center of the universal enveloping algebra of the complexification $\mathfrak{g}_{\mathbb{C}}$ of the Lie algebra $\mathfrak{g}$ of G, and let V be a finite-dimensional complex vector space. A slowly increasing analytic function $f : G \to V$ is an automorphic form for Γ if it is left Γ-invariant, right K-finite, and $Z(\mathfrak{g})$-finite. Let G' be another semisimple Lie group with the corresponding objects K', Γ' and V', and let $\varphi : G \to G'$ be a homomorphism such that $\varphi(K) \subset K'$ and $\varphi(\Gamma) \subset \Gamma'$. Then the associated mixed automorphic forms may be described as linear combinations of functions of the form $f \otimes (f' \circ \varphi) : G \to V \otimes V'$, where $f : G \to V$ is an automorphic form for Γ and $f' : G' \to V'$ is an automorphic form for Γ' (see Section 5.2.

The equivariant pair (τ, ρ) considered in the previous paragraph also determines a family of abelian varieties parametrized by a locally symmetric space if G' is a symplectic group. Let $\mathcal{H}_n$ be the Siegel upper half space of degree n on which the symplectic group $Sp(n, \mathbb{R})$ acts as usual. Then the semidirect product $Sp(n, \mathbb{R}) \ltimes \mathbb{R}^{2n}$ operates on the space $\mathcal{H}_n \times \mathbb{C}^n$ by

$$\left(\left(\begin{smallmatrix} A & B \\ C & D \end{smallmatrix} \right), (\mu, \nu) \right) \cdot (Z, \zeta) = ((AZ + B)(CZ + D)^{-1}, (\zeta + \mu Z + \nu)(CZ + D)^{-1})$$

for $\left(\begin{smallmatrix} A & B \\ C & D \end{smallmatrix} \right) \in Sp(n, \mathbb{R})$, $(\mu, \nu) \in \mathbb{R}^{2n}$ and $(Z, \zeta) \in \mathcal{H}_n \times \mathbb{C}^n$, where elements of $\mathbb{R}^{2n}$ and $\mathbb{C}^n$ are considered as row vectors. We consider the discrete subgroup $\Gamma_0 = Sp(n, \mathbb{Z})$ of $Sp(n, \mathbb{R})$, and set

$$X_0 = \Gamma_0 \backslash \mathcal{H}_n, \quad Y_0 = \Gamma_0 \ltimes \mathbb{Z}^{2n} \backslash \mathcal{H}_n \times \mathbb{C}^n.$$

Then the map $\pi_0 : Y_0 \to X_0$ induced by the natural projection map $\mathcal{H}_n \times \mathbb{C}^n \to \mathcal{H}_n$ has the structure of a fiber bundle over the Siegel modular space X_0 whose fibers are complex tori of dimension n. In fact, each fiber of this bundle has the structure of a principally polarized abelian variety, and therefore the Siegel modular variety $X_0 = \Gamma_0 \backslash \mathcal{H}_n$ can be regarded as the parameter space of the family of principally polarized abelian varieties (cf. [63]). In order to consider a more general family of abelian varieties, we need to consider an equivariant holomorphic map of a Hermitian symmetric domain into a Siegel upper half space. Let G be a semisimple Lie group of Hermitian type, and let $\mathcal{D}$ be the associated Hermitian symmetric domain, which can be identified with the quotient G/K of G by a maximal compact subgroup K. We assume that there are a homomorphism $\rho : G \to Sp(n, \mathbb{R})$ of Lie groups and a holomorphic map $\tau : \mathcal{D} \to \mathcal{H}_n$ that is equivariant with respect to ρ. If Γ is a torsion-free discrete subgroup of G with $\rho(\Gamma) \subset \Gamma_0$ and if we set $X = \Gamma \backslash \mathcal{D}$, then τ induces a map $\tau_X : X \to X_0$ of the locally symmetric space X into the Siegel modular variety X_0. By pulling the bundle $\pi_0 : Y_0 \to X_0$ back via τ_X we obtain a fiber bundle $\pi : Y \to X$ over X whose fibers are n-dimensional complex tori. As in the case of π_0, each fiber is a polarized abelian variety, so that the total space Y of the bundle may be regarded as a family of abelian varieties parametrized by the locally symmetric space X. Such a family Y is known as a Kuga fiber variety, and various arithmetic and geometric aspects of Kuga fiber varieties have been studied in numerous papers over the years (see e.g. [1, 2, 31, 61, 62, 69, 74, 84, 96, 108, 113]). A Kuga fiber variety is also an example of a mixed Shimura variety in more modern language (cf. [94]). Holomorphic forms of the highest degree on the Kuga fiber variety Y can be identified with mixed automorphic forms on the symmetric domain $\mathcal{D}$ (see Section 6.3).

Equivariant holomorphic maps of symmetric domains and Kuga fiber varieties are also closely linked to Jacobi forms of several variables. Jacobi forms on the Poincaré upper half plane $\mathcal{H}$, or on $SL(2, \mathbb{R})$, share properties in common with both elliptic functions and modular forms in one variable, and they were systematically developed by Eichler and Zagier in [23]. They are functions defined on the space $\mathcal{H} \times \mathbb{C}$ which satisfy certain transformation formulas with respect to the action of a discrete subgroup Γ of $SL(2, \mathbb{R})$, and important examples of Jacobi forms include theta functions and Fourier coefficients of Siegel modular forms. Numerous papers have been devoted to the study of such Jacobi forms in connection with various topics in number theory (see e.g. [7, 9, 54, 116]). In the mean time, Jacobi forms of several variables have been studied mostly for symplectic groups of the form $Sp(m, \mathbb{R})$, which are defined on the product of a Siegel upper half space and a complex vector space. Such Jacobi forms and their relations with Siegel modular forms and theta functions have also been studied extensively over the years (cf. [25, 49, 50, 59, 123, 124]). Jacobi forms for more general semisimple Lie groups were in fact considered more than three decades ago by Piatetskii-Shapiro in [102, Chapter 4]. Such Jacobi forms occur as coefficients of Fourier-Jacobi series of automorphic forms on symmetric domains. Since then, there have not been many investigations about such Jacobi forms. In recent years, however, a number of papers which deal with Jacobi forms for orthogonal groups have appeared, and one notable such paper was written by Borcherds [12] (see also [11, 55]). Borcherds gave a highly interesting construction of Jacobi forms and modular forms for an orthogonal group of the form $O(n+2, 2)$ and investigated their connection with generalized Kac-Moody algebras. Such a Jacobi form appears as a denominator function for an affine Lie algebra and can be written as an infinite product. The denominator function for the fake monster Lie algebra on the other hand is a modular form for an orthogonal group, which can also be written as an infinite product. Thus many new examples of generalized Kac-Moody algebras may be constructed from modular or Jacobi forms for $O(n+2, 2)$, and conversely examples of modular or Jacobi forms may be obtained from generalized Kac-Moody algebras. In this book we consider Jacobi forms associated to an equivariant holomorphic map of symmetric domains of the type that is used in the construction of a Kuga fiber variety (see Chapter 7). Such Jacobi forms can be used to construct an embedding of a Kuga fiber variety into a complex projective space. They can also be identified with sections of a certain line bundle on the corresponding Kuga fiber variety. Similar identifications have been studied by Kramer and Runge for $SL(2, \mathbb{R})$ and $Sp(n, \mathbb{R})$ (see [57, 58, 105]).

The construction of Kuga fiber varieties can be extended to the one of more general complex torus bundles by using certain cocycles of discrete groups. Let (τ, ρ) be the equivariant pair that was used above for the construction of a Kuga fiber variety. Thus $\tau : \mathcal{D} \to \mathcal{H}_n$ is a holomorphic map that is equivariant with respect to the homomorphism $\rho : G \to Sp(n, \mathbb{R})$ of Lie groups. Let L be a lattice in $\mathbb{R}^{2n}$, and let Γ be a torsion-free discrete subgroup of G such that $\ell \cdot \rho(\gamma) \in L$ for all $\ell \in L$ and $\gamma \in \Gamma$, where we regarded elements of L as row vectors. If L denotes the lattice $\mathbb{Z}^{2n}$ in $\mathbb{Z}^{2n}$, the multiplication operation for the semidirect product $\Gamma \ltimes L$ is given by

$$(\gamma_1, \ell_1) \cdot (\gamma_2, \ell_2) = (\gamma_1 \gamma_2, \ell_1 \rho(\gamma_2) + \ell_2) \tag{0.6}$$

for all $\gamma_1, \gamma_2 \in \Gamma$ and $\ell_1, \ell_2 \in L$, and $\Gamma \ltimes L$ acts on $\mathcal{D} \times \mathbb{C}^n$ by

$$(\gamma, (\mu, \nu)) \cdot (z, w) = (\gamma z, (w + \mu \tau(z) + \nu)(C_\rho \tau(z) + D_\rho)^{-1}), \tag{0.7}$$

for all $(z, w) \in \mathcal{D} \times \mathbb{C}^n$, $(\mu, \nu) \in L \subset \mathbb{R}^n \times \mathbb{R}^n$ and $\gamma \in \Gamma$ with $\rho(\gamma) = \begin{pmatrix} A_\rho & B_\rho \\ C_\rho & D_\rho \end{pmatrix} \in Sp(n, \mathbb{R})$. Then the associated Kuga fiber variety is given by the quotient

$$Y = \Gamma \ltimes L \backslash \mathcal{D} \times \mathbb{C}^n,$$

which is a fiber bundle over the locally symmetric space $X = \Gamma \backslash \mathcal{D}$. We now consider a 2-cocycle $\psi : \Gamma \times \Gamma \to L$ define the generalized semidirect product $\Gamma \ltimes_\psi L$ by replacing the multiplication operation (0.6) with

$$(\gamma_1, \ell_1) \cdot (\gamma_2, \ell_2) = (\gamma_1 \gamma_2, \ell_1 \rho(\gamma_2) + \ell_2 + \psi(\gamma_1, \gamma_2)).$$

We denote by $\mathcal{A}(\mathcal{D}, \mathbb{C}^n)$ the space of $\mathbb{C}^n$-valued holomorphic functions on $\mathcal{D}$, and let ξ be a 1-cochain for the cohomology of Γ with coefficients in $\mathcal{A}(\mathcal{D}, \mathbb{C}^n)$ satisfying

$$\delta \xi(\gamma_1, \gamma_2)(z) = \psi(\gamma_1, \gamma_2) \begin{pmatrix} \tau(z) \\ 1 \end{pmatrix}$$

for all $z \in \mathcal{D}$ and $\gamma_1, \gamma_2 \in \Gamma$, where δ is the coboundary operator on 1-cochains. Then an action of $\Gamma \ltimes_\psi L$ on $\mathcal{D} \times \mathbb{C}^n$ can be defined by replacing (0.7) with

$$(\gamma, (\mu, \nu)) \cdot (z, w) = (\gamma z, (w + \mu \tau(z) + \nu + \xi(\gamma)(z))(C_\rho \tau(z) + D_\rho)^{-1}).$$

If the quotient of $\mathcal{D} \times \mathbb{C}^n$ by $\Gamma \ltimes_\psi L$ with respect to this action is denoted by $Y_{\psi, \xi}$, the map $\pi : Y_{\psi, \xi} \to X = \Gamma \backslash \mathcal{D}$ induced by the natural projection $\mathcal{D} \times \mathbb{C}^n \to \mathcal{D}$ is a torus bundle over X which may be called a twisted torus bundle (see Chapter 8). As in the case of Kuga fiber varieties, holomorphic forms of the highest degree on $Y_{\psi,\xi}$ can also be identified with mixed automorphic forms for Γ of certain type.

This book is organized as follows. In Chapter 1 we discuss basic properties of mixed automorphic and cusp forms of one variable including the construction of Eisenstein and Poincaré series. We also study some cusp forms associated to mixed cusp forms and describe mixed automorphic forms associated to a certain class of linear ordinary differential equations. Geometric aspects of mixed automorphic forms of one variable are presented in Chapter 2. We construct elliptic varieties and interpret holomorphic forms of the highest degree on such a variety as mixed automorphic forms. Discussions of modular symbols and Eichler-Shimura relations for mixed automorphic forms are also included. In Chapter 3 we investigate connections between parabolic cohomology and mixed automorphic forms and discuss a generalization of the Eichler-Shimura isomorphism. In order to consider mixed automorphic forms of several variables we introduce mixed Hilbert modular forms and mixed Siegel modular forms in Chapter 4 and show that certain types of such forms occur as holomorphic forms on certain families of abelian varieties parametrized by Hilbert or Siegel modular varieties. In Chapter 5 we describe mixed automorphic forms on Hermitian symmetric domains associated to equivariant holomorphic maps of symmetric domains. We then introduce a representation-theoretic description of mixed automorphic forms on semisimple Lie groups and real reductive groups. We also construct the associated Poincaré and Eisenstein series as well as Whitaker vectors. In Chapter 6 we describe Kuga fiber varieties associated to an equivariant holomorphic map of a symmetric domain into a Siegel upper half space and show that holomorphic forms of the highest degree on a Kuga fiber variety can be identified with mixed automorphic forms on a symmetric domain. Jacobi forms on symmetric domains and their relations with bundles over Kuga fiber varieties are discussed in Chapter 7. In Chapter 8 we are concerned with complex torus bundles over a locally symmetric space which generalize Kuga fiber varieties. Such torus bundles are constructed by using certain 2-cocycles and 1-cochains of a discrete group. We discuss their connection with mixed automorphic forms and determine certain cohomology of such a bundle.

Mixed Automorphic Forms

Classical automorphic or cusp forms of one variable are holomorphic functions on the Poincaré upper half plane $\mathcal{H}$ satisfying a transformation formula with respect to a discrete subgroup Γ of $SL(2,\mathbb{R})$ as well as certain regularity conditions at the cusps (see e.g. [14, 95, 114]). Given a nonnegative integer k, the transformation formula for an automorphic form f for Γ of weight k is of the form

$$f(\gamma z) = j(\gamma, z)^k f(z)$$

for $z \in \mathcal{H}$ and $\gamma \in \Gamma$, where $j(\gamma, z) = cz + d$ with c and d being the $(2,1)$ and $(2,2)$ entries of the matrix γ.

Mixed automorphic forms generalize automorphic forms, and they are associated with a holomorphic map $\omega : \mathcal{H} \to \mathcal{H}$ that is equivariant with respect to a homomorphism $\chi : \Gamma \to SL(2,\mathbb{R})$. Indeed, the transformation formula for mixed automorphic forms is of the form

$$f(\gamma z) = j(\gamma, z)^k j(\chi(\gamma), \omega(z))^\ell f(z)$$

for some nonnegative integers k and ℓ. Such equivariant pairs (ω, χ) occur naturally in the theory of elliptic surfaces (see Chapter 2) or in connection with certain linear ordinary differential equations. For example, an equivariant pair is obtained by using the period map ω of an elliptic surface E and the monodromy representation χ of E. In this case, a holomorphic form on E of degree two can be interpreted as a mixed automorphic form (cf. [18, 43, 68]). Similarly, the period map and the monodromy representation of a certain type of second order linear ordinary differential equation also provide us an equivariant pair (cf. [79]; see also [83]). In this chapter we introduce mixed automorphic and mixed cusp forms of one variable and discuss some their properties.

In Section 1.1 we describe the definition of mixed automorphic forms as well as mixed cusp forms of one variable associated to an equivariant holomorphic map of the Poincaré upper half plane. As examples of mixed automorphic forms, Eisenstein series and Poincaré series for mixed automorphic forms are constructed in Section 1.2. In Section 1.3 we consider certain cusp forms associated to pairs of mixed cusp forms and discuss relations among the Fourier coefficients of the cusp forms and those of the mixed cusp forms. Section 1.4 is about mixed automorphic forms associated to a certain class of linear ordinary differential equation. We relate the monodromy of such a differential equations with the periods of the associated mixed automorphic forms.

1.1 Mixed Automorphic Forms of One Variable

In this section we introduce mixed automorphic forms associated to an equivariant pair, which generalize elliptic modular forms. In particular, we discuss cusp conditions for mixed automorphic and cusp forms.

Let $\mathcal{H}$ denote the Poincaré upper half plane

$$\{z \in \mathbb{C} \mid \operatorname{Im} z > 0\}$$

on which $SL(2, \mathbb{R})$ acts by linear fractional transformations. Thus, if $z \in \mathcal{H}$ and $\gamma = \left(\begin{smallmatrix} a & b \\ c & d \end{smallmatrix}\right) \in SL(2, \mathbb{R})$, we have

$$\gamma z = \frac{az + b}{cz + d}.$$

For the same z and γ, we set

$$j(\gamma, z) = cz + d. \tag{1.1}$$

Then the resulting map $j : SL(2, \mathbb{R}) \times \mathcal{H} \to \mathbb{C}$ is an automorphy factor, meaning that it satisfies the cocycle condition

$$j(\gamma\gamma', z) = j(\gamma, \gamma'z)j(\gamma', z) \tag{1.2}$$

for all $z \in \mathcal{H}$ and $\gamma, \gamma' \in SL(2, \mathbb{R})$.

We fix a discrete subgroup Γ of $SL(2, \mathbb{R})$ and extend the action of Γ on $\mathcal{H}$ continuously to the set $\mathcal{H} \cup \mathbb{R} \cup \{\infty\}$. An element $s \in \mathbb{R} \cup \{\infty\}$ is a cusp for Γ if it is fixed under an infinite subgroup, called a parabolic subgroup, of Γ. Elements of a parabolic subgroup of Γ are parabolic elements of Γ. We assume that Γ is a Fuchsian group of the first kind, which means that the volume of the quotient space $\Gamma \backslash \mathcal{H}$ is finite. Let $\chi : \Gamma \to SL(2, \mathbb{R})$ be a homomorphism of groups such that its image $\chi(\Gamma)$ is a Fuchsian group of the first kind, and let $\omega : \mathcal{H} \to \mathcal{H}$ be a holomorphic map that is equivariant with respect to χ. Thus (ω, χ) is an equivariant pair satisfying the condition

$$\omega(\gamma z) = \chi(\gamma)\omega(z) \tag{1.3}$$

for all $\gamma \in \Gamma$ and $z \in \mathcal{H}$. We assume that the inverse image of the set of parabolic elements of $\chi(\Gamma)$ under χ consists of the parabolic elements of Γ. In particular, χ carries parabolic elements to parabolic elements. Given a pair of nonnegative integers k and ℓ, we set

$$J_{k,\ell}(\gamma, z) = j(\gamma, z)^k j(\chi(\gamma), \omega(z))^\ell \tag{1.4}$$

for all $\gamma \in \Gamma$ and $z \in \mathcal{H}$.

Lemma 1.1 *The map $J_{k,\ell} : \Gamma \times \mathcal{H} \to \mathbb{C}$ determined by (1.4) is an automorphy factor, that is, it satisfies the cocycle condition*

$$J_{k,\ell}(\gamma\gamma', z) = J_{k,\ell}(\gamma, \gamma'z) \cdot J_{k,\ell}(\gamma', z) \tag{1.5}$$

for all $\gamma, \gamma' \in \Gamma$ and $z \in \mathcal{H}$.

Proof. This follows easily from (1.2), (1.3), and (1.4). ☐

If $f : \mathcal{H} \to \mathbb{C}$ is a function and $\gamma \in \Gamma$, we denote by $f \mid_{k,\ell} \gamma$ the function on $\mathcal{H}$ given by

$$(f \mid_{k,\ell} \gamma)(z) = J_{k,\ell}(\gamma, z)^{-1} f(\gamma z) \tag{1.6}$$

for all $z \in \mathcal{H}$. Using (1.5), we see easily that

$$((f \mid_{k,\ell} \gamma) \mid_{k,\ell} \gamma' = f \mid_{k,\ell} (\gamma\gamma')$$

for all $\gamma, \gamma' \in \Gamma'$.

Let $s \in \mathbb{R} \cup \{\infty\}$ be a cusp for Γ, so that we have

$$\alpha s = s = \sigma\infty \tag{1.7}$$

for some $\sigma \in SL(2, \mathbb{R})$ and a parabolic element α of Γ. If Γ_s denotes the subgroup

$$\Gamma_s = \{\gamma \in \Gamma \mid \gamma s = s\} \tag{1.8}$$

of Γ consisting of the elements fixing s, then we have

$$\sigma^{-1}\Gamma_s\sigma \cdot \{\pm 1\} = \left\{ \pm \begin{pmatrix} 1 & h \\ 0 & 1 \end{pmatrix}^n \;\middle|\; n \in \mathbb{Z} \right\} \tag{1.9}$$

for some positive real number h. Since $\chi(\alpha)$ is a parabolic element of $\chi(\Gamma)$, there is a cusp s_χ for $\chi(\Gamma)$ and an element $\chi(\sigma) \in SL(2, \mathbb{R})$ such that

$$\chi(\alpha)s_\chi = s_\chi = \chi(\sigma)\infty. \tag{1.10}$$

We assume that

$$\omega(\sigma z) = \chi(\sigma)\omega(z) \tag{1.11}$$

for all $z \in \mathcal{H}$. Given an element $z \in \mathcal{H}$ and a holomorphic function f on $\mathcal{H}$, we extend the maps $\gamma \mapsto J_{k,\ell}(\gamma, z)$ and $\gamma \mapsto f \mid_{k,\ell} \gamma$ given by (1.4) and (1.6), respectively, to $\Gamma \cup \{\sigma\}$. In particular, we may write

$$J_{k,\ell}(\sigma, z) = j(\sigma, z)^k j(\chi(\sigma), \omega(z))^\ell, \tag{1.12}$$

$$(f \mid_{k,\ell} \sigma)(z) = J_{k,\ell}(\sigma, z)^{-1} f(\sigma z) \tag{1.13}$$

for all $z \in \mathcal{H}$.

In order to discuss Fourier series, let $f : \mathcal{H} \to \mathbb{C}$ be a holomorphic function that satisfies

$$f \mid_{k,\ell} \gamma = f \qquad (1.14)$$

for all $\gamma \in \Gamma$. Then we can consider the Fourier expansion of f at the cusps of Γ as follows. Suppose first that ∞ is a cusp of Γ. Then the subgroup Γ_∞ of Γ that fixes ∞ is generated by an element of the form $\left(\begin{smallmatrix} 1 & h \\ 0 & 1 \end{smallmatrix}\right)$ for a positive real number h. Since χ carries a parabolic element of Γ to a parabolic element of $\chi(\Gamma)$, we may assume

$$\chi\begin{pmatrix} 1 & h \\ 0 & 1 \end{pmatrix} = \pm\begin{pmatrix} 1 & h_\chi \\ 0 & 1 \end{pmatrix}$$

for some positive real number h_χ. Thus, using (1.1) and (1.4), we see that

$$J^{k,\ell}_{\omega,\chi}\left(\left(\begin{smallmatrix} 1 & h \\ 0 & 1 \end{smallmatrix}\right), z\right) = 1,$$

and hence we obtain $f(z + h) = f(z)$ for all $z \in \mathcal{H}$. This leads us to the Fourier expansion of f at ∞ of the form

$$f(z) = \sum_{n \geq n_0} a_n e^{2\pi i n z / h}$$

for some $n_0 \in \mathbb{Z}$.

We now consider an arbitrary cusp s of Γ with $\sigma(\infty) = s$. If Γ_s is as in (1.8) and if $\Gamma^\sigma = \sigma^{-1}\Gamma\sigma$, then we see that $\gamma \in \Gamma_s$ if and only if

$$(\sigma^{-1}\gamma\sigma)\infty = \sigma^{-1}\gamma s = \sigma^{-1}s = \infty;$$

hence $\sigma^{-1}\Gamma_s\sigma = (\Gamma^\sigma)_\infty$. In particular, ∞ is a cusp for Γ^σ.

Lemma 1.2 *If f satisfies the functional equation (1.14), then the function $f \mid_{k,\ell} \sigma : \mathcal{H} \to \mathbb{C}$ in (1.13) satisfies the relation*

$$(f \mid_{k,\ell} \sigma)(gz) = (f \mid_{k,\ell} \sigma)(z)$$

for all $g \in (\Gamma^\sigma)_\infty = \sigma^{-1}\Gamma_s\sigma$ and $z \in \mathcal{H}$.

Proof. Let $g = \sigma^{-1}\gamma\sigma \in \Gamma^\sigma$ with $\gamma \in \Gamma_s$. Then by (1.13) we have

$$(f \mid_{k,\ell} \sigma)(gz) = j(\sigma, gz)^{-k} j(\chi(\sigma), \omega(gz))^{-\ell} f(\gamma\sigma z) \qquad (1.15)$$

for all $z \in \mathcal{H}$. Since both g and $\chi(\sigma)^{-1}\chi(\gamma)\chi(\sigma)$ fix ∞, we have

$$j(g, z) = 1 = j(\chi(\sigma)^{-1}\chi(\gamma)\chi(\sigma), \omega(z)).$$

Using this, (1.2) and (1.11), we see that

$$j(\sigma, gz) = j(\sigma, gz)j(g, z) = j(\sigma g, z) = j(\gamma\sigma, z), \qquad (1.16)$$

$$j(\chi(\sigma), \omega(gz)) = j(\chi(\sigma), \chi(\sigma)^{-1}\chi(\gamma)\chi(\sigma)\omega(z)) \qquad (1.17)$$
$$\times j(\chi(\sigma)^{-1}\chi(\gamma)\chi(\sigma), \omega(z))$$
$$= j(\chi(\sigma)\chi(\sigma)^{-1}\chi(\gamma)\chi(\sigma), \omega(z))$$
$$= j(\chi(\gamma)\chi(\sigma), \omega(z)).$$

Thus, by combining this with (1.15), (1.16), and (1.17), we obtain

$$(f \mid_{k,\ell} \sigma)(gz) = j(\gamma\sigma, z)^{-k} j(\chi(\gamma)\chi(\sigma), \omega(z))^{-\ell} f(\gamma\sigma z)$$
$$= j(\gamma, \sigma z)^{-k} j(\sigma, z)^{-k} j(\chi(\gamma), \chi(\sigma)\omega(z))^{-\ell}$$
$$\times j(\chi(\sigma), \omega(z))^{-\ell} f(\gamma\sigma z)$$
$$= j(\sigma, z)^{-k} j(\chi(\sigma), \omega(z))^{-\ell} (f \mid_{k,\ell} \gamma)(\sigma z)$$
$$= (f \mid_{k,\ell} \sigma)(z);$$

hence the lemma follows. □

By Lemma 1.2 the Fourier expansion of $f \mid_{k,\ell} \sigma$ at ∞ can be written in the form

$$(f \mid_{k,\ell} \sigma)(z) = \sum_{n \geq n_0} a_n e^{2\pi i n z / h},$$

which is called the Fourier expansion of f at s.

Definition 1.3 *Let Γ, ω, and χ as above, and let k and ℓ be nonnegative integers. A* mixed automorphic form *of type (k, ℓ) associated to Γ, ω and χ is a holomorphic function $f : \mathcal{H} \to \mathbb{C}$ satisfying the following conditions:*

(i) $f \mid_{k,\ell} \gamma = f$ for all $\gamma \in \Gamma$.

(ii) The Fourier coefficients a_n of f at each cusp s satisfy the condition that $n \geq 0$ whenever $a_n \neq 0$.

The holomorphic function f is a mixed cusp form *of type (k, ℓ) associated to Γ, ω and χ if (ii) is replaced with the following condition:*

(ii)′ The Fourier coefficients a_n of f at each cusp s satisfy the condition that $n > 0$ whenever $a_n \neq 0$.

We shall denote by $M_{k,\ell}(\Gamma, \omega, \chi)$ (resp. $S_{k,\ell}(\Gamma, \omega, \chi)$) the space of mixed automorphic (resp. cusp) forms associated to Γ, ω and χ.

Remark 1.4 *If $\ell = 0$ in Definition 1.3(i), then f is a classical automorphic form or a cusp form (see e.g. [95, 114]). Thus, if $M_k(\Gamma)$ denotes the space of automorphic forms of weight k for Γ, we see that*

$$M_{k,0}(\Gamma, \omega, \chi) = M_k(\Gamma), \quad M_{k,\ell}(\Gamma, 1_{\mathcal{H}}, 1_{\Gamma}) = M_{k+\ell}(\Gamma),$$

where $1_{\mathcal{H}}$ is the identity map on $\mathcal{H}$ and 1_{Γ} is the inclusion map of Γ into $SL(2, \mathbb{R})$. On the other hand for $k = 0$ the elements of $M_{0,\ell}(\Gamma, \omega, \chi)$ are generalized automorphic forms of weight ℓ in the sense of Hoyt and Stiller (see e.g. [120, p. 31]). In addition, if $f \in M_{k,\ell}(\Gamma, \omega, \chi)$ and $g \in M_{k',\ell'}(\Gamma, \omega, \chi)$, then we see that $fg \in M_{k+k',\ell+\ell'}(\Gamma, \omega, \chi)$.

1.2 Eisenstein Series and Poincaré Series

In this section we construct Eisenstein series and Poincaré series, which provide examples of mixed automorphic forms. We shall follow closely the descriptions in [70] and [80].

Let $\omega : \mathcal{H} \to \mathcal{H}$ and $\chi : \Gamma \to SL(2,\mathbb{R})$ be as in Section 1.1, and let s be a cusp for Γ. Let $\sigma \in SL(2,\mathbb{R})$ and $\alpha \in \Gamma$ be the elements associated to s satisfying (1.7), and assume that Γ_s in (1.8) satisfies (1.9). We consider the corresponding parabolic element $\chi(\alpha)$ of $\chi(\Gamma)$ and the element $\chi(\sigma) \in SL(2,\mathbb{R})$ satisfying (1.10) and (1.11). We fix a positive integer k and a nonnegative integer m. For each nonnegative integer ν, we define the holomorphic function $\phi_\nu : \mathcal{H} \to \mathbb{C}$ associated to the cusp s by

$$\phi_\nu(z) = J_{2k,2m}(\sigma, z)^{-1} \exp(2\pi i \nu \sigma z/h) \tag{1.18}$$
$$= j(\sigma, z)^{-2k} j(\chi(\sigma), \omega(z))^{-2m} \exp(2\pi i \nu \sigma z/h)$$

for all $z \in \mathcal{H}$, where we used the notation in (1.12).

Lemma 1.5 *If s is a cusp of Γ described above, then the associated function ϕ_ν given by (1.18) satisfies*

$$\phi_\nu \mid_{2k,2m} \gamma = \phi_\nu \tag{1.19}$$

for all $\gamma \in \Gamma_s$.

Proof. Given $z \in \mathcal{H}$ and $\gamma \in \Gamma_s$, using (1.18), we have

$$\phi_\nu(\gamma z) = j(\sigma, \gamma z)^{-2k} j(\chi(\sigma), \omega(\gamma z))^{-2m} \exp(2\pi i \nu \sigma \gamma z/h)$$
$$= j(\sigma, \gamma z)^{-2k} j(\chi(\sigma), \chi(\gamma)\omega(z))^{-2m} \exp(2\pi i \nu \sigma \gamma z/h)$$
$$= j(\sigma\gamma, z)^{-2k} j(\gamma, z)^{2k} j(\chi(\sigma)\chi(\gamma), \omega(z))^{-2m}$$
$$\times j(\chi(\gamma), \omega(z))^{2m} \exp(2\pi i \nu (\sigma\gamma\sigma^{-1})\sigma z/h),$$

where we used (1.2). Since $\sigma\gamma\sigma^{-1}$ and $\chi(\sigma)\chi(\gamma)\chi(\sigma)^{-1}$ stabilize ∞, we have

$$j(\sigma\gamma\sigma^{-1}, w) = j(\chi(\sigma)\chi(\gamma)\chi(\sigma)^{-1}, \chi(\sigma)w) = 1$$

for all $w \in \mathcal{H}$, and hence we see that

$$j(\sigma\gamma, z) = j(\sigma\gamma\sigma^{-1}, \sigma z) \cdot j(\sigma, z) = j(\sigma, z),$$

$$j(\chi(\sigma)\chi(\gamma), \omega(z)) = j(\chi(\sigma)\chi(\gamma)\chi(\sigma)^{-1}, \chi(\sigma)\omega(z)) \cdot j(\chi(\sigma), \omega(z))$$
$$= j(\chi(\sigma), \omega(z)),$$

and $\sigma\gamma z/h = (\sigma\gamma\sigma^{-1})\sigma z/h = \sigma z/h + d$ for some integer d. Thus we obtain

$$\phi_\nu(\gamma z) = j(\sigma, z)^{-2k} j(\gamma, z)^{2k}$$
$$\times j(\chi(\sigma), \omega(z))^{-2m} j(\chi(\gamma), \omega(z))^{2m} \exp(2\pi i \nu \sigma z / h)$$
$$= j(\gamma, z)^{2k} j(\chi(\gamma), \omega(z))^{2m} \phi_\nu(z),$$

and therefore the lemma follows. $\square$

Let s be a cusp of Γ considered above, and set

$$P^\nu_{2k,2m}(z) = \sum_{\gamma \in \Gamma_s \backslash \Gamma} (\phi_\nu \mid_{2k,2m} \gamma)(z) \tag{1.20}$$

for all $z \in \mathcal{H}$. Note that by Lemma 1.5 the summation is well-defined.

Definition 1.6 *The function $P^\nu_{2k,2m}(z)$ is called a* Poincaré series *for mixed automorphic forms if $\nu \geq 1$, and the function $P^0_{2k,2m}(z)$ is called an* Eisenstein series *for mixed automorphic forms.*

We shall show below that the series in (1.20) defining the function $P^\nu_{2k,2m}(z)$ converges and is holomorphic on $\mathcal{H}$.

Lemma 1.7 *Let $z_0 \in \mathcal{H}$, and let ε be a positive real number such that*

$$N_{3\varepsilon} = \{z \in \mathbb{C} \mid |z - z_0| \leq 3\varepsilon\} \subset \mathcal{H},$$

and let k and m be nonnegative integers. If ψ is a continuous function on $N_{3\varepsilon}$ that is holomorphic on the interior of $N_{3\varepsilon}$, then there exists a positive real number C such that

$$|\psi(z_1)| \leq C \int_{N_{3\varepsilon}} |\psi(z)| (\operatorname{Im} z)^k (\operatorname{Im} \omega(z))^m dV$$

for all $z_1 \in N_\varepsilon = \{z \in \mathbb{C} \mid |z - z_0| \leq \varepsilon\}$, where $dV = dx dy / y^2$ with $x = \operatorname{Re} z$ and $y = \operatorname{Im} z$.

Proof. Let z_1 be an element of N_ε, and consider the Taylor expansion of $\psi(z)$ about z_1 of the form

$$\psi(z) = \sum_{n=0}^\infty a_n (z - z_1)^n.$$

We set $N'_\varepsilon = \{z \in \mathbb{C} \mid |z - z_1| < \varepsilon\}$. Then $N'_\varepsilon \subset N_{3\varepsilon}$, and we have

$$\int_{N'_\varepsilon} \psi(z) dx dy = \int_0^{2\pi} \int_0^\varepsilon \sum_{n=0}^\infty a_n r^{n+1} e^{in\theta} dr d\theta = \pi \varepsilon^2 a_0 = \pi \varepsilon^2 \psi(z_1).$$

Hence we obtain

$$
\begin{aligned}
|\psi(z_1)| &\leq (\pi\varepsilon^2)^{-1} \int_{N_{3\varepsilon}} |\psi(z)| dx dy \\
&= (\pi\varepsilon^2)^{-1} \int_{N_{3\varepsilon}} \frac{|\psi(z)|(\operatorname{Im} z)^k (\operatorname{Im}\omega(z))^m}{(\operatorname{Im} z)^{k-2}(\operatorname{Im}\omega(z))^m} dV \\
&\leq (\pi\varepsilon^2 C_1)^{-1} \int_{N_{3\varepsilon}} |\psi(z)|(\operatorname{Im} z)^k (\operatorname{Im}\omega(z))^m dV,
\end{aligned}
$$

where

$$
C_1 = \inf\{(\operatorname{Im} z)^{k-2}(\operatorname{Im}\omega(z))^m \mid z \in N_{3\varepsilon}\}.
$$

Thus the lemma follows by setting $C = (\pi\varepsilon^2 C_1)^{-1}$.   □

If U is a connected open subset of $\mathcal{H}$, then we define the norm $\|\cdot\|_U$ on the space of holomorphic functions on U by

$$
\|\psi\|_U = \int_U |\psi(z)|(\operatorname{Im} z)^k (\operatorname{Im}\omega(z))^m dV,
$$

where ψ is a holomorphic function on U.

Lemma 1.8 *Let $\{f_n\}$ be a Cauchy sequence of holomorphic functions on U with respect to the norm $\|\cdot\|_U$. Then the sequence $\{f_n\}$ converges absolutely to a holomorphic function on U uniformly on any compact subsets of U.*

Proof. Let $\{f_n\}$ be a Cauchy sequence of holomorphic functions on an open set $U \subset \mathcal{H}$. Then by Lemma 1.7, for each $z \in U$, there is a constant C such that

$$
|f_n(z) - f_m(z)| \leq C\|f_n - f_m\|_U
$$

for all $n, m \geq 0$. Thus the sequence $\{f_n(z)\}$ of complex numbers is also a Cauchy sequence, and therefore it converges. We set $f(z) = \lim_{n\to\infty} f_n(z)$ for all $z \in U$. Let $z_0 \in U$, and choose $\delta > 0$ such that

$$
N_{3\delta} = \{z \in \mathbb{C} \mid |z - z_0| \leq 3\delta\} \subset U.
$$

Using Lemma 1.7 again, we have

$$
|f_n(z) - f_m(z)| \leq C'\|f_n - f_m\|_U
$$

for all $z \in N_\delta = \{z \in \mathbb{C} \mid |z - z_0| \leq \delta\}$. Given $\varepsilon > 0$, let N be a positive integer such that $\|f_n - f_m\|_U < \varepsilon/(2C')$ whenever $m, n > N$. For each $z \in N_\delta$, if we choose an integer $n' > N$ so that $|f_{n'}(z) - f(z)| < \varepsilon/2$, then we obtain

$$
|f_n(z) - f(z)| \leq |f_n(z) - f_{n'}(z)| + |f_{n'}(z) - f(z)| < \varepsilon
$$

for all $n > N$. Thus the sequence $\{f_n\}$ converges to f uniformly on N_δ and therefore on any compact subsets of U. Hence it follows that f is holomorphic function on U.   □

Let ϕ_ν be as in (1.18), and let $\{s_1, \ldots, s_\mu\}$ be the set of all Γ-inequivalent cusps of Γ. We choose a neighborhood U_i of s_i for each $i \in \{1, \ldots, \mu\}$, and set

$$\mathcal{H}' = \mathcal{H} - \bigcup_{i=1}^{\mu} \bigcup_{\gamma \in \Gamma} \gamma U_i. \tag{1.21}$$

Then, using the relations

$$\operatorname{Im} \gamma w = |j(\gamma, w)|^{-2} \cdot \operatorname{Im} w, \quad \operatorname{Im} \omega(\gamma w) = |j(\chi(\gamma), \omega(w))|^{-2} \cdot \operatorname{Im} \omega(w) \tag{1.22}$$

for $\gamma \in \Gamma$ and $w \in \mathcal{H}$ and the fact that ϕ_ν satisfies (1.19) for $\gamma \in \Gamma_s$, it can be shown that

$$\int_{\Gamma_s \backslash \mathcal{H}'} |\phi_\nu(z)| (\operatorname{Im} z)^k (\operatorname{Im} \omega(z))^m dV < \infty. \tag{1.23}$$

Theorem 1.9 *The series in* (1.20) *defining* $P_{2k,2m}^\nu(z)$ *converges absolutely on* $\mathcal{H}$ *and uniformly on compact subsets, and, in particular, the function* $P_{2k,2m}^\nu(z)$ *is holomorphic on* $\mathcal{H}$.

Proof. Let $s_1, \ldots, s_\mu$ be the Γ-inequivalent cusps of Γ as above, and let z_0 be an element of $\mathcal{H}$. We choose neighborhoods W of z_0 and U_i of s_i for $1 \leq i \leq \mu$ such that

$$\{\gamma \in \Gamma \mid \gamma W \cap W \neq \emptyset\} = \Gamma_{z_0}, \quad \gamma W \cap U_i = \emptyset \tag{1.24}$$

for all $\gamma \in \Gamma$ and $1 \leq i \leq \mu$, where Γ_{z_0} is the stabilizer of z_0 in Γ. Then, using (1.18) and (1.22), we have

$$\|P_{2k,2m}^\nu\|_W = \int_W \left| \sum_{\gamma \in \Gamma_s \backslash \Gamma} (\phi_\nu|_{2k,2m}\gamma)(z) \right| (\operatorname{Im} z)^k (\operatorname{Im} \omega(z))^m dV$$

$$\leq \int_W \sum_{\gamma \in \Gamma_s \backslash \Gamma} |(\phi_\nu|_{2k,2m}\gamma)(z)| (\operatorname{Im} z)^k (\operatorname{Im} \omega(z))^m dV$$

$$= \sum_{\gamma \in \Gamma_s \backslash \Gamma} \int_W |\phi_\nu(\gamma z)| (\operatorname{Im} \gamma z)^k (\operatorname{Im} \omega(\gamma z))^m dV$$

$$= \sum_{\gamma \in \Gamma_s \backslash \Gamma} \int_{\gamma W} |\phi_\nu(z)| (\operatorname{Im} z)^k (\operatorname{Im} \omega(z))^m dV.$$

In order to estimate the number of terms in the above sum, let $\gamma' \in \Gamma$ and set

$$\Xi = \{\gamma \in \Gamma \mid \gamma'' \gamma W \cap \gamma' W \neq \emptyset \text{ for some } \gamma'' \in \Gamma_s\}.$$

Then by (1.24) we see that $\gamma' W \in \mathcal{H}'$ and

$$|\Gamma_s \backslash \Xi| \leq |\Gamma_s \backslash \Gamma_s \gamma' \Gamma_{z_0}| \leq |\Gamma_{z_0}|,$$

where $|\cdot|$ denotes the cardinality. Thus, using this and (1.23), we have

$$\sum_{\gamma \in \Gamma_s \backslash \Gamma} \int_{\gamma W} |\phi_\nu(z)| (\operatorname{Im} z)^k (\operatorname{Im} \omega(z))^m dV$$

$$\leq |\Gamma_{z_0}| \int_{\Gamma_s \backslash \mathcal{H}'} |\phi_\nu(z)| (\operatorname{Im} z)^k (\operatorname{Im} \omega(z))^m dV < \infty.$$

Hence we obtain $\|P_{2k,2m}^\nu\|_W < \infty$, and by Lemma 1.8 we see that $P_{2k,2m}^\nu(z)$ converges absolutely on W and uniformly on compact subsets of W. Thus it follows that the function $P_{2k,2m}^\nu(z)$ is holomorphic on W, and therefore is holomorphic on $\mathcal{H}$ as well. $\square$

Now we need to show that the function $P_{2k,2m}^\nu(z)$ is holomorphic at each cusp for all nonnegative integers ν and that it vanishes at each cusp for all positive integers ν.

Lemma 1.10 *Let s' be a cusp of Γ such that $\sigma' s' = \infty$ with $\sigma' \in SL(2,\mathbb{R})$, and let $\sigma'_\chi \in SL(2,\mathbb{R})$ be an element with $\sigma'_\chi \omega(s) = \infty$. Using the notation in (1.6), the function ϕ_ν given in (1.18) satisfies the following conditions.*

(i) If s' is not Γ-equivalent to s, then there exist positive real numbers M and λ such that

$$|(\phi_\nu|_{2k,2m}\sigma'^{-1})(z)| \leq M|z|^{-2k} \tag{1.25}$$

whenever $\operatorname{Im} z > \lambda$.

(ii) If s' is Γ-equivalent to s, then there exist positive real numbers M and λ such that

$$|(\phi_\nu|_{2k,2m}\sigma'^{-1})(z)| \leq M \tag{1.26}$$

whenever $\operatorname{Im} z > \lambda$. If in addition $\nu > 0$, then we have

$$(\phi_\nu|_{2k,2m}\sigma'^{-1})(z) \to 0 \tag{1.27}$$

as $\operatorname{Im} z \to \infty$.

Proof. Using (1.6) and (1.18), for $z \in \mathcal{H}$ we have

$$(\phi_\nu|_{2k,2m}\sigma'^{-1})(z) = j(\sigma'^{-1}, z)^{-2k} j(\sigma'^{-1}_\chi, \omega(z))^{-2m} j(\sigma, \sigma'^{-1}z)^{-2k}$$
$$\times j(\chi(\sigma), \omega(\sigma'^{-1}z))^{-2m} \exp(2\pi i \nu \sigma \sigma'^{-1}/h).$$

If $\sigma\sigma'^{-1} = \begin{pmatrix} a & b \\ c & d \end{pmatrix}$ and if $\operatorname{Im} z > 2|d|/|c|$, then we have

$$|j(\sigma, \sigma'^{-1}z) \cdot j(\sigma'^{-1}, z)| = |j(\sigma\sigma'^{-1}, z)| = |cz + d|$$
$$\geq |c||z| - |d| \geq |c||z| - (|c|/2)\operatorname{Im} z$$
$$= |c||z| - (|c|/2)|z| = |c||z|/2.$$

On the other hand, if $\sigma'^{-1}_\chi = \begin{pmatrix} a' & b' \\ c' & d' \end{pmatrix}$ and $\chi(\sigma) = \begin{pmatrix} a'' & b'' \\ c'' & d'' \end{pmatrix}$, then we obtain

$$|j(\chi(\sigma)'^{-1}, \omega(z))||j(\chi(\sigma), \omega(\sigma'^{-1}z))| = |c'\omega(z) + d'||c''\omega(\sigma'^{-1}z) + d''|.$$

Since $\operatorname{Im}\omega(z) \to \infty$ and $\omega(\sigma'^{-1}z) \to \omega(s')$ as $\operatorname{Im} z \to \infty$, there exist real numbers $A, \lambda' > 0$ such that

$$|j(\chi(\sigma)'^{-1}, \omega(z))||j(\chi(\sigma), \omega(\sigma'^{-1}z))| \geq A$$

whenever $\operatorname{Im} z > \lambda'$. We set $\lambda = \max(\lambda', 2|d|/|c|)$. Then, whenever $\operatorname{Im} z > \lambda$, we have

$$|(\phi_\nu \mid_{2k,2m} \sigma'^{-1})(z)| \leq (|c||z|/2)^{-2k} A^{-2m} \exp(-2\pi\nu\sigma\sigma'(\operatorname{Im} z)/h).$$

Thus (1.25) holds for $M = (|c|/2)^{-2k} A^{-2m} \exp(-2\pi\nu\sigma\sigma'\lambda/h)$, and therefore (i) follows. As for (ii), if s' is equivalent to s, we may assume that $\sigma = \sigma'$. Thus we have

$$(\phi_\nu \mid_{2k,2m} \sigma'^{-1})(z) = j(1, z)^{-2k} j(\chi(\sigma)^{-1}, \omega(z))^{-2m}$$
$$\times j(\chi(\sigma), \omega(\sigma^{-1}z))^{-2m} \exp(2\pi i\nu z/h).$$

Since $j(1, z) = 1$, we obtain (1.26) by arguing as in the case of (i). $\square$

Theorem 1.11 *Let s_0 be a cusp of Γ. Then the function $P^\nu_{2k,2m}(z)$ is holomorphic at s_0 for all nonnegative integers ν. Furthermore, $P^\nu_{2k,2m}(z)$ vanishes at s_0 if $\nu > 0$.*

Proof. Let $\Gamma_{s_0} \subset \Gamma$ be the stabilizer of the cusp s_0, and let $\{\delta\}$ be a complete set of representatives of $\Gamma_s \backslash \Gamma / \Gamma_{s_0}$. Given δ, let $\{\eta\}$ be a complete set of representatives of $\delta^{-1}\Gamma_s\delta \cap \Gamma_{s_0} \backslash \Gamma_{s_0}$, so that we have $\Gamma = \coprod_{\delta,\eta} \Gamma_s\delta\eta$. We set

$$\phi_{\nu,\delta}(z) = \sum_\eta (\phi_\nu \mid_{2k,2m} \delta\eta)(z)$$

for all $z \in \mathcal{H}$. Then we have

$$P^\nu_{2k,2m}(z) = \sum_\delta \sum_\eta (\phi_\nu \mid_{2k,2m} \delta\eta)(z) = \sum_\delta \phi_{\nu,\delta}(z).$$

By Theorem 1.9 there is a neighborhood U of s_0 in $\mathcal{H}$ such that $P^\nu_{2k,2m}(z)$ converges uniformly on any compact subset of U. Hence, if $\sigma_0 s_0 = \infty$ with $\sigma_0 \in SL(2, \mathbb{R})$, then the function

$$P^\nu_{2k,2m} \mid_{2k,2m} \sigma_0^{-1} = \sum_\delta \phi_{\nu,\delta} \mid_{2k,2m} \sigma_0^{-1}$$

converges uniformly on any compact subset of $\{z \in \mathcal{H} \mid \operatorname{Im} z > d\}$ for some positive real number d. Therefore it suffices to show that each $\phi_{\nu,\delta} \mid_{2k,2m} \sigma_0^{-1}$ is holomorphic at ∞ and that it has zero at ∞ if $\nu > 0$. First, suppose that δs_0 is not a cusp of Γ_s. Then $\delta^{-1}\Gamma_s\delta \cap \Gamma_{s_0}$ coincides with $\{1\}$ or $\{\pm 1\}$, and hence we have

$$\phi_{\nu,\delta} \,|_{2k,2m}\, \sigma_0^{-1} = C \cdot \sum_{\eta \in \Gamma_{s_0}} (\phi_\nu \,|_{2k,2m}\, \delta\sigma_0^{-1}\sigma_0\eta\sigma_0^{-1})$$

with $C = 1$ or $1/2$, respectively. Applying (1.25) for $s = \delta s_0$, $\sigma = \sigma_0\delta^{-1}$, we obtain

$$|(\phi_\nu \,|_{2k,2m}\, \delta\sigma_0^{-1})(z)| \leq M|z|^{-2k}$$

for all z with $\operatorname{Im} z > \lambda$ for some $M, \lambda > 0$. Thus we obtain

$$|(\phi_{\nu,\delta} \,|_{2k,2m}\, \sigma_0^{-1})(z)| \leq 2M \sum_{\alpha \in \mathbb{Z}} |z + \alpha b|^{-2k}, \qquad (1.28)$$

where b is a positive real number such that

$$\sigma_0 \Gamma_{s_0} \sigma_0^{-1} \cdot \{\pm 1\} = \{\pm \left(\begin{smallmatrix} 1 & b \\ 0 & 1 \end{smallmatrix}\right)^\alpha \mid \alpha \in \mathbb{Z}\}.$$

By comparing the series on the right hand side of (1.28) with the series $\sum_{\alpha \in \mathbb{Z}} \alpha^{-2k}$, we see that it converges uniformly on any compact subset of the domain $\operatorname{Im} z > \lambda$. Hence it follows that $\phi_{\nu,\delta} \,|_{2k,2m}\, \sigma_0^{-1}$ is holomorphic at ∞. Furthermore, $\phi_{\nu,\delta} \,|_{2k,2m}\, \sigma_0^{-1}$ vanishes at ∞ because the right hand side of (1.28) approaches zero as $z \to \infty$. Next, suppose δs_0 is a cusp of Γ_s. Then $\delta^{-1}\Gamma_s\delta \cap \Gamma_{s_0}$ is a subgroup of Γ_{s_0} of finite index; hence the sum on the right hand side of

$$\phi_{\nu,\delta} \,|_{2k,2m}\, \sigma_0^{-1} = \sum_{\eta}(\phi_\nu \,|_{2k,2m}\, \delta\sigma_0^{-1}\sigma_0\eta\sigma_0^{-1}),$$

where the summation is over $\eta \in \delta^{-1}\Gamma_s\delta \cap \Gamma_{s_0}\backslash\Gamma_{s_0}$, is a finite sum. Using (1.26) for $s = \delta s_0$ and $\sigma = \sigma_0\delta^{-1}$, for each δ we obtain

$$|(\phi_\nu \,|_{2k,2m}\, \delta\sigma_0^{-1})(z)| \leq M$$

for all $\operatorname{Im} z > \lambda$ for some $M, \lambda > 0$. For each $\eta \in \Gamma_{s_0}$ we have

$$\sigma_0\eta\sigma_0^{-1} = \pm \left(\begin{smallmatrix} 1 & \beta b \\ 0 & 1 \end{smallmatrix}\right)$$

for some $\beta \in \mathbb{Z}$; hence we have

$$|(\phi_{\nu,\delta} \,|_{2k,2m}\, \sigma_0^{-1})(z)| \leq M$$

for all $\operatorname{Im} z > \lambda$, and it follows that $\phi_{\nu,\delta} \,|_{2k,2m}\, \sigma_0^{-1}$ is holomorphic at ∞. Furthermore, if $\nu > 0$, then by (1.27) we have

$$(\phi_\nu \,|_{2k,2m}\, \delta\sigma_0^{-1})(z) \to 0$$

as $\operatorname{Im} z \to \infty$; hence we see that $\phi_{\nu,\delta} \,|_{2k,2m}\, \sigma_0^{-1}$ vanishes at ∞. $\qquad\square$

Theorem 1.12 *The Eisenstein series* $P^0_{2k,2m}(z)$ *is a mixed automorphic form and the Poincaré series* $P^\nu_{2k,2m}(z)$ *is a mixed cusp form for* Γ *of type* $(2k, 2m)$.

Proof. Using the relations

$$j(\gamma, \gamma' z) = j(\gamma', z)^{-1} j(\gamma\gamma', z),$$

$$j(\chi(\gamma), \chi(\gamma')\omega(z)) = j(\chi(\gamma'), \omega(z))^{-1} j(\chi(\gamma\gamma'), \omega(z))$$

for $\gamma, \gamma' \in \Gamma$ and $z \in \mathcal{H}$, we obtain

$$
\begin{aligned}
P_{2k,2m}^\nu(\gamma' z) &= \sum_{\gamma \in \Gamma_s \backslash \Gamma} (\phi_\nu \mid_{2k,2m} \gamma)(\gamma' z) \\
&= \sum_{\gamma \in \Gamma_s \backslash \Gamma} j(\gamma, \gamma' z)^{-2k} j(\chi(\gamma), \omega(\gamma' z))^{-2m} \phi_\nu(\gamma\gamma' z) \\
&= j(\gamma', z)^{2k} j(\chi(\gamma'), \omega(z))^{2m} \\
&\quad \times \sum_{\gamma \in \Gamma_s \backslash \Gamma} j(\gamma\gamma', z)^{-2k} j(\chi(\gamma\gamma'), \omega(z))^{-2m} \phi_\nu(\gamma\gamma' z) \\
&= j(\gamma', z)^{2k} j(\chi(\gamma'), \omega(z))^{2m} P_{2k,2m}^\nu(z)
\end{aligned}
$$

for all $\gamma' \in \Gamma$ and $z \in \mathcal{H}$; hence we see that $P_{2k,2m}^\nu$ satisfies the condition (i) in Definition 1.3. Therefore the theorem follows from the cusp conditions given in Theorem 1.11. □

Remark 1.13 *If ω is the identity map on $\mathcal{H}$ and if χ is the inclusion map of Γ into $SL(2,\mathbb{R})$, then $P_{2k,2m}^0(z)$ and $P_{2k,2m}^\nu(z)$ for $\nu > 0$ are the Eisenstein series and the Poincaré series, respectively, for elliptic modular forms for Γ of weight $2(k+m+1)$. Poincaré series were also considered in [70] for mixed cusp forms of type $(2, 2m)$.*

1.3 Cusp Forms Associated to Mixed Cusp Forms

Let $S_{\ell,k}(\Gamma, \omega, \chi)$ be the space of mixed cusp forms of type (ℓ, k) associated to Γ, ω and χ as in Definition 1.3, and let $S_m(\Gamma)$ be the space of cusp forms of weight m for Γ. If ω is the identity map on $\mathcal{H}$ and χ is the inclusion map of Γ into $SL(2,\mathbb{R})$, then a mixed cusp form of type (ℓ, k) associated to Γ, ω and χ becomes a cusp form of weight $\ell + k$ for Γ. Similarly, a mixed cusp form of type $(\ell, 0)$ is a cusp form of weight ℓ. Given a mixed cusp form g belonging to $S_{k,\ell}(\Gamma, \omega, \chi)$, denote by $\mathcal{L}_g^* : S_{\ell+m,k}(\Gamma, \omega, \chi) \to S_m(\Gamma)$ be the map that is the adjoint of the linear map $\mathcal{L}_g : S_m(\Gamma) \to S_{\ell+m,k}(\Gamma, \omega, \chi)$ sending $h \in S_m(\Gamma)$ to gh. In this section we express the Fourier coefficients of the cusp form $\mathcal{L}_g^*(f)$ associated to a mixed cusp form f in terms of series involving Fourier coefficients of f and g by following the method of W. Kohnen [53] who treated the case of classical cusp forms.

If $g \in S_{\ell,k}(\Gamma, \omega, \chi)$ and $h \in S_m(\Gamma)$, then we see easily that $gh \in S_{\ell+m,k}(\Gamma, \omega, \chi)$. Thus, given an element $g \in S_{\ell,k}(\Gamma, \omega, \chi)$, we can consider a linear map $\mathcal{L}_g : S_m(\Gamma) \to S_{\ell+m,k}(\Gamma, \omega, \chi)$ defined by

$$\mathcal{L}_g(h) = gh \tag{1.29}$$

for all $h \in S_m(\Gamma)$. The space $S_m(\Gamma)$ is equipped with the Petersson inner product, and the Petersson inner product can also be defined on the space $S_{\ell+m,k}(\Gamma, \omega, \chi)$ of mixed cusp forms (see [70, Proposition 2.1]). Thus we can consider the adjoint map $\mathcal{L}_g^* : S_{\ell+m,k}(\Gamma, \omega, \chi) \to S_m(\Gamma)$ of $\mathcal{L}_g$ satisfying the condition

$$\langle \mathcal{L}_g^* f, h \rangle = \langle\langle f, \mathcal{L}_g h \rangle\rangle \tag{1.30}$$

for all $f \in S_{\ell+m,k}(\Gamma, \omega, \chi)$ and $h \in S_m(\Gamma)$, where $\langle\,,\,\rangle$ (resp. $\langle\langle\,,\,\rangle\rangle$) is the Petersson inner product on the space $S_m(\Gamma)$ (resp. $S_{\ell+m,k}(\Gamma, \omega, \chi)$).

Throughout the rest of this section we shall assume that the Fuchsian group Γ is a congruence subgroup of $SL(2, \mathbb{R})$. If Γ_∞ is the subgroup of Γ consisting of the elements of Γ fixing ∞, then Γ_∞ is an infinite cyclic group generated by the translation map $z \mapsto z + b$ for some $b \in \mathbb{Z}$. Let $P_{m,n}$ be the n-th Poincaré series in $S_m(\Gamma)$ given by

$$P_{m,n}(z) = \sum_{\gamma \in \Gamma_\infty \backslash \Gamma} e^{2\pi i n \gamma z / b} j(\gamma, z)^{-m}, \tag{1.31}$$

where $j(\gamma, z)$ is as in (1.1) (see e.g. [33]). If h is a cusp form in $S_m(\Gamma)$ whose Fourier expansion is of the form

$$h(z) = \sum_{p=1}^\infty A_p(h) e^{2\pi i p z / b}$$

and if $\langle\,,\,\rangle$ is the Petersson inner product on $S_m(\Gamma)$, then we have

$$\langle h, P_{m,n} \rangle = \frac{b^m \Gamma(m-1)}{(4\pi n)^{m-1}} A_n(h) \tag{1.32}$$

(see Theorem 5 in [33, Section 11]), where Γ is the Gamma function. Thus, if $\mathcal{L}_g^* : S_{\ell+m,k}(\Gamma, \omega, \chi) \to S_m(\Gamma)$ is the adjoint of $\mathcal{L}_g$ as before, for each $f \in S_{\ell+m,k}(\Gamma, \omega, \chi)$ and a positive integer n by using (1.29), (1.30) and (1.32), we obtain

$$\frac{b^m \Gamma(m-1)}{(4\pi n)^{m-1}} A_n(\mathcal{L}_g^* f) = \langle \mathcal{L}_g^* f, P_{m,n} \rangle = \langle\langle f, \mathcal{L}_g P_{m,n} \rangle\rangle = \langle\langle f, g P_{m,n} \rangle\rangle$$

$$= \int_{\Gamma \backslash \mathcal{H}} f(z) \overline{g(z) P_{m,n}(z)} y^{m+\ell} (\mathrm{Im}\,\omega(z))^k dV$$

$$= \int_{\Gamma \backslash \mathcal{H}} \Phi(z) \overline{P_{m,n}(z)} y^m dV,$$

where $\langle\langle\,,\,\rangle\rangle$ is the Petersson inner product on $S_{\ell+m,k}(\Gamma, \omega, \chi)$ (cf. [70, Proposition 2.1]), $z = x + iy$, $\Phi(z) = f(z)\overline{g(z)} y^\ell (\mathrm{Im}\,\omega(z))^k$, and $dV = y^{-2} dx dy$.

Lemma 1.14 *The function* $\Phi(z) = f(z)\overline{g(z)}(\operatorname{Im} z)^{\ell}(\operatorname{Im}\omega(z))^{k}$ *satisfies the relation*

$$\Phi(\gamma z) = j(\gamma, z)^{m}\Phi(z) \tag{1.33}$$

for all $\gamma \in \Gamma$ *and* $z \in \mathcal{H}$.

Proof. Since $f \in S_{\ell+m,k}(\Gamma, \omega, \chi)$ and $g \in S_{\ell,k}(\Gamma, \omega, \chi)$, for $\gamma \in \Gamma$ and $z \in \mathcal{H}$, we have

$$
\begin{aligned}
\Phi(\gamma z) &= f(\gamma z)\overline{g(\gamma z)}(\operatorname{Im}\gamma z)^{\ell}(\operatorname{Im}\omega(\gamma z))^{k}\\
&= f(\gamma z)\overline{g(\gamma z)}(\operatorname{Im}\gamma z)^{\ell}(\operatorname{Im}\chi(\gamma)\omega(z))^{k}\\
&= j(\gamma, z)^{\ell+m}j(\chi(\gamma), \omega(z))^{k}f(z)\,\overline{j(\gamma, z)^{\ell}}\,\overline{j(\chi(\gamma), \omega(z))^{k}}\,\overline{g(z)}\\
&\quad \times |j(\gamma, z)|^{-2\ell}(\operatorname{Im} z)^{\ell}|j(\chi(\gamma), \omega(z))|^{-2k}(\operatorname{Im}\omega(z))^{k}\\
&= j(\gamma, z)^{m}f(z)\overline{g(z)}(\operatorname{Im} z)^{\ell}(\operatorname{Im}\omega(z))^{k}\\
&= j(\gamma, z)^{m}\Phi(z).
\end{aligned}
$$

Hence the lemma follows.   □

By Lemma 1.14 the function Φ satisfies $\Phi(z + kb) = \Phi(b)$ for all $k \in \mathbb{Z}$. Note that Φ is not a holomorphic function. However, if $z = x + iy$, then Φ is periodic as a function of x with period b and therefore has a Fourier expansion of the form

$$\Phi(z) = \sum_{p \in \mathbb{Z}} A_p(\Phi; y)e^{2\pi i p x/b}, \tag{1.34}$$

where the $A_p(\Phi; y)$ are functions of y. On the other hand, given $f \in S_{\ell+m,k}(\Gamma, \omega, \chi)$, the Fourier expansion of the cusp form $\mathcal{L}_g^* f$ in $S_m(\Gamma)$ can be written in the form

$$\mathcal{L}_g^* f(z) = \sum_{p=1}^{\infty} A_p(\mathcal{L}_g^* f)e^{2\pi i p z/b} \tag{1.35}$$

for some constants $A_p(\mathcal{L}_g^* f) \in \mathbb{C}$.

Theorem 1.15 *Given* $g \in S_{\ell,k}(\Gamma, \omega, \chi)$, *let* f *be a mixed cusp form in* $S_{\ell+m,k}(\Gamma, \omega, \chi)$, *and let* $A_n(\mathcal{L}_g^* f)$ *be the n-th Fourier coefficient of the cusp form* $\mathcal{L}_g^* f$ *in* $S_m(\Gamma)$ *as in* (1.35). *Then we have*

$$A_n(\mathcal{L}_g^* f) = \frac{(4\pi n)^{m-1}}{\Gamma(m-1)b^{m-1}}\int_0^{\infty} A_n(\Phi; y)e^{-2\pi n y/b}y^{m-2}dy, \tag{1.36}$$

where $A_n(\Phi; y)$ *is the n-th Fourier coefficient of* $\Phi(z) = f(z)\overline{g(z)}y^{\ell}(\operatorname{Im}\omega(z))^{k}$ *regarded as a function of* y *as in* (1.34).

Proof. Let $\varPhi(z)$ be as above, and let $P_{m,n}$ be the Poincaré series in (1.31). We set

$$I = \int_D \varPhi(z)\overline{P_{m,n}(z)}y^m dV = \frac{b^m \varGamma(m-1)}{(4\pi n)^{m-1}} A_n(\mathcal{L}_g^* f),\qquad(1.37)$$

where $D \subset \mathcal{H}$ is a fundamental domain of $\varGamma$. Using the fact that

$$dV = y^{-2}dxdy = \mathrm{Im}(z)^{-2}(i/2)dz \wedge d\bar{z},$$

we have

$$I = \sum_{\gamma\in\varGamma_\infty\backslash\varGamma} \int_D \varPhi(z)e^{-2\pi in\overline{\gamma z}/b}\,\overline{j(\gamma,z)}^{-m}\,\mathrm{Im}(z)^{m-2}(i/2)dz \wedge d\bar{z}.$$

In terms of the variable $w = \gamma z = u + iv$, a typical term in the above sum becomes

$$\int_D \varPhi(z)e^{-2\pi in\overline{\gamma z}/b}\,\overline{j(\gamma,z)}^{-m}\,\mathrm{Im}(z)^{m-2}(i/2)\,dz \wedge d\bar{z}\qquad(1.38)$$

$$= \int_{\gamma D} \varPhi(\gamma^{-1}w)e^{-2\pi in\overline{w}/b}\,\overline{j(\gamma,\gamma^{-1}w)}^{-m}$$

$$\times \mathrm{Im}(\gamma^{-1}w)^{m-2}(i/2)\,d(\gamma^{-1}w) \wedge d(\overline{\gamma^{-1}w}).$$

However, using (1.2) and (1.38), we see that

$$\varPhi(\gamma^{-1}w) = j(\gamma^{-1},w)^m\varPhi(w), \quad \overline{j(\gamma,\gamma^{-1}w)}^{-m} = \overline{j(\gamma^{-1},w)}^{-m},$$

$$\mathrm{Im}(\gamma^{-1}w)^{m-2} = |j(\gamma^{-1},w)|^{-2m+4}\,\mathrm{Im}(w)^{m-2}$$

$$= j(\gamma^{-1},w)^{-m+2}\,\overline{j(\gamma^{-1},w)}^{-m+2}\,\mathrm{Im}(w)^{m-2},$$

and

$$(i/2)\,d(\gamma^{-1}w) \wedge d(\overline{\gamma^{-1}w}) = (i/2)\,j(\gamma^{-1},w)^{-2}dw \wedge \overline{j(\gamma^{-1},w)}^{-2}d\overline{w}$$

$$= j(\gamma^{-1},w)^{-2}\,\overline{j(\gamma^{-1},w)}^{-2}dudv.$$

Thus by substituting these into (1.38) we obtain

$$\int_D \varPhi(z)e^{-2\pi in\overline{\gamma z}/b}\,\overline{j(\gamma,z)}^{-m}\,\mathrm{Im}(z)^{m-2}(i/2)\,dz \wedge d\bar{z}$$

$$= \int_{\gamma D} \varPhi(w)e^{-2\pi in\overline{w}/b}\,\mathrm{Im}(w)^{m-2}dudv,$$

and hence the integral I in (1.37) can be written in the form

$$I = \int_{D'} \varPhi(z)e^{-2\pi in\bar{z}/b}y^{m-2}dxdy,$$

where

$$D' = \bigcup \{\gamma D \mid \gamma \in \Gamma_\infty \backslash \Gamma\}. \tag{1.39}$$

From (1.39) we see easily that D' is a fundamental domain of Γ_∞, and hence it can be written as

$$D' = \{z \in \mathcal{H} \mid 0 \le \operatorname{Re} z \le b\}.$$

Therefore it follows that

$$I = \int_0^\infty \int_0^b \Phi(z) e^{-2\pi i n \bar{z}/b} y^{m-2} dx dy.$$

Now by using the Fourier expansion of $\Phi(z)$ in (1.34) we obtain

$$I = \sum_{k=0}^\infty \int_0^b e^{2\pi i (k-n) x/b} dx \int_0^\infty A_k(\Phi; y) e^{-2\pi n y/b} y^{m-2} dy$$

$$= b \int_0^\infty A_n(\Phi; y) e^{-2\pi n y/b} y^{m-2} dy.$$

Thus, using this and (1.37), we see that

$$A_n(\mathcal{L}_g^* f) = \frac{(4\pi n)^{m-1}}{b^m \Gamma(m-1)} I$$

$$= \frac{(4\pi n)^{m-1}}{\Gamma(m-1) b^{m-1}} \int_0^\infty A_n(\Phi; y) e^{-2\pi n y/b} y^{m-2} dy,$$

and hence the proof of the theorem is complete.   □

We now want to establish relations among the Fourier coefficients of f and g and those of the image $\mathcal{L}_g^* f \in S_m(\Gamma)$ of a mixed cusp form f in $S_{\ell+m,k}(\Gamma, \omega, \chi)$ under the map $\mathcal{L}_g^*$ associated to $g \in S_{\ell,k}(\Gamma, \omega, \chi)$. First, we assume that the mixed cusp forms f and g have Fourier expansions of the form

$$f(z) = \sum_{\rho=1}^\infty B(\rho) e^{2\pi i \rho z/b}, \qquad g(z) = \sum_{\mu=1}^\infty C(\mu) e^{2\pi i \mu z/b} \tag{1.40}$$

(see [18]). Note that the homomorphism χ maps parabolic elements to parabolic elements. Therefore we may assume that

$$\chi \begin{pmatrix} 1 & b \\ 0 & 1 \end{pmatrix} = \begin{pmatrix} 1 & b_\chi \\ 0 & 1 \end{pmatrix}$$

for some $b_\chi \in \mathbb{R}$. Thus, if $\psi(z) = (\operatorname{Im} \omega(z))^k$ and if $T_b = \begin{pmatrix} 1 & b \\ 0 & 1 \end{pmatrix}$, then $j(\chi(T_b), \omega(z)) = 1$ and

$$\psi(z+b) = (\operatorname{Im} \omega(T_b z))^k = (\operatorname{Im} \chi(T_b) \omega(z))^k$$

$$= |j(\chi(T_b), \omega(z))|^{-2k} (\operatorname{Im} \omega(z))^k = \psi(z)$$

for all $z \in \mathcal{H}$. Hence the real-valued function $\psi(z)$ has a Fourier expansion of the form

$$(\operatorname{Im} \omega(z))^k = \sum_{\nu \in \mathbb{Z}} F(\nu; y) e^{2\pi i \nu x/b}, \tag{1.41}$$

where the $F(\nu; y)$ are functions of y.

Theorem 1.16 *Given $f \in S_{\ell+m,k}(\Gamma, \omega, \chi)$ and $g \in S_{\ell,k}(\Gamma, \omega, \chi)$, the n-th Fourier coefficient $A_n(\mathcal{L}_g^* f)$ of the cusp form $\mathcal{L}_g^* f$ in $S_m(\Gamma)$ can be written in the form*

$$A_n(\mathcal{L}_g^* f) = \frac{n^{m-1} b^\ell}{\Gamma(m-1)(4\pi)^\ell} \times \sum_{\mu,\nu} \frac{B(\mu+n-\nu)\overline{C(\mu)}}{(\mu+n-\nu/2)^{\ell+m-1}}$$

$$\times \int_0^\infty F\left(\nu; \frac{bt}{4\pi(\mu+n-\nu/2)}\right) t^{\ell+m-2} e^{-t} dt,$$

where B, C and F are as in (1.40) and (1.41).

Proof. In terms of the Fourier coefficients of f, g and $(\operatorname{Im} \omega(z))^k$ in (1.40) and (1.41) the function $\Phi(z) = f(z) \overline{g(z)} y^\ell (\operatorname{Im} \omega(z))^k$ can be written in the form

$$\Phi(z) = y^\ell \sum_{\rho,\mu,\nu} B(\rho) \overline{C(\mu)} F(\nu; y) e^{2\pi i(\nu+\rho-\mu)x/b} e^{-2\pi(\rho+\mu)y/b}.$$

Thus, using $\nu + \rho - \mu = n$, we have $\rho = \mu + n - \nu$, and by (1.34) the n-th Fourier coefficient of $\Phi(z)$ is given by

$$A_n(\Phi; y) = y^\ell \sum_{\mu,\nu} B(\mu+n-\nu) \overline{C(\mu)} F(\nu; y) e^{-2\pi(2\mu+n-\nu)y/b}.$$

Substituting this into (1.36), we obtain

$$A_n(\mathcal{L}_g^* f) = \frac{(4\pi n)^{m-1}}{\Gamma(m-1)b^{m-1}}$$

$$\times \int_0^\infty y^{\ell+m-2} \sum_{\mu,\nu} B(\mu+n-\nu) \overline{C(\mu)} F(\nu; y) e^{-4\pi(\mu+n-\nu/2)y/b} dy.$$

Now the the theorem follows by expressing the above integral in terms of the new variable $t = 4\pi(\mu+n-\nu/2)y/b$. $\qquad \square$

Remark 1.17 *If $k = 0$, the map $\mathcal{L}_g^* : S_{\ell+m,k}(\Gamma, \omega, \chi) \to S_m(\Gamma)$ becomes a map from $S_{\ell+m}(\Gamma)$ to $S_m(\Gamma)$. If in addition $\Gamma = SL(2, \mathbb{Z})$, then $b = 1$ and the formula for $A_n(\mathcal{L}_g^* f)$ given in Theorem 1.16 reduces to the one obtained by Kohnen in [53].*

1.4 Mixed Automorphic Forms and Differential Equations

In this section we discuss mixed automorphic forms associated to a certain class of linear ordinary differential equations and establish a relation between the monodromy of such a differential equation and the periods of the corresponding mixed automorphic forms.

Let X be a Riemann surface regarded as a smooth complex algebraic curve over $\mathbb{C}$, and consider a linear ordinary differential operator

$$\Lambda^n = \frac{d^n}{dx^n} + P_{n-1}(x)\frac{d^{n-1}}{dx^{n-1}} + \cdots + P_1(x)\frac{d}{dx} + P_0(x)$$

of order n on X, where x is a nonconstant element of the function field $K(X)$ of X and $P_i \in K(X)$ for $0 \le i \le n-1$. We assume that Λ^n has only regular singular points. Let $U \subset X$ be a Zariski open set on which the functions P_i are all regular, and choose a base point $x_0 \in U$. Let $\omega_1, \ldots, \omega_n$ be holomorphic functions which form a basis of the space of solutions of the equation $\Lambda^n f = 0$ near x_0. Given a closed path γ that determines an element of the fundamental group $\pi_1(U, x_0)$, there is a matrix $M(\gamma) \in GL_n(\mathbb{C})$ such that the analytic continuation of the solution ω_i becomes $\sum_{j=1}^n m_{ij}\omega_j$ for each $i \in \{1, \ldots, n\}$. Thus we obtain a representation $M : \pi_1(U, x_0) \to GL_n(\mathbb{C})$ of the fundamental group of U called the *monodromy representation*.

Let V_{x_0} be the space of local solutions of the equation $\Lambda^n f = 0$ near $x_0 \in U$, and consider an element ψ of $K(X)$. By shrinking the Zariski open set U if necessary, we may assume that ψ is regular on U. Then the monodromy representation M determines an action of the fundamental group $\pi_1(U, x_0)$ on V_{x_0}. We shall determine an element of the cohomology $H^1(\pi_0(U, x_0), V_{x_0})$ of $\pi_1(U, x_0)$ with coefficients in V_{x_0} associated to $\psi \in K(X)$. Let f^ψ be a solution of the nonhomogeneous equation $\Lambda^n f = \psi$, and suppose that the solution of $\Lambda^n f = \psi$ obtained by the analytic continuation of f^ψ around a closed path $\gamma \in \pi_1(U, x_0)$ is given by

$$f^\psi + a_{\gamma,1}^\psi \omega_1 + \cdots + a_{\gamma,n}^\psi \omega_n = f^\psi + {}^t\mathbf{a}_\gamma^\psi \boldsymbol{\omega},$$

where

$$ {}^t\mathbf{a}_\gamma^\psi = (a_{\gamma,1}^\psi, \ldots, a_{\gamma,n}^\psi), \quad {}^t\boldsymbol{\omega} = \omega_1, \ldots, \omega_n) \tag{1.42}$$

with $a_{\gamma,1}^\psi, \ldots, a_{\gamma,n}^\psi \in \mathbb{C}$. If $\tau \in \pi_1(U, x_0)$ is another closed path, then the analytic continuation of $f^\psi + {}^t\mathbf{a}_\gamma^\psi \boldsymbol{\omega}$ around τ becomes

$$f^\psi + {}^t\mathbf{a}_\tau^\psi \boldsymbol{\omega} + {}^t\mathbf{a}_\gamma^\psi M(\tau).$$

On the other hand, since the analytic continuation of f^ψ around $\gamma\tau$ is $f^\psi + {}^t\mathbf{a}_{\gamma\tau}^\psi \boldsymbol{\omega}$, it follows that

$$^t\mathbf{a}_{\gamma\tau}^{\psi} = \,^t\mathbf{a}_{\tau}^{\psi} + \,^t\mathbf{a}_{\gamma}^{\psi} M(\tau).$$

Thus the map $\gamma \mapsto \,^t\mathbf{a}_{\gamma}^{\psi}$ is a cocycle, and this cocycle is independent of the choice of the solution f^{ψ} up to coboundary (cf. [120]). Note that each $^t\mathbf{a}_{\gamma}^{\psi}$ determines an element $^t\mathbf{a}_{\gamma}^{\psi}\omega$ of V_{x_0}. Consequently $^t\mathbf{a}_{\gamma}^{\psi}$ can be regarded as an element of V_{x_0}, and hence it defines an element of the cohomology $H^1(\pi_1(U, x_0), V_{x_0})$.

Let $\Gamma \subset SL(2, \mathbb{R})$ be a Fuchsian group of the first kind that does not contain elements of finite order. Then the quotient $X = \Gamma \backslash \mathcal{H}^*$ is a compact Riemann surface, where $\mathcal{H}^*$ is the union of $\mathcal{H}$ and the cusps of Γ, and can be regarded as a smooth algebraic curve over $\mathbb{C}$. Let x be a nonconstant element of the function field $K(X)$ of X when X is regarded as an algebraic curve over $\mathbb{C}$. We consider a second order linear differential equation $\Lambda_X^2 f = 0$ with

$$\Lambda_X^2 = \frac{d^2}{dx^2} + P_X(x)\frac{d}{dx} + Q_X(x), \qquad (1.43)$$

where $P_X(x)$ and $Q_X(x)$ are elements of $K(X)$. We assume that the differential equation $\Lambda_X^2 f = 0$ has only regular singular points and that the set of singular points coincides with the set of cusps of Γ. Thus, if X_0 is the set of regular points of $\Lambda_X^2 f = 0$ and if $x_0 \in X_0$, then X_0 can be regarded as the quotient space $\Gamma \backslash \mathcal{H}$ and the fundamental group $\pi_1(X_0, x_0)$ of X_0 can be identified with Γ. Let $\chi : \Gamma \to GL(2, \mathbb{C})$ be the monodromy representation for the differential equation $\Lambda_X^2 f = 0$, and assume that $\chi(\Gamma) \subset SL(2, \mathbb{R})$.

Let ω_1 and ω_2 be linearly independent solutions of $\Lambda_X^2 f = 0$, and for each positive integer m let $S^m(\Lambda^2)$ be the linear ordinary differential operator of order $m + 1$ such that the $m + 1$ functions

$$\omega_1^m, \omega_1^{m-1}\omega_2, \dots, \omega_1\omega_2^{m-1}, \omega_2^m$$

are linearly independent solutions of the corresponding homogeneous equation $S^m(\Lambda_X^2)f = 0$.

By pulling back the operator in (1.43) via the natural projection $\mathcal{H}^* \to X = \Gamma \backslash \mathcal{H}^*$ we obtain a differential operator

$$\Lambda^2 = \frac{d^2}{dz^2} + P(z)\frac{d}{dz} + Q(z) \qquad (1.44)$$

such that $P(z)$ and $Q(z)$ are meromorphic functions on $\mathcal{H}^*$. Let $\omega_1(z)$ and $\omega_2(z)$ for $z \in \mathcal{H}$ be the two linearly independent solutions of $\Lambda^2 f = 0$ corresponding to ω_1 and ω_2 above. Then the monodromy representation for the differential equation $\Lambda^2 f = 0$ is the group homomorphism $\chi : \Gamma \to SL(2, \mathbb{R})$ defined as follows. Given elements $\gamma \in \Gamma$ and $z \in \mathcal{H}$, we assume that the elements $\omega_1(\gamma z), \omega_2(\gamma z) \in \mathcal{H}$ can be written in the form

$$\omega_1(\gamma z) = a_\chi \omega_1(z) + b_\chi \omega_2(z), \quad \omega_2(\gamma z) = c_\chi \omega_1(z) + d_\chi \omega_2(z). \qquad (1.45)$$

Then the image of $\gamma \in \Gamma$ under the monodromy representation χ is given by

$$\chi(\gamma) = \begin{pmatrix} a_\chi & b_\chi \\ c_\chi & d_\chi \end{pmatrix} \in SL(2, \mathbb{R}). \tag{1.46}$$

We now set

$$\omega(z) = \omega_1(z)/\omega_2(z), \tag{1.47}$$

and assume that $\omega(z) \in \mathcal{H}$ for all $z \in \mathcal{H}$. Then the resulting map $\omega : \mathcal{H} \to \mathcal{H}$ is the period map for the differential equation $\Lambda^2 f = 0$, and by (1.45) it satisfies

$$\omega(\gamma z) = \frac{a_\chi \omega(z) + b_\chi}{c_\chi \omega(z) + d_\chi} = \chi(\gamma)\omega(z)$$

for all $z \in \mathcal{H}$ and $\gamma \in \Gamma$. Thus the maps ω and χ form an equivariant pair, and we may consider the associated mixed automorphic or cusp forms as in Section 1.1. Note, however, in this section we deal with meromorphic functions rather than holomorphic functions on $\mathcal{H}$, so we need to modify Definition 1.3 by replacing holomorphic functions with meromorphic functions.

Definition 1.18 Let $f : \mathcal{H} \to \mathbb{C}$ be a mixed automorphic form of type (μ, ν) associated to Γ, ω and χ with $\mu \geq 2$. Then for $z_0 \in \mathcal{H}$ and $\gamma \in \Gamma$ the integrals of the form

$$\int_{z_0}^{\gamma z_0} f(z)\omega'(z)^{1-\mu/2}\omega(z)^i dz, \qquad i = 0, 1, \dots, \mu + \nu - 2$$

are called the periods of f.

If $S^m(\Lambda^2)$ is the differential operator acting on the functions on $\mathcal{H}$ obtained by pulling back $S^m(\Lambda_X^2)$ via the projection $\mathcal{H}^* \to X$, then the solutions of the equation $S^m(\Lambda^2)f = 0$ are of the form

$$\sum_{i=0}^{m} c_i \omega_1(z)^{m-i} \omega_2(z)^i \tag{1.48}$$

for some constants $c_0, \dots, c_m$. Let ψ be a meromorphic function on $\mathcal{H}^*$ corresponding to an element ψ_X in $K(X)$, and let f^ψ be a solution of the non-homogeneous equation

$$S^m(\Lambda^2)f = \psi. \tag{1.49}$$

Given a nonnegative integer k, we set

$$\Phi_k^\psi(z) = \omega'(z)^k \frac{d^{m+1}}{d\omega(z)^{m+1}} \left(\frac{f^\psi(z)}{\omega_2(z)^m} \right) \tag{1.50}$$

for all $z \in \mathcal{H}$.

Lemma 1.19 The function Φ_k^ψ in (1.50) is independent of the choice of the solution f^ψ of the differential equation in (1.49).

Proof. Let f^ψ and h^ψ of the differential equation in (1.49). Then, using (1.48), we see that

$$h^\psi(z) = f^\psi(z) + \sum_{i=0}^{m} c_i \omega_1(z)^{m-i} \omega_2(z)^i$$

for some $c_0, \dots, c_m \in \mathbb{C}$. Thus by using (1.47) we have

$$\omega'(z)^k \frac{d^{m+1}}{d\omega(z)^{m+1}} \left(\frac{h^\psi(z)}{\omega_2(z)^m} \right) = \omega'(z)^k \frac{d^{m+1}}{d\omega(z)^{m+1}} \left(\frac{f^\psi(z)}{\omega_2(z)^m} + \sum_{i=0}^{m} c_i \omega(z) \right)$$

$$= \omega'(z)^k \frac{d^{m+1}}{d\omega(z)^{m+1}} \left(\frac{f^\psi(z)}{\omega_2(z)^m} \right),$$

which proves the lemma. $\square$

Theorem 1.20 *(i) The function $\Phi_k^\psi(z)$ is a mixed automorphic form of type $(2k, m - 2k + 2)$ associated to Γ, ω and the monodromy representation χ.*

(ii) If z_0 is a fixed point in $\mathcal{H}$, then a solution f^ψ of the equation $S^m(\Lambda^2)f = \psi$ is of the form

$$f^\psi(z) = \frac{\omega_2(z)^m}{m!} \int_{z_0}^{z} \Phi_k^\psi(z) \omega'(z)^{-k} (\omega(z) - \omega(t))^m \, d\omega(t) \qquad (1.51)$$

$$+ \sum_{i=1}^{m+1} c_i \omega_1(z)^{m+1-i} \omega_2(z)^{i-1}$$

for some constants $c_1, \dots, c_{m+1}$.

Proof. It is known that the function

$$\frac{d^{m+1}}{d\omega(z)^{m+1}} \left(\frac{f^\psi(z)}{\omega_2(z)^m} \right)$$

is a mixed automorphic form of type $(0, m + 2)$ associated to Γ, ω and χ (cf. [120, p. 32]). On the other hand, for $\gamma = \left(\begin{smallmatrix} a & b \\ c & d \end{smallmatrix} \right) \in \Gamma$ and $\chi(\gamma)$ as in (1.46) we have

$$\frac{d(\gamma z)}{dz} = \frac{d}{dz} \left(\frac{az + d}{cz + d} \right) = \frac{1}{(cz + d)^2},$$

$$\frac{d\omega(\gamma z)}{d\omega(z)} = \frac{d(\chi(\gamma)\omega(z))}{d\omega(z)} = \frac{1}{(c_\chi z + d_\chi)^2}.$$

Thus we obtain

$$\Phi_k^\psi(\gamma z) = \left(\frac{d\omega(\gamma z)}{d\omega(z)} \right)^k \left(\frac{d(\gamma z)}{dz} \right)^{-k} \left(\frac{d\omega(z)}{dz} \right)^k \frac{d^{m+1}}{d\omega(z)^{m+1}} \left(\frac{f^\psi(z)}{\omega_2(z)^m} \right)(z)$$

$$= (cz + d)^{2k} (c_\chi z + d_\chi)^{m-2k+2} \Phi_k^\psi(z),$$

and therefore (i) follows. As for (ii), we have

$$\int_{z_0}^z \Phi_k^\psi(z)\omega'(z)^{-k}(\omega(z) - \omega(t))^m d\omega(t)$$

$$= \int_{z_0}^z \frac{d^{m+1}}{d\omega^{m+1}} \left(\frac{f^\psi(z)}{\omega_2(z)^m} \right) (\omega(z) - \omega(t))^m d\omega(t)$$

$$= m! \left(\frac{f^\psi(z)}{\omega_2(z)^m} \right)^{m+1} + \sum_{i=1}^{m+1} c_i' \omega(z)^{m+1-i}$$

for some constants $c_1', \ldots, c_{m+1}'$ by applying the integration by parts m times. Now using $c_i = -c_i'/(m!)$ and $\omega = \omega_1/\omega_2$, we obtain the desired formula in (1.51). $\square$

Let Λ^2 be the second-order differential operator in (1.44) with monodromy representation $\chi : \Gamma \to SL(2, \mathbb{R})$ described above. Then the monodromy representation of $S^m(\Lambda^2)$ is given by $S^m\chi = \mathrm{Sym}^m \circ \chi$, where $\mathrm{Sym}^m : SL(2, \mathbb{R}) \to SL(m + 1, \mathbb{R})$ is the m-th symmetric power representation.

Let V_{x_0} be the space of local solutions of the equation $S^m(\Lambda^2)f = 0$ near $x_0 \in X_0$, and for $\psi \in K(X)$ let ${}^t\mathbf{a}_\gamma^\psi$ be the cocycle in $H^1(\Gamma, V_{x_0})$ associated to a solution f^ψ of $S^m(\Lambda^2)f = \psi$ in (1.42). Then we can express the cocycle ${}^t\mathbf{a}_\gamma^\psi$ in terms of the periods of the mixed automorphic form Φ_k^ψ and the monodromy representation $S^m\chi$ of $S^m(\Lambda^2)f = 0$ as follows:

Theorem 1.21 Let Φ_k^ψ be the mixed automorphic form of type $(2k, m - 2k + 2)$ associated to Γ, ω and χ determined by a solution of f^ψ of $S^m(\Lambda^2)f = \psi$, and let z_0 be a fixed point in $\mathcal{H}$. Then we have

$$\left[{}^t\mathbf{a}_\gamma^\psi \right] = \left[(\Xi_{\gamma,1}^\psi, \ldots, \Xi_{\gamma,m+1}^\psi) \cdot S^m\chi(\gamma) \right]$$

for each $\gamma \in \Gamma$, where

$$\Xi_{\gamma,\nu}^\psi = (-1)^{\nu-1}(m!)^{-1} \binom{m}{\nu-1} \int_{z_0}^{\gamma z_0} \Phi_k^\psi(z)\omega'(z)^{1-k}\omega(z)^{\nu-1} dz$$

for $1 \leq \nu \leq m + 1$; here the square brackets denote the cohomology class in $H^1(\Gamma, V_{x_0})$.

Proof. Using [120, Proposition 3 bis. 10] and the proof of [120, Theorem 3 bis. 17], we obtain

$${}^t\mathbf{a}_\gamma^\psi = (\Xi_{\gamma,1}^\psi, \ldots, \Xi_{\gamma,m+1}^\psi) \cdot S^m\chi(\gamma) + (c_1, \ldots, c_{m+1}) \cdot (S^m\chi(\gamma) - I_{m+1})$$

for some constants $c_1, \ldots, c_{m+1}$, where I_{m+1} is the $(m+1) \times (m+1)$ identity matrix and

$$\Xi_{\gamma,\nu}^{\psi} = (-1)^{\nu-1}(m!)^{-1} \binom{m}{\nu-1}$$
$$\times \int_{z_0}^{\gamma z_0} \frac{d^{m+1}}{d\omega(z)^{m+1}} \left(\frac{f^{\psi}(z)}{\omega_2(z)^m} \right) \omega'(z)^{k-1} \omega(z)^{\nu-1} d\omega(z)$$

for $1 \leq \nu \leq m+1$. However, we have

$$\frac{d^{m+1}}{d\omega(z)^{m+1}} \left(\frac{f^{\psi}(z)}{\omega_2(z)^m} \right) d\omega(z) = \Phi_k^{\psi}(z)\omega'(z)^{-k} d\omega(z) = \Phi_k^{\psi}(z)\omega'(z)^{1-k} dz.$$

Now the theorem follows from the fact that

$$(c_1, \ldots, c_{m+1}) \cdot (S^m \chi(\gamma) - I_{m+1})$$

is a coboundary in $H^1(\Gamma, V_{x_0})$.   □

Line Bundles and Elliptic Varieties

An elliptic surface is the total space of a fiber bundle over a Riemann surface whose generic fiber is an elliptic curve, and it was Hunt and Meyer [43] who observed that a holomorphic form of degree two on an elliptic surface can be interpreted as a mixed cusp form of type $(2,1)$. An elliptic variety, on the other hand, can be constructed by considering a fiber bundle whose generic fiber is the product of a finite number of elliptic curves, and a holomorphic form of the highest degree on an elliptic variety can be identified with a mixed cusp form of more general type (cf. [68, 18]). In this chapter we discuss certain aspects of mixed automorphic forms of one variable that are related to their geometric connections with elliptic varieties.

It is well-known that automorphic forms for a discrete subgroup $\Gamma \subset SL(2,\mathbb{R})$ can be identified with sections of a line bundle over the Riemann surface $X = \Gamma \backslash \mathcal{H}$. Such an interpretation can be extended to the case of mixed automorphic forms. Thus a mixed automorphic form associated to Γ and an equivariant pair (ω, χ) can be regarded as a section of a line bundle over X determined by the given equivariant pair. This identification can in turn be used to establish a correspondence between mixed automorphic forms and holomorphic forms of the highest degree on an elliptic variety over X.

If f is a cusp form of weight w for a discrete subgroup $\Gamma \subset SL(2,\mathbb{R})$, the periods of f are given by the integrals

$$\int_0^{i\infty} f(z) z^k dz$$

with $0 \leq k \leq w - 2$, and it is well-known that such periods of cusp forms are closely related to the values at the integer points in the critical strip of the Hecke L-series (see e.g. [51, 104]). In [22] Eichler discovered certain relations among the periods of cusp forms, which were extended later by Shimura [112]; these relations are called Eichler-Shimura relations. More explicit connections between the Eichler-Shimura relations and the Fourier coefficients of cusp forms were found by Manin [91]. On the other hand, if f is a mixed cusp form of type $(2, m)$ associated to Γ and an equivariant pair (ω, χ), then the periods of f are the integrals

$$\int_0^{i\infty} f(z) \omega(z)^k dz$$

with $0 \leq k \leq m$. The Eichler-Shimura relations for these periods can be obtained by regarding mixed cusp forms as holomorphic forms on an elliptic variety.

In Section 2.1 we construct line bundles over a Riemann surface whose sections can be identified with mixed automorphic forms. Such an identification is used in Section 2.2 to establish a correspondence between mixed cusp forms and holomorphic forms of the highest degree on an elliptic variety. Section 2.3 describes some of the properties of the modular symbols and periods of mixed cusp forms, which are used in Section 2.4 to obtain Eichler-Shimura relations for mixed cusp forms.

2.1 Mixed Cusp Forms and Line Bundles

We shall adopt the notations used in Section 1.1. In particular, $\Gamma \subset SL(2,\mathbb{R})$ is a Fuchsian group of the first kind, and the holomorphic map $\omega : \mathcal{H} \to \mathcal{H}$ and the homomorphism $\chi : \Gamma \to SL(2,\mathbb{R})$ form an equivariant pair. Recall also that $\mathcal{H}^* = \mathcal{H} \cup \Sigma$ with Σ being the set of cusps for Γ and that the quotient $X = \Gamma \backslash \mathcal{H}^*$ has the structure of a compact Riemann surface. In this section we establish an isomorphism between the space $S_{k,\ell}(\Gamma, \omega, \chi)$ of mixed cusp forms of type (k, ℓ) associated to Γ, ω, and χ and the space of sections of a certain line bundle over the Riemann surface X.

If $\mathcal{S}_{\mathcal{H}^*}$ is a sheaf on $\mathcal{H}^*$ on which Γ acts on the right, the associated Γ-fixed sheaf $(\mathcal{S}_{\mathcal{H}^*})^\Gamma$ can be constructed by defining $(\mathcal{S}_{\mathcal{H}^*})^\Gamma$ for each open subset $U \subset X$ to be the space of Γ-invariant elements of $\mathcal{S}_{\mathcal{H}^*}(\pi^{-1}(U))$, where $\pi : \mathcal{H}^* \to X$ is the natural projection map. Thus we may write

$$(\mathcal{S}_{\mathcal{H}^*})^\Gamma(U) = (\mathcal{S}_{\mathcal{H}^*}(\pi^{-1}(U)))^\Gamma. \tag{2.1}$$

If $s \in \Sigma \subset \mathcal{H}^*$ is a cusp of Γ and if $\Gamma_s = \{\gamma \in \Gamma \mid \gamma s = s\}$, then the stalk $(\mathcal{S}_{\mathcal{H}^*})^\Gamma_{\pi(s)}$ of $(\mathcal{S}_{\mathcal{H}^*})^\Gamma$ over $\pi(s) \in X$ can be identified with the Γ_s-invariant elements of the stalk $\mathcal{S}_{\mathcal{H}^*,s}$ of $\mathcal{S}_{\mathcal{H}^*}$ over s, that is,

$$(\mathcal{S}_{\mathcal{H}^*})^\Gamma_{\pi(s)} = (\mathcal{S}_{\mathcal{H}^*,s})^\Gamma_s \tag{2.2}$$

(see [6, Proposition 0.2]).

Let $\mathcal{O}_\mathcal{H}$ be the sheaf of germs of holomorphic functions on $\mathcal{H}$, and let $\eta : \mathcal{H} \to \mathcal{H}^*$ denote the natural inclusion map. Then the direct image sheaf $\eta_* \mathcal{O}_\mathcal{H}$ is an extension of $\mathcal{O}_\mathcal{H}$ to $\mathcal{H}^*$ such that its stalk at each $s \in \Sigma$ is given by

$$(\eta_* \mathcal{O}_\mathcal{H})_s = \varinjlim_U \mathcal{O}_\mathcal{H}(U \cap \mathcal{H}),$$

where U runs through the set of open neighborhoods of s in $\mathcal{H}^*$. Thus an element of $(\eta_* \mathcal{O}_\mathcal{H})_s$ with $s \in \Sigma$ is the germ of a section $f : (U - \{s\}) \to \mathbb{C}$ of $\mathcal{O}_\mathcal{H}$ on a punctured neighborhood $U - \{s\}$ of s. We denote by $\mathcal{O}_{\mathcal{H}^*}$ the subsheaf of $\eta_* \mathcal{O}_\mathcal{H}$ such that the stalk at each $s \in \Sigma$ is given by

$$\mathcal{O}_{\mathcal{H}^*,s} = \{f \in (\eta_* \mathcal{O}_{\mathcal{H}})_s \mid f(\sigma^{-1}z) = O(|z|^k) \quad \text{for some} \quad k \in \mathbb{Z}\}, \quad (2.3)$$

where σ is an element of $SL(2, \mathbb{R})$ with $\sigma s = \infty$. Thus the germs at s belonging to $\mathcal{O}_{\mathcal{H}^*}$ are the elements of $\eta_* \mathcal{O}_{\mathcal{H}}$ that are meromorphic at s. The left action of Γ on $\mathcal{H}^*$ induces a right action of Γ on $\mathcal{O}_{\mathcal{H}^*}$ given by

$$(f \cdot \gamma)(z) = f(\gamma z)$$

for all $z \in U$ and $\gamma \in \Gamma$, where $f : U \to \mathbb{C}$ represents a germ belonging to $\mathcal{O}_{\mathcal{H}^*}$. We denote by $(\mathcal{O}_{\mathcal{H}^*})^\Gamma$ the Γ-fixed sheaf on X associated to $\mathcal{O}_{\mathcal{H}^*}$.

Lemma 2.1 *The sheaf $(\mathcal{O}_{\mathcal{H}^*})^\Gamma$ can be identified with the sheaf $\mathcal{O}_X$ of holomorphic functions on X.*

Proof. Given an open subset $U \subset X$, by (2.1) a section $f \in (\mathcal{O}_{\mathcal{H}^*})^\Gamma(U)$ of $(\mathcal{O}_{\mathcal{H}^*})^\Gamma$ on U is a Γ-invariant function on $\pi^{-1}(U)$. Thus f determines a function $\tilde{f}$ on U. Since f is holomorphic on $\pi^{-1}(U) \cap \mathcal{H}$, we see that $\tilde{f}$ is holomorphic on $U \cap X_0$, where $X_0 = \Gamma \backslash \mathcal{H}$. If $s \in \pi^{-1}(U)$ is a cusp of Γ with $\sigma s = \infty$ and if the subgroup $\sigma \Gamma_s \sigma^{-1} \subset SL(2, \mathbb{R})$ is generated by $\left(\begin{smallmatrix} 1 & h \\ 0 & 1 \end{smallmatrix}\right)$, then the function $z \mapsto f(\sigma^{-1}z)$ has a Fourier expansion of the form

$$f(\sigma^{-1}z) = \sum_{n=-\infty}^{\infty} a_n e^{2\pi i n z/h}.$$

By (2.3) we see that f satisfies

$$f(\sigma^{-1}z) = O(|z|^k)$$

for some $k \in \mathbb{Z}$. However, we have

$$e^{2\pi i z/h} = O(1), \quad e^{-2\pi i z/h} \neq O(|z|^\ell)$$

for every $\ell \in \mathbb{Z}$; hence it follows that $a_n = 0$ for $n < 0$. Thus f is holomorphic at s, and the proof of the lemma is complete. $\square$

Let $\Omega^1_{\mathcal{H}^*}$ be the sheaf of holomorphic 1-forms on $\mathcal{H}^*$. Then $\Omega^1_{\mathcal{H}^*}$ is an $\mathcal{O}_{\mathcal{H}^*}$-module, and we may write

$$\Omega^1_{\mathcal{H}^*} = d\mathcal{O}_{\mathcal{H}^*} = \mathcal{O}_{\mathcal{H}^*} dz.$$

The group Γ acts on $\Omega^1_{\mathcal{H}^*}$, and therefore we can consider the associated Γ-fixed sheaf $(\Omega^1_{\mathcal{H}^*})^\Gamma$ on X. We denote by $\Omega^1(\Sigma)$ the sheaf of meromorphic 1-forms on X that are holomorphic on $X_0 = \Gamma \backslash \mathcal{H}$ and have a pole of order at most 1 at the cusps.

Lemma 2.2 *The sheaf $(\Omega^1_{\mathcal{H}^*})^\Gamma$ on X coincides with $\Omega^1(\Sigma)$.*

Proof. Let $s \in \Sigma \subset \mathcal{H}^*$ be a cusp of Γ with $\sigma s = \infty$. If $q_s = e^{2\pi i \sigma z/h}$, then we have

$$dq_s = (2\pi i/h)q_s d(\sigma z), \quad d(\sigma z) = \frac{h}{2\pi i}q_s^{-1}dq_s.$$

Using this and (2.2), we see that

$$(\Omega^1_{\mathcal{H}^*})^\Gamma_{\pi(s)} = (\Omega^1_{\mathcal{H}^*, s})^{\Gamma_s} = (\mathcal{O}_{\mathcal{H}^*, s}d(\sigma z))^{\Gamma_s} = q_s^{-1}(\mathcal{O}_{\mathcal{H}^*, s})^{\Gamma_s}dq_s;$$

hence the lemma follows.   □

We now consider the sheaf $\mathcal{O}_{\mathcal{H}^*} \otimes_{\mathbb{C}} \mathbb{C}^2$ of $\mathbb{C}^2$-valued meromorphic functions of the form described above. Thus elements of this sheaf are germs of functions of the form $f_1(z)e_1 + f_2(z)e_2$, where f_1, f_2 are functions representing germs belonging to $\mathcal{O}_{\mathcal{H}^*}$ and $e_1 = (1,0), e_2 = (0,1) \in \mathbb{C}^2$. Note that in this section we are considering elements of $\mathbb{C}^2$ as row vectors. The group Γ acts on the sheaf $\mathcal{O}_{\mathcal{H}^*}$ on the right by

$$(f_1(z)e_1 + f_2(z)e_2) \cdot \gamma = (f_1(\gamma z), f_2(\gamma z)) \begin{pmatrix} a & b \\ c & d \end{pmatrix} \tag{2.4}$$
$$= (af_1(\gamma z) + cf_2(\gamma z))e_1$$
$$+ (bf_1(\gamma z) + df_2(\gamma z))e_2$$

for $\gamma = \left(\begin{smallmatrix} a & b \\ c & d \end{smallmatrix}\right) \in \Gamma$. We define the sheaf $\mathcal{F}_{\mathcal{H}^*}$ on $\mathcal{H}^*$ to be the subsheaf of $\mathcal{O}_{\mathcal{H}^*} \otimes_{\mathbb{C}} \mathbb{C}^2$ generated by the global section $e_1 - ze_2$, that is,

$$\mathcal{F}_{\mathcal{H}^*} = \{f(z)(e_1 - ze_2) \mid f \in \mathcal{O}_{\mathcal{H}^*}\}. \tag{2.5}$$

Lemma 2.3 *The sheaf $\mathcal{F}_{\mathcal{H}^*}$ on $\mathcal{H}^*$ is Γ-invariant.*

Proof. Given $\gamma = \left(\begin{smallmatrix} a & b \\ c & d \end{smallmatrix}\right) \in \Gamma$ and $f \in \mathcal{O}_{\mathcal{H}^*}$, by (2.4) we have

$$f(z)(e_1 - ze_2) \cdot \gamma = f(\gamma z)(a - c(\gamma z))e_1 + (b - d(\gamma z))e_2).$$

Using the relations $ad - bc = 1$ and $\gamma z = (az + b)/(cz + d)$, we see that

$$a - c(\gamma z) = \frac{a(cz + d) - c(az + d)}{cz + d} = \frac{1}{cz + d},$$
$$b - d(\gamma z) = \frac{b(cz + d) - d(az + d)}{cz + d} = \frac{-z}{cz + d}.$$

Hence we obtain

$$f(z)(e_1 - ze_2) \cdot \gamma = (cz + d)^{-1}f(\gamma z)(e_1 - ze_2) \in \mathcal{F}_{\mathcal{H}^*}, \tag{2.6}$$

which proves the lemma.   □

By Lemma 2.3 we can consider the Γ-fixed sheaf $(\mathcal{F}_{\mathcal{H}^*})^\Gamma$ on X. Given a positive integer m, we can also consider the sheaf $(\mathcal{F}^m_{\mathcal{H}^*})^\Gamma$ on X, where $\mathcal{F}^m_{\mathcal{H}^*}$ is the m-th tensor power of $\mathcal{F}_{\mathcal{H}^*}$.

Proposition 2.4 *The space* $(\mathcal{F}_{\mathcal{H}*}^m)^\Gamma(X)$ *of global sections of* $(\mathcal{F}_{\mathcal{H}*}^m)^\Gamma$ *is canonically isomorphic to the space* $M_m(\Gamma)$ *of automorphic forms of weight* m *for* Γ.

Proof. By (2.1) and (2.5) we have

$$(\mathcal{F}_{\mathcal{H}*}^m)^\Gamma(X) = (\mathcal{F}_{\mathcal{H}*}^m(\mathcal{H}^*))^\Gamma = (\mathcal{O}_{\mathcal{H}*}(\mathcal{H}^*)(e_1 - ze_2)^m)^\Gamma.$$

Thus an element $\phi \in (\mathcal{F}_{\mathcal{H}*}^m)^\Gamma(X)$ can be regarded as a Γ-invariant function of the form

$$\phi(z) = f(z)(e_1 - ze_2)^m$$

for all $z \in \mathcal{H}^*$ with $f \in \mathcal{O}_{\mathcal{H}*}(\mathcal{H}^*)$. Hence, in order to prove the proposition, it suffices to show that f is an automorphic form for Γ of weight m. Since ϕ is Γ-invariant, by using (2.6) we see that

$$f(\gamma z) = j(\gamma, z)^m f(z) = (f \mid_m \gamma)(z)$$

for all $z \in \mathcal{H}$ and $\gamma \in \Gamma$. Now let $s \in \Sigma$ be a cusp of Γ with $\sigma s = \infty$. Then $f \mid_m \sigma^{-1}$ has a Fourier expansion of the form

$$(f \mid_m \sigma^{-1})(z) = \sum_{-\infty}^{\infty} a_n e^{2\pi i z/h}.$$

However, as in the proof of Lemma 2.1, the condition $f(\sigma^{-1}z) = O(|z|^k)$ implies that $a_n = 0$ for $n < 0$. Thus $f \in M_m(\Gamma)$, and therefore the proof of the proposition is complete. $\qquad\square$

Let $\mathcal{O}_{\mathcal{H}*}(-\Sigma)$ be the sheaf of functions on $\mathcal{H}^*$ which are holomorphic on $\mathcal{H}$ and zero on Σ. For each positive integer m we set

$$\mathcal{F}_{\mathcal{H}*}^m(-\Sigma) = \mathcal{F}_{\mathcal{H}*}^m \otimes \mathcal{O}_{\mathcal{H}*}(-\Sigma),$$

and denote by

$$\mathcal{F}_\Gamma^m = (\mathcal{F}_{\mathcal{H}*}^m(-\Sigma))^\Gamma \tag{2.7}$$

the Γ-fixed sheaf of $\mathcal{F}_{\mathcal{H}*}^m(-\Sigma)$ on $X = \Gamma\backslash\mathcal{H}^*$.

Proposition 2.5 *The space* $\mathcal{F}_\Gamma^m(X)$ *of global sections of* $\mathcal{F}_\Gamma^m$ *is canonically isomorphic to the space* $S_m(\Gamma)$ *of cusp forms of weight* m *for* Γ.

Proof. Using (2.1) and (2.5), we see that

$$\mathcal{F}_\Gamma^m(X) = (\mathcal{F}_{\mathcal{H}*}^m(-\Sigma))^\Gamma(X) = (\mathcal{F}_{\mathcal{H}*}^m(-\Sigma)(\mathcal{H}^*))^\Gamma$$
$$= (\mathcal{O}_{\mathcal{H}*}(-\Sigma)(\mathcal{H}^*)(e_1 - ze_2)^m)^\Gamma.$$

Thus an element $\phi \in \mathcal{F}_\Gamma^m(X)$ can be regarded as a Γ-invariant function of the form

$$\phi(z) = f(z)(e_1 - ze_2)^m$$

for all $z \in \mathcal{H}^*$ with $f \in \mathcal{O}_{\mathcal{H}^*}(-\Sigma)(\mathcal{H}^*)$. Hence, as in the proof of Proposition 2.4, we see that f satisfies $f \mid_m \gamma = f$ for all $\gamma \in \Gamma$ on $\mathcal{H}$. Now the fact that the f is zero at each cusp follows from the definition of the sheaf $\mathcal{O}_{\mathcal{H}^*}(-\Sigma)$. Thus it follows that f is a cusp form of weight m for Γ. $\square$

Let $\Gamma' = \chi(\Gamma)$ be the image of Γ under χ, and let Σ' be the set of cusps of Γ'. Then for each positive integer m we can define the sheaf $\mathcal{F}_{\Gamma'}^m$ over the Riemann surface $X_{\Gamma'} = \Gamma' \backslash \mathcal{H}^\#$ with $\mathcal{H}^\# = \mathcal{H} \cup \Sigma'$ given by

$$\mathcal{F}_{\Gamma'}^m = (\mathcal{F}_{\mathcal{H}^\#}^m)^{\Gamma'} = (\mathcal{F}_{\mathcal{H}^\#}^m(-\Sigma'))^{\Gamma'}, \tag{2.8}$$

and its sections can be identified with the cusp forms of weight m for Γ'. Let $\omega_X : X \to X_{\Gamma'}$ be the morphism of complex algebraic curves induced by the holomorphic map $\omega : \mathcal{H} \to \mathcal{H}$, and denote by $\omega_X^* \mathcal{F}_{\Gamma'}^m$ the sheaf on X obtained by pulling the sheaf $\mathcal{F}_{\Gamma'}^m$ on $X_{\Gamma'}$ via the map ω_X.

Proposition 2.6 *The space $S_{k,\ell}(\Gamma, \omega, \chi)$ of mixed cusp forms associated to Γ, ω and χ is canonically isomorphic to the space $H^0(X, \mathcal{F}_\Gamma^k \otimes \omega_X^* \mathcal{F}_{\Gamma'}^\ell)$ of sections of the sheaf $\mathcal{F}_\Gamma^k \otimes \omega_X^* \mathcal{F}_{\Gamma'}^\ell$ over X.*

Proof. Each global section ϕ of the sheaf $\mathcal{F}_\Gamma^k \otimes \omega_X^* \mathcal{F}_{\Gamma'}^\ell$ is of the form

$$\phi = \sum_{i=1}^r \alpha_i \beta_i,$$

where $\alpha_i \in \mathcal{F}_\Gamma^k(X)$ and $\beta_i \in \omega_X^* \mathcal{F}_{\Gamma'}^\ell(X)$ for each i. As in the proof of Proposition 2.5 the sections α_i and β_i may be regarded as Γ-invariant functions of the form

$$\alpha_i(z) = \sum_{i=1}^r f_i(z)(e_1 - ze_2)^k, \quad \beta_i(z) = h_i(\omega(z))(e_1 - \omega(z)e_2)^\ell$$

for all $z \in \mathcal{H}^*$, so that we have

$$\phi(z) = \sum_{i=1}^r f_i(z)h_i(\omega(z))(e_1 - ze_2)^k(e_1 - \omega(z)e_2)^\ell.$$

Thus we see that the function

$$F(z) = \sum_{i=1}^r f_i(z)h_i(\omega(z))$$

for $z \in \mathcal{H}^*$ satisfies

$$F(\gamma z) = j(\gamma, z)^k j(\chi(\gamma), \omega(z))^\ell F(z)$$

for all $\gamma \in \Gamma$ and $z \in \mathcal{H}$. Thus ϕ has the same transformation property as the one for an element in $S_{k,\ell}(\Gamma, \omega, \chi)$. Now the proposition follows from the fact that the Γ-cusps and Γ'-cusps correspond via ω and χ, since χ maps parabolic elements to parabolic elements. $\square$

2.2 Elliptic Varieties

In this section we describe the interpretation of mixed cusp forms as holomorphic forms on certain families of abelian varieties called elliptic varieties. An abelian variety belonging to such a family is the product of a finite number of elliptic curves.

Let E be an elliptic surface in the sense of Kodaira [52]. Thus E is a compact smooth surface over $\mathbb{C}$, and it is the total space of an elliptic fibration $\pi : E \to X$ over a Riemann surface X whose generic fiber is an elliptic curve. Let E_0 be the union of the regular fibers of π, and let $\Gamma \subset PSL(2,\mathbb{R})$ be the fundamental group of $X_0 = \pi(E_0)$. Then Γ acts on the universal covering space $\mathcal{H}$ of X_0 by linear fractional transformations, and we have

$$X = \Gamma \backslash \mathcal{H} \cup \{\Gamma\text{-cusps}\}.$$

For $z \in X_0$, let Φ be a holomorphic 1-form on the fiber $E_z = \pi^{-1}(z)$, and choose an ordered basis $\{\gamma_1(z), \gamma_2(z)\}$ for $H_1(E_z, \mathbb{Z})$ that depends on the parameter z in a continuous manner. Consider the periods ω_1 and ω_2 of E given by

$$\omega_1(z) = \int_{\gamma_1(z)} \Phi, \quad \omega_2(z) = \int_{\gamma_2(z)} \Phi.$$

Then the imaginary part of the quotient $\omega_1(z)/\omega_2(z)$ is nonzero for each z, and therefore we may assume that $\omega_1(z)/\omega_2(z) \in \mathcal{H}$. In fact, ω_1/ω_2 is a many-valued holomorphic function from X_0 to $\mathcal{H}$ which can be lifted to a single-valued function $\omega : \mathcal{H} \to \mathcal{H}$ on the universal cover of X_0 such that

$$\omega(\gamma z) = \chi(\gamma)\omega(z)$$

for all $\gamma \in \Gamma$ and $z \in \mathcal{H}$; here $\chi : \Gamma \to SL(2,\mathbb{R})$ is the monodromy representation of the elliptic fibration $\pi : E \to X$.

In order to discuss connections of elliptic varieties with mixed cusp forms we shall regard Γ as a subgroup of $SL(2,\mathbb{R})$. As in Section 1.1 we denote by $S_{j+2,k}(\Gamma, \chi, \omega)$ the space of mixed cusp forms of type $(j+2, k)$ associated to Γ, ω and χ. Let $E(\chi)$ (resp. $E(1)$) be an elliptic surface over X whose monodromy representation is χ (resp. the inclusion map), and let $\pi(\chi) : E(\chi) \to X$ (resp. $\pi(1) : E(1) \to X$) be the associated elliptic fibration. We set

$$E(\chi)_0 = \pi(\chi)^{-1}(\Gamma \backslash \mathcal{H}), \qquad E(1)_0 = \pi(1)^{-1}(\Gamma \backslash \mathcal{H}),$$

and denote by $(E_{\chi,1}^{j,k})_0$ the fiber product of j-copies of $E(1)_0$ and k copies of $E(\chi)_0$ over X corresponding to the maps $\pi(1)$ and $\pi(\chi)$, respectively. The space $(E_{\chi,1}^{j,k})_0$ can also be constructed as below. Consider the semidirect product $\Gamma \ltimes_{1,\chi} \mathbb{Z}^{2j} \times \mathbb{Z}^{2k}$ consisting of the triples $(\gamma, \boldsymbol{\mu}, \boldsymbol{\nu})$ in $\Gamma \times \mathbb{Z}^{2j} \times \mathbb{Z}^{2k}$ whose multiplication law is defined as follows. Let

$$(\gamma, \boldsymbol{\mu}, \boldsymbol{\nu}), (\gamma', \boldsymbol{\mu}', \boldsymbol{\nu}') \in \Gamma \times \mathbb{Z}^{2j} \times \mathbb{Z}^{2k}$$

with

$$\boldsymbol{\mu}' = (\boldsymbol{\mu}_1', \boldsymbol{\mu}_2') = (\mu_{11}', \dots, \mu_{j1}', \mu_{12}', \dots, \mu_{j2}') \in \mathbb{Z}^{2j},$$

$$\boldsymbol{\nu}' = (\boldsymbol{\nu}_1', \boldsymbol{\nu}_2') = (\nu_{11}', \dots, \nu_{k1}', \nu_{12}', \dots, \nu_{k2}') \in \mathbb{Z}^{2k},$$

$$\gamma = \begin{pmatrix} a & b \\ c & d \end{pmatrix} \in \Gamma, \qquad \chi(\gamma) = \begin{pmatrix} a_\chi & b_\chi \\ c_\chi & d_\chi \end{pmatrix} \in SL(2, \mathbb{Z}).$$

Then we have

$$(\gamma, \boldsymbol{\mu}, \boldsymbol{\nu}) \cdot (\gamma', \boldsymbol{\mu}', \boldsymbol{\nu}') = (\gamma\gamma', \gamma \cdot (\boldsymbol{\mu}', \boldsymbol{\nu}') + (\boldsymbol{\mu}, \boldsymbol{\nu})),$$

where $\gamma \cdot (\boldsymbol{\mu}', \boldsymbol{\nu}') = (\boldsymbol{\mu}'', \boldsymbol{\nu}'')$ with

$$\boldsymbol{\mu}'' = (a\mu_{11}' + b\mu_{12}', \dots, a\mu_{j1}' + b\mu_{j2}',$$
$$c\mu_{11}' + d\mu_{12}', \dots, c\mu_{j1}' + d\mu_{j2}') \in \mathbb{Z}^{2j},$$

$$\boldsymbol{\nu}'' = (a_\chi\nu_{11}' + b_\chi\nu_{12}', \dots, a_\chi\nu_{k1}' + b_\chi\nu_{k2}',$$
$$c_\chi\nu_{11}' + d_\chi\nu_{12}', \dots, c_\chi\nu_{k1}' + d_\chi\nu_{k2}') \in \mathbb{Z}^{2k}.$$

The group $\Gamma \ltimes_{1,\chi} \mathbb{Z}^{2j} \times \mathbb{Z}^{2k}$ acts on the space $\mathcal{H} \times \mathbb{C}^j \times \mathbb{C}^k$ by

$$(\gamma, \boldsymbol{\mu}, \boldsymbol{\nu}) \cdot (z, \boldsymbol{\xi}, \boldsymbol{\zeta}) \tag{2.9}$$
$$= \big(\gamma z, (cz + d)^{-1}(\boldsymbol{\xi} + z\boldsymbol{\mu}_1 + \boldsymbol{\mu}_2),$$
$$(c_\chi \omega(z) + d_\chi)^{-1}(\boldsymbol{\zeta} + \omega(z)\boldsymbol{\nu}_1 + \boldsymbol{\nu}_2)\big)$$

for $\gamma \in \Gamma$, $z \in \mathcal{H}$, $\boldsymbol{\xi} \in \mathbb{C}^j$, $\boldsymbol{\zeta} \in \mathbb{C}^k$, $\boldsymbol{\mu} = (\boldsymbol{\mu}_1, \boldsymbol{\mu}_2) \in \mathbb{Z}^{2j}$, and $\boldsymbol{\nu} = (\boldsymbol{\nu}_1, \boldsymbol{\nu}_2) \in \mathbb{Z}^{2k}$. Then we have

$$(E_{1,\chi}^{j,k})_0 = \Gamma \ltimes_{1,\chi} \mathbb{Z}^{2j} \times \mathbb{Z}^{2k} \backslash \mathcal{H} \times \mathbb{C}^j \times \mathbb{C}^k. \tag{2.10}$$

Now we obtain the *elliptic variety* $E_{1,\chi}^{k,m}$ by resolving the singularities of the compactification of $(E_{1,\chi}^{j,k})_0$ (cf. [117]). The elliptic fibration π induces a fibration $\pi_{1,\chi}^{j,k} : E_{1,\chi}^{j,k} \to X$ whose generic fiber is the product of $(j + k)$ elliptic curves.

Theorem 2.7 *Let $E_{1,\chi}^{j,k}$ be an elliptic variety described above. Then there is a canonical isomorphism*

$$H^0(E_{1,\chi}^{j,k}, \Omega^{j+k+1}) \cong S_{j+2,k}(\Gamma, \omega, \chi)$$

between the space of holomorphic $(j + k + 1)$-forms on $E_{\chi,1}^{j,k}$ and the space of mixed cusp forms of type $(j + 2, k)$.

Proof. Let $(E_{1,\chi}^{j,k})_0$ be as in (2.10). Then a holomorphic $(j+k+1)$-form on $E_{1,\chi}^{j,k}$ can be regarded as a holomorphic $(j+k+1)$-form on $\mathcal{H} \times \mathbb{C}^j \times \mathbb{C}^k$ that is invariant under the operation of $\Gamma \ltimes_{1,\chi} \mathbb{Z}^{2j} \times \mathbb{Z}^{2k}$. Since the complex dimension of the space $\mathcal{H} \times \mathbb{C}^j \times \mathbb{C}^k$ is $j+k+1$, a holomorphic $(j+k+1)$-form on $\mathcal{H} \times \mathbb{C}^j \times \mathbb{C}^k$ is of the form

$$\Theta = \widetilde{f}(z, \boldsymbol{\xi}, \boldsymbol{\zeta}) dz \wedge d\boldsymbol{\xi} \wedge d\boldsymbol{\zeta},$$

where $\widetilde{f}$ is holomorphic. For $z_0 \in \mathcal{H}$, the holomorphic form Θ descends to a holomorphic $(j+k)$-form on the fiber $(\pi_{1,\chi}^{j,k})^{-1}(z_0)$. However, the dimension of the fiber is $j+k$, and therefore the space of holomorphic $(j+k)$-forms on $(\pi_{1,\chi}^{j,k})^{-1}(z_0)$ is one. Hence the map $(\boldsymbol{\xi}, \boldsymbol{\zeta}) \mapsto \widetilde{f}(z, \boldsymbol{\xi}, \boldsymbol{\zeta})$ is a holomorphic function on a compact complex manifold, and consequently is a constant function. Thus we have $\widetilde{f}(z, \boldsymbol{\xi}, \boldsymbol{\zeta}) = f(z)$, where f is a holomorphic function on $\mathbb{C}^j \times \mathbb{C}^k$. From (2.9) the action of $(\gamma, \boldsymbol{\mu}, \boldsymbol{\nu}) \in \Gamma \ltimes_{1,\chi} \mathbb{Z}^{2j} \times \mathbb{Z}^{2k}$ on the form $\Theta = f(z) dz \wedge d\boldsymbol{\xi} \wedge d\boldsymbol{\zeta}$ is given by

$$\begin{aligned}
\Theta \cdot (\gamma, \boldsymbol{\mu}, \boldsymbol{\nu}) &= f(gz) d(gz) \wedge d\big((cz+d)^{-1}(\boldsymbol{\xi} + z\boldsymbol{\mu}_1 + \boldsymbol{\mu}_2)\big) \\
&\quad \wedge d\big((c_\chi \omega(z) + d_\chi)^{-1}(\boldsymbol{\zeta} + \omega(z)\boldsymbol{\nu}_1 + \boldsymbol{\nu}_2)\big) \\
&= f(gz)(cz+d)^{-2}(cz+d)^{-j}(c_\chi \omega(z) + d_\chi)^{-k} dz \wedge d\boldsymbol{\xi} \wedge d\boldsymbol{\zeta}.
\end{aligned}$$

Thus it follows that $f(z)$ satisfies the condition (i) of Definition 1.3, and it remains to show that f satisfies the cusp condition. Using Theorem 3.1 in [67], we see that the differential form Θ can be extended to $E_{1,\chi}^{j,k}$ if and only if

$$\int_{(E_{1,\chi}^{j,k})_0} \Theta \wedge \overline{\Theta} < \infty.$$

From (2.9) it follows that a fundamental domain F in $\mathcal{H} \times \mathbb{C}^j \times \mathbb{C}^k$ for the action of $\Gamma \ltimes_{1,\chi} \mathbb{Z}^{2j} \times \mathbb{Z}^{2k}$ can be chosen in the form

$$\begin{aligned}
F = \{(z, \boldsymbol{\xi}, \boldsymbol{\zeta}) \in \mathcal{H} \times \mathbb{C}^j \times \mathbb{C}^k \mid z \in F_0, \quad \boldsymbol{\xi} = \mathbf{s} + \mathbf{t}z, \quad \boldsymbol{\zeta} = \mathbf{u} + \mathbf{v}\omega(z), \\
\mathbf{s}, \mathbf{t} \in I^j, \quad \mathbf{u}, \mathbf{v} \in I^k\},
\end{aligned}$$

where $F_0 \subset \mathcal{H}$ is a fundamental domain of Γ and I is the closed interval $[0,1] \subset \mathbb{R}$. Thus we have

$$\begin{aligned}
\int_{(E_{1,\chi}^{k,m})_0} \Theta \wedge \overline{\Theta} &= \int_F \Theta \wedge \overline{\Theta} = \int_F |f(z)|^2 dz \wedge d\boldsymbol{\xi} \wedge d\boldsymbol{\zeta} \wedge d\overline{z} \wedge d\overline{\boldsymbol{\xi}} \wedge d\overline{\boldsymbol{\zeta}} \\
&= K \int_F |f(z)|^2 (\operatorname{Im} z)^j (\operatorname{Im} \omega(z))^k dz \wedge d\overline{z},
\end{aligned}$$

where K is a nonzero constant. Thus the integral $\int_F \Theta \wedge \overline{\Theta}$ is a nonzero constant multiple of the Petersson inner product $\langle f, f \rangle$ described in Proposition 2.1 in [70]; hence it is finite if and only if f satisfies the cusp condition, and the proof of the theorem is complete. $\qquad \square$

Remark 2.8 *Theorem 2.7 is an extension of the results of [43, Theorem 1.6] and [68, Theorem 3.2], where mixed cusp forms of types (2,1) and (2,m), respectively, were considered.*

2.3 Modular Symbols

Modular symbols for automorphic forms were first introduced by Birch [10] and were developed systematically by Manin [89, 90]. More general modular symbols were introduced by Šokurov [119] by using the fact that cusp forms can be identified with holomorphic forms on an elliptic variety. In this section we extend the notion of modular symbols of Šokurov to include the ones associated to mixed cusp forms.

Let (ω, χ) be the equivariant pair associated to an elliptic surface E over $X = \Gamma \backslash \mathcal{H}^*$ considered in Section 2.2. Thus $\omega : \mathcal{H} \to \mathcal{H}$ is the period map and $\chi : \Gamma \to SL(2, \mathbb{R})$ is the monodromy representation for the elliptic fibration $\pi : E \to X$ satisfying

$$\omega(\gamma z) = \chi(\gamma)\omega(z)$$

for all $z \in \mathcal{H}$ and $\gamma \in \Gamma$.

Let $R_1 \pi_* \mathbb{Q}$ be the sheaf on X corresponding to the presheaf determined by the map

$$U \mapsto H_1(\pi^{-1}(U), \mathbb{Q})$$

for each open subset $U \subset X$. We fix a positive integer m and denote by $(R_1 \pi_* \mathbb{Q})^m$ the m-th symmetric tensor power of the sheaf $R_1 \pi_* \mathbb{Q}$. We shall construct below a map

$$\{,,\}_{\omega,\chi} : \widetilde{\mathbb{Q}} \times \mathbb{Z}^m \times \mathbb{Z}^m \longrightarrow H_0(\Sigma, (R_1 \pi_* \mathbb{Q})^m) \qquad (2.11)$$

which assigns to each triple $(\alpha, p, q) \in \widetilde{\mathbb{Q}} \times \mathbb{Z}^m$ the element $\{\alpha, p, q\}_{\omega,\chi} \in H_0(\Sigma, (R_1 \pi_* \mathbb{Q})^m)$ called a *boundary modular symbol*, where Σ is the set of cusps for Γ and $\widetilde{\mathbb{Q}} = \mathbb{Q} \cup \{\infty\}$. If ω is the identity map on $\mathcal{H}$ and χ is the inclusion map, then $\{\alpha, p, q\}_{\omega,\chi}$ is the boundary modular symbol $\{\alpha, p, q\}_\Gamma$ of Šokurov defined in [118, Section 1].

Note that the elliptic surface E can be written in the form

$$E = \Gamma \ltimes_\chi \mathbb{Z}^2 \backslash \mathcal{H} \times \mathbb{C},$$

where the quotient is taken with respect to the action of the semidirect product $\Gamma \ltimes_\chi \mathbb{Z}^2$ on $\mathcal{H} \times \mathbb{C}$ given by

$$(\gamma, (\mu_1, \mu_2)) \cdot (z, \zeta) = (\gamma z, (c_\chi z + d_\chi)^{-1}(\zeta + \mu_1 \omega(z) + \mu_2))$$

for $(\mu_1, \mu_2) \in \mathbb{Z}^2$, $(z, \zeta) \in \mathcal{H} \times \mathbb{C}$ and $\gamma \in \Gamma$ with $\chi(\gamma) = \left(\begin{smallmatrix} a_\chi & b_\chi \\ c_\chi & d_\chi \end{smallmatrix} \right)$. Thus, if $\varpi : \mathcal{H}^* \to X$ is the natural projection map and $z \in \mathcal{H}$, the fiber E_z of the elliptic fibration $\pi : E \to X$ over $\varpi(z) \in X$ is the elliptic curve of the form

$$E_z = \mathbb{C}/(\mathbb{Z}\omega(z) + \mathbb{Z}).$$

We choose a basis $\{\eta_1(z), \eta_2(z)\}$ of $H_1(E_z, \mathbb{Z})$ represented by the cycles $\eta_1(z)$ and $\eta_2(z)$ corresponding to the paths $\{t\omega(z) \mid 0 \le t \le 1\}$ and $\{t \mid 0 \le t \le 1\}$, respectively, in the complex plane $\mathbb{C}$. Thus we may write

$$\eta_1(z) = \left[\tilde{\pi}\{t\omega(z) \mid 0 \le t \le 1\}\right], \quad \eta_2(z) = \left[\{t \mid 0 \le t \le 1\}\right], \qquad (2.12)$$

where $\tilde{\pi} : \mathbb{C} \to E_z$ is the natural projection map. The codomain of the map in (2.11) to be constructed has a direct sum decomposition of the form

$$H_0(\Sigma, (R_1\pi_*\mathbb{Q})^m) = \bigoplus_{s \in \Sigma} H_0(s, (R_1\pi_*\mathbb{Q})^m).$$

Let $s_0 \in \Sigma$ be a cusp for Γ determined by $\alpha \in \tilde{\mathbb{Q}}$, and let $(p, q) \in \mathbb{Z}^m \times \mathbb{Z}^m$. Then we define the element $\{\alpha, p, q\}_{\omega,\chi} \in H_0(\Sigma, (R_1\pi_*\mathbb{Q})^m)$ to be trivial on the summands $H_0(s, (R_1\pi_*\mathbb{Q})^m)$ for $s \ne s_0$ so that we may write

$$\{\alpha, p, q\}_{\omega,\chi} \in H_0(s_0, (R_1\pi_*\mathbb{Q})^m).$$

Now we choose a small disk D of s_0, and let $\tilde{D}$ be the corresponding neighborhood of α that covers D. If $z_D \in \tilde{D} \subset \mathcal{H}^*$, $p = (p_1, \ldots, p_m) \in \mathbb{Z}^m$ and $q = (q_1, \ldots, q_m) \in \mathbb{Z}^m$, then we define the element

$$\{z_D, p, q\}_{\omega,\chi}^D \in H_0(D, (R_1\pi_*\mathbb{Q})^m)$$

to be the homology class of the cycle

$$\prod_{i=1}^{m}(p_i\eta_1 + q_i\eta_2)v_D,$$

where $v_D \in X$ is the point corresponding to $z_D \in \mathcal{H}^*$. Then the boundary modular symbol for the triple (α, p, q) is defined by

$$\{\alpha, p, q\}_{\omega,\chi} = \varprojlim_{D}\{z_D, p, q\}_{\omega,\chi}^D, \qquad (2.13)$$

where the inverse limit is taken over the set of open disks D containing s_0.

In this section we assume that Γ is a subgroup of $SL(2, \mathbb{Z})$ of finite index. If s is a Γ-cusp in X, then it is of type I_b or I_b^* in the sense of Kodaira [52] where b is the ramification index of the canonical map $\mu : X \to SL(2, \mathbb{Z})\backslash\mathcal{H} \cup \tilde{\mathbb{Q}}$ induced by χ at s. If $\tau = e^{2\pi i z}$ is the canonical local parameter at the unique cusp in $SL(2, \mathbb{Z})\backslash\mathcal{H} \cup \tilde{\mathbb{Q}}$, we can take a local parameter τ_s at s to be a fixed branch of the root $(\tau_s\mu)^{1/b}$. We define the subsets E_s^ε and F^ε by

$$E_s^\varepsilon = \{\tau_s \in X \mid \mid \tau_s \mid \le \varepsilon\}, \quad F^\varepsilon = \bigcup_{s \in \Sigma} E_s^\varepsilon$$

for each small $\varepsilon > 0$. For $\gamma \in SL(2,\mathbb{Z})$, let s_1 and s_2 be the cusps corresponding to $\gamma(0)$ and $\gamma(i\infty)$ respectively. Let $[\gamma] = \Gamma\gamma \in SL(2,\mathbb{Z})/\Gamma$ be the Γ-coset containing γ, and denote by $\overline{0\,i\infty}$ the nonnegative part of the imaginary axis in $\mathcal{H} \cup \overline{\mathbb{Q}}$ oriented from 0 to $i\infty$. We define the 1-cell $\mathfrak{c}[\gamma]_\varepsilon$ to be the image of the oriented 1-cell $\gamma(\overline{0\,i\infty})$ that lies outside E_{s_1} and E_{s_2}, that is,

$$\mathfrak{c}[\gamma]_\varepsilon = \varpi(\overline{0\,i\infty}) - E_{s_1} \cap E_{s_2}, \tag{2.14}$$

where $\varpi : \mathcal{H}^* \to X = \Gamma\backslash\mathcal{H}^*$ is the natural projection map. The basis $\{\eta_1(z), \eta_2(z)\}$ of $H_1(E_z, \mathbb{Z})$ for each fiber E_z considered before induces the basis $\{\eta_1, \eta_2\}$ of the group

$$(R_1\pi_*\mathbb{Q})\,|_{\mathfrak{c}[\gamma]_\varepsilon}\,.$$

If k is an integer with $0 \le k \le m$, we set

$$1_k = (\underbrace{1,\ldots,1}_{k},\underbrace{0,\ldots,0}_{m-k}), \tag{2.15}$$

and define the element

$$\{\gamma(0),\gamma(i\infty),1_k,1_m - 1_k\}^\varepsilon_{\omega,\chi} \in H_1(X, F^\varepsilon, (R_1\pi_*\mathbb{Q})^m)$$

to be the homology class of the cycle

$$\eta_1^k \eta_2^{m-k} \mathfrak{c}[\gamma]_\varepsilon = \prod_{i=1}^{m}(p_i\eta_1 + q_i\eta_2)\mathfrak{c}[\gamma]_\varepsilon, \tag{2.16}$$

where $p = (p_1,\ldots,p_m) = 1_k$ and $q = (q_1,\ldots,q_m) = 1_m - 1_k$. Noting that the pairs (X, F^ε) with $\varepsilon > 0$ form a cofinal system, we define the modular symbol

$$\{\gamma(0),\gamma(i\infty),1_k,1_m - 1_k\}_{\omega,\chi} \in H_1(X, \Sigma, (R_1\pi_*\mathbb{Q})^m)$$

to be the inverse limit

$$\varprojlim_{F^\varepsilon}\{\gamma(0),\gamma(i\infty),1_k,1_m - 1_k\}^\varepsilon_{\omega,\chi}. \tag{2.17}$$

We now denote by E^m the elliptic variety $E_{1,\chi}^{0,m}$ over $X = \Gamma\backslash\mathcal{H}^*$ constructed in Section 2.2 associated to the elliptic fibration $\pi : E \to X$.

Lemma 2.9 *There is a canonical pairing*

$$\langle\,,\,\rangle : H_1(X, \Sigma, (R_1\pi_*\mathbb{Q})^m) \times H^0(E^m, \Omega^{m+1} \oplus \overline{\Omega}^{m+1}) \longrightarrow \mathbb{C} \tag{2.18}$$

that is nondegenerate on the right.

Proof. See [119, Section 4.1].   □

Note that the pairing in (2.18) is given by

$$\langle [\delta], \Phi \rangle = \int_\delta \Phi$$

for all $[\delta] \in H_1(X, \Sigma, (R_1\pi_*\mathbb{Q})^m)$ and $\Phi \in H^0(E^m, \Omega^{m+1} \oplus \overline{\Omega}^{m+1})$, where δ is a cycle representing the cohomology class $[\delta]$. Using (2.18) and the canonical isomorphism

$$S_{2,m}(\Gamma, \omega, \chi) \cong H^0(E^m, \Omega^{m+1}) \qquad (2.19)$$

given in Theorem 2.7, we obtain the canonical pairing

$$\langle \, , \, \rangle : H_1(X, \Sigma, (R_1\pi_*\mathbb{Q})^m) \times (S_{2,m}(\Gamma, \omega, \chi) \oplus \overline{S_{2,m}(\Gamma, \omega, \chi)}) \longrightarrow \mathbb{C}. \quad (2.20)$$

Lemma 2.10 *Let* $\partial : H_1(X, \Sigma, (R_1\pi_*\mathbb{Q})^m) \rightarrow H_0(\Sigma, (R_1\pi_*\mathbb{Q})^m)$ *be the boundary map for the homology sequence of the pair* (X, Σ)*. Then for* $1 \leq k \leq m$ *we have*

$$\partial\{\gamma(0), \gamma(i\infty), 1_k, 1_m - 1_k\}_{\omega,\chi} = \{\gamma(i\infty), 1_k, 1_m - 1_k\}_{\omega,\chi}$$
$$- \{\gamma(0), 1_k, 1_m - 1_k\}_{\omega,\chi}$$

for all $\gamma \in SL(2, \mathbb{Z})$.

Proof. This follows easily from the definitions in (2.13) and (2.17). $\qquad \square$

Proposition 2.11 *Let* $\langle \, , \, \rangle$ *be the canonical pairing in* (2.20), *and let* $1 \leq k \leq m$*. Then we have*

$$\langle \{\gamma(0), \gamma(i\infty), 1_k, 1_m - 1_k\}_{\omega,\chi}, (f_1, \overline{f}_2) \rangle$$
$$= \int_{\gamma(0)}^{\gamma(i\infty)} f_1 \omega(z)^k dz + \int_{\gamma(0)}^{\gamma(i\infty)} \overline{f}_2 \overline{\omega(z)}^k d\overline{z}$$
$$= \int_{\gamma(0)}^{\gamma(i\infty)} f_1 \prod_{i=1}^m (p_i \omega(z) + q_i) dz$$
$$+ \int_{\gamma(0)}^{\gamma(i\infty)} \overline{f}_2 \prod_{i=1}^m (p_i \overline{\omega(z)} + q_i) d\overline{z}$$

for all $\gamma \in SL(2, \mathbb{Z})$ *and* $f_1, f_2 \in S_{2,m}(\Gamma, \omega, \chi)$*, where* $p = (p_1, \ldots, p_m) = 1_k$ *and* $q = (q_1, \ldots, q_m) = 1_m - 1_k$*.*

Proof. We note first that by (2.17) the element

$$\{\gamma(0), \gamma(i\infty), 1_k, 1_m - 1_k\}_{\omega,\chi} \in H_1(X, \Sigma, (R_1\pi_*\mathbb{Q})^m)$$

is obtained by taking the inverse limit of

$$[\delta_\varepsilon] = \{\gamma(0), \gamma(i\infty), 1_k, 1_m - 1_k\}^\varepsilon_{\omega,\chi} \in H_1(X, F^\varepsilon, (R_1\pi_*\mathbb{Q})^m)$$

over the set $\{F^\varepsilon \mid \varepsilon > 0\}$. However, by (2.16) the homology class $[\delta_\varepsilon]$ can be represented by the cycle

$$\delta_\varepsilon = \eta_1^k \eta_2^{m-k} \mathfrak{c}[\gamma]_\varepsilon. \qquad (2.21)$$

Given $f_1, f_2 \in S_{2,m}(\Gamma, \omega, \chi)$, let Φ_{f_1} and Φ_{f_2} be the holomorphic $(m+1)$-forms on E^m corresponding to f_1 and f_2, respectively, under the canonical isomorphism (2.19). Then we see that

$$\langle \{\gamma(0), \gamma(i\infty), 1_k, 1_m - 1_k\}_{\omega, \chi}, (f_1, \overline{f}_2) \rangle = \int_{m! \delta_\varepsilon} \left(\Phi_{f_1} + \overline{\Phi}_{f_2} \right). \qquad (2.22)$$

Given $\varepsilon > 0$, if we identify $(R_1 \pi_* \mathbb{Q})^m$ with its embedded image in $(R_1 \pi_* \mathbb{Q})^{\otimes m}$ and if $\mathfrak{c}[\gamma]_\varepsilon$ is the 1-cell in (2.14), then the chain

$$\left[m! \eta_1^k \eta_2^{m-k} \mathfrak{c}[\gamma]_\varepsilon \right]$$

corresponds to the set $\widetilde{\delta} \subset \mathcal{H} \times \mathbb{C}^m$ whose restriction to $\{z\} \times \mathbb{C}^m \cong \mathbb{C}^m$ is given by

$$\widetilde{\delta}\,|_z = \bigcup_{(p,q)} \prod_{i=1}^m \{(t\, p_i \omega(z), t\, q_i) \mid 0 \leq t \leq 1\} \subset \mathbb{C}^m;$$

here $z \in \mathcal{H}$ belongs to the set corresponding to $\mathfrak{c}[\gamma]_\varepsilon$, and (p, q) runs through all permutations $(\sigma(1_k), \sigma(1_m - 1_k))$ for $\sigma \in \mathfrak{S}_m$. Now recall that there is a canonical isomorphism

$$E_0^m \cong \Gamma \ltimes_\chi \mathbb{Z}^{2m} \backslash \mathcal{H} \times \mathbb{C}^m,$$

where E_0^m is the elliptic variety $(E_{1,\chi}^{0,m})_0$ in (2.10). Let $\Phi_{f_i}^0$ for $i \in \{1, 2\}$ be the lifting of $\Phi_{f_i} |_{X_0}$ to $\mathcal{H} \times \mathbb{C}^m$. Then we have

$$\Phi_{f_i}^0 = f_i(z) dz \wedge d\zeta_1 \wedge \cdots \wedge d\zeta_m, \quad \overline{\Phi}_{f_i}^0 = \overline{f_i(z)} d\overline{z} \wedge d\overline{\zeta}_1 \wedge \cdots \wedge d\overline{\zeta}_m,$$

where z and $(\zeta_1, \ldots, \zeta_m)$ are the standard coordinate systems for $\mathcal{H} \subset \mathbb{C}$ and $\mathbb{C}^m$, respectively. On the other hand, for each $j \in \{1, \ldots, m\}$ we have

$$\int_{\{t\,\omega(z) \mid 0 \leq t \leq 1\}} d\zeta_j = \omega(z), \quad \int_{\{t\,\omega(z) \mid 0 \leq t \leq 1\}} d\overline{\zeta}_j = \overline{\omega(z)},$$

$$\int_{[0,1]} d\zeta_j = \int_{[0,1]} d\overline{\zeta}_j = 1.$$

Using this, (2.12) and (2.14), we see that

$$\int_{m! \eta_1^k \eta_2^{m-k} \mathfrak{c}[\gamma]_\varepsilon} \left(\Phi_{f_1} + \overline{\Phi}_{f_2} \right) = m! \left(\int_{z_\varepsilon}^{z_\varepsilon'} f_1 \omega(z)^k dz + \int_{z_\varepsilon}^{z_\varepsilon'} \overline{f_2} \, \overline{\omega(z)}^k d\overline{z} \right)$$

where $\partial \mathfrak{c}[\gamma]_\varepsilon = z_\varepsilon' - z_\varepsilon$ on $\mathcal{H}$. However, by (2.21) we have

$$m!\eta_1^k \eta_2^{m-k} \mathfrak{c}[\gamma]_\varepsilon = m!\delta_\varepsilon;$$

hence we obtain

$$\int_{\delta_\varepsilon} \left(\Phi_{f_1} + \overline{\Phi}_{f_2} \right) = \int_{z_\varepsilon}^{z'_\varepsilon} f_1 \omega(z)^k dz + \int_{z_\varepsilon}^{z'_\varepsilon} \overline{f_2} \overline{\omega(z)}^k d\overline{z}.$$

Thus the proposition follows by taking the limit of this relation as $\varepsilon \to 0$ and then using (2.22). □

Now in order to consider more general modular symbols

$$\{\alpha, \beta, p, q\}_{\omega, \chi} \in H_1(X, \Sigma, (R_1\pi_*\mathbb{Q})^m)$$

for $\alpha, \beta \in \widetilde{\mathbb{Q}}$ and $p, q \in \mathbb{Z}^m$, we first state the following theorem.

Theorem 2.12 *There exists a unique map*

$$\{,,,\}_{\omega, \chi} : \widetilde{\mathbb{Q}} \times \widetilde{\mathbb{Q}} \times \mathbb{Z}^m \times \mathbb{Z}^m \longrightarrow H_1(X, \Sigma, (R_1\pi_*\mathbb{Q})^m)$$

satisfying the following properties:
(i) If $\partial : H_1(X, \Sigma, (R_1\pi_*\mathbb{Q})^m) \to H_0(\Sigma, (R_1\pi_*\mathbb{Q})^m)$ is the boundary map for the homology sequence of the pair (X, Σ), then we have

$$\partial\{\alpha, \beta, p, q\}_{\omega, \chi} = \{\beta, p, q\}_{\omega, \chi} - \{\alpha, p, q\}_{\omega, \chi}. \tag{2.23}$$

for all $\alpha, \beta \in \widetilde{\mathbb{Q}}$ and $p, q \in \mathbb{Z}^m$.
(ii) If $\langle , \rangle$ denotes the canonical pairing in (2.20), then we have

$$\langle \{\alpha, \beta, p, q\}_{\omega, \chi}, (f_1, \overline{f}_2) \rangle \tag{2.24}$$

$$= \int_\alpha^\beta f_1 \prod_{i=1}^m (p_i \omega(z) + q_i) dz + \int_\alpha^\beta \overline{f}_2 \prod_{i=1}^m (p_i \overline{\omega(z)} + q_i) d\overline{z} \tag{2.25}$$

for all $\alpha, \beta \in \widetilde{\mathbb{Q}}$, $p, q \in \mathbb{Z}^m$ and $f_1, f_2 \in S_{2,m}(\Gamma, \omega, \chi)$.

Definition 2.13 *The values $\{\alpha, \beta, p, q\}_{\omega, \chi}$ of the map in Theorem 2.12 are called* modular symbols.

Before we prove Theorem 2.12, we shall first verify a few properties satisfied by modular symbols.

Lemma 2.14 *Suppose that $\{\alpha, \beta, p, q\}_{\omega, \chi}$ and $\{\beta, \gamma, p, q\}_{\omega, \chi}$ are modular symbols, and set*

$$\{\alpha, \gamma, p, q\}_{\omega, \chi} = \{\alpha, \beta, p, q\}_{\omega, \chi} + \{\beta, \gamma, p, q\}_{\omega, \chi}. \tag{2.26}$$

Then $\{\alpha, \gamma, p, q\}_{\omega, \chi}$ is also a modular symbol.

Proof. We need to verify (i) and (ii) in Theorem 2.12 for $\{\alpha, \gamma, p, q\}_{\omega,\chi}$ defined by the relation in (2.26). However, both of these properties follow easily from the definitions and the uniqueness of the modular symbols.   □

Lemma 2.15 *Consider the modular symbols*

$$\{\alpha, \beta, p(j, 1), q(j, 0)\}_{\omega,\chi}, \quad \{\alpha, \beta, p(j, 0), q(j, 1)\}_{\omega,\chi},$$

where $p(j, *)$ (resp. $q(j, *)$) denotes the element in $\mathbb{Z}^m$ obtained from $p = (p_1, \ldots, p_m)$ (resp. $q = (q_1, \ldots, q_m)$) by replacing p_j (resp. q_j) by $*$. If we set

$$\{\alpha, \beta, p, q\}_{\omega,\chi} = p_j\{\alpha, \beta, p(j, 1), q(j, 0)\}_{\omega,\chi} + q_j\{\alpha, \beta, p(j, 0), q(j, 1)\}_{\omega,\chi},$$

then $\{\alpha, \beta, p, q\}_{\omega,\chi}$ is a modular symbol.

Proof. Using the definition of $\{\,,\,\}_{\omega,\chi}$, we see that

$$\{\xi, p, q\}_{\omega,\chi} = p_j\{\xi, p(j, 1), q(j, 0)\}_{\omega,\chi} + q_j\{\xi, p(j, 0), q(j, 1)\}_{\omega,\chi}$$

for all $\xi \in \widetilde{\mathbb{Q}}$ The fact that $\{\alpha, \beta, p, q\}_{\omega,\chi}$ satisfies (2.23) follows easily from this relation. On the other hand, we have

$$
\begin{aligned}
&\langle\{\alpha, \beta, p, q\}_{\omega,\chi}, (f_1, \overline{f}_2)\rangle \\
&= p_j \int_\alpha^\beta f_1\omega(z) \prod_{i\neq j}(p_i\omega(z)q_i)dz + p_j \int_\alpha^\beta \overline{f_2\omega(z)} \prod_{i\neq j}(p_i\overline{\omega(z)} + q_i)d\overline{z} \\
&\quad + q_j \int_\alpha^\beta f_1 \prod_{i\neq j}(p_i\omega(z) + q_i)dz + q_j \int_\alpha^\beta \overline{f}_2 \prod_{i\neq j}(p_i\overline{\omega(z)} + q_i)d\overline{z} \\
&= \int_\alpha^\beta f_1 \prod_{i=1}^m (p_i\omega(z) + q_i)dz + \int_\alpha^\beta \overline{f}_2 \prod_{i=1}^m (p_i\overline{\omega(z)} + q_i)d\overline{z}.
\end{aligned}
$$

for all $f_1, f_2 \in S_{2,m}(\Gamma, \omega, \chi)$; hence we see that $\{\alpha, \beta, p, q\}_{\omega,\chi}$ satisfies (2.24).

□

Lemma 2.16 *Let $\sigma \in \mathfrak{S}_m$ be a permutation of the set $\{1, \ldots m\}$. If $\{\alpha, \beta, p, q\}_{\omega,\chi}$ is a modular symbol, then $\{\alpha, \beta, \sigma p, \sigma q\}_{\omega,\chi}$ is also a modular symbol. Furthermore, we have*

$$\{\alpha, \beta, p, q\}_{\omega,\chi} = \{\alpha, \beta, \sigma p, \sigma q\}_{\omega,\chi},$$

where $\sigma(p) = (p_{\sigma(1)}, \ldots, p_{\sigma(m)})$ and $\sigma(q) = (q_{\sigma(1)}, \ldots, q_{\sigma(m)})$.

Proof. If $\xi \in \widetilde{\mathbb{Q}}$, by using the definition of $\{\,,\,\}_{\omega,\chi}$it can be easily shown that

$$\{\xi, p, q\}_{\omega,\chi} = \{\xi, \sigma p, \sigma q\}_{\omega,\chi},$$

hence it follows that

$$\partial\{\alpha, \beta, \sigma p, \sigma q\}_{\omega,\chi} = \partial\{\alpha, \beta, p, q\}_{\omega,\chi}.$$

We also see that

$$\langle\{\alpha, \beta, \sigma p, \sigma q\}_{\omega,\chi}, (f_1, \overline{f}_2)\rangle = \langle\{\alpha, \beta, p, q\}_{\omega,\chi}, (f_1, \overline{f}_2)\rangle$$

for $f_1, f_2 \in S_{2,m}(\Gamma, \omega, \chi)$. Therefore the lemma follows from these results and the uniqueness of modular symbols. □

Now we are ready to prove Theorem 2.12.

Proof. (Theorem 2.12) Given $\alpha, \beta \in \widetilde{\mathbb{Q}}$ and $p, q \in \mathbb{Z}^m$, in order to prove uniqueness suppose we have two modular symbols $\{\alpha, \beta, p, q\}_{\omega,\chi}$ and $\{\alpha, \beta, p, q\}'_{\omega,\chi}$. If we set

$$\mu = \{\alpha, \beta, p, q\}_{\omega,\chi} - \{\alpha, \beta, p, q\}'_{\omega,\chi},$$

then $\partial\mu = 0$ by (2.23), which implies that $\mu \in H_1(X, (R_1\pi_*\mathbb{Q})^m)$. On the other hand, using (2.24), we have

$$\langle\mu, (f_1, \overline{f}_2)\rangle = 0$$

for all $f_1, f_2 \in S_{2,m}(\Gamma, \omega, \chi)$. Since the pairing $\langle\, , \rangle$ is nondegenerate, we obtain $\mu = 0$. As for the existence, we first note that by Proposition 2.11 there are modular symbols of the form

$$\{\gamma(0), \gamma(i\infty), 1_k, 1_m - 1_k\}_{\omega,\chi}.$$

By combining this with Lemma 2.15 and Lemma 2.16, we see that there exist modular symbols of the form

$$\{\gamma(0), \gamma(i\infty), p, q\}_{\omega,\chi}$$

for arbitrary $p, q \in \mathbb{Z}^m$. Now for arbitrary $\alpha, \beta \in \widetilde{\mathbb{Q}}$, we consider a finite sequence of points $\eta_1, \ldots, \eta_l \in \widetilde{\mathbb{Q}}$ such that

$$(\alpha, \eta_1) = (\gamma_1(0), \gamma_1(i\infty)), \ldots, (\eta_i, \eta_{i+1}) = (\gamma_{i+1}(0), \gamma_{i+1}(i\infty)), \ldots$$

$$\ldots, (\eta_l, \beta) = (\gamma_{l+1}(0), \gamma_{l+1}(i\infty))$$

for $\gamma_1, \ldots, \gamma_{l+1} \in SL(2, \mathbb{Z})$ (see [90, Theorem 1.6] for the proof of the existence of such a sequence). Using this and Lemma 2.15 provides us the existence of the modular symbol $\{\alpha, \beta, p, q\}_{\omega,\chi}$, and therefore the proof of the theorem is complete. □

2.4 Eichler-Shimura Relations

In this section we discuss periods of mixed cusp forms in connection with modular symbols described in Section 2.3. In particular, we establish relations among such periods, which generalize Eichler-Shimura relations for classical cusp forms studied by Šokuov [118].

As in Section 2.3, we consider an equivariant pair (ω, χ) associated to an elliptic surface E over $X = \Gamma \backslash \mathcal{H}^*$, so that the period map $\omega : \mathcal{H} \to \mathcal{H}$ is equivariant with respect to the monodromy representation $\chi : \Gamma \to SL(2, \mathbb{Z})$ for E. In this section we assume that χ can be extended to a homomorphism defined on $SL(2, \mathbb{Z})$. We shall use the same symbol χ for the extension. Thus we have a homomorphism $\chi : SL(2, \mathbb{Z}) \to SL(2, \mathbb{Z})$ satisfying

$$\omega(\gamma z) = \chi(\gamma)\omega(z)$$

for all $z \in \mathcal{H}$ and $\gamma \in \Gamma$. Then $SL(2, \mathbb{Z})$ acts on the modular symbols $\{\alpha, \beta, p, q\}_{\omega, \chi}$ described in Section 2.3 by

$$\gamma \cdot \{\alpha, \beta, p, q\}_{\omega, \chi} = \{\gamma\alpha, \gamma\beta, (p, q)\chi(\gamma)^{-1}\}_{\omega, \chi} \tag{2.27}$$
$$= \{\gamma\alpha, \gamma\beta, d_\chi p - c_\chi q, -b_\chi p + a_\chi q\}_{\omega, \chi}$$

for $\alpha, \beta \in \widetilde{\mathbb{Q}}$, $p, q \in \mathbb{Z}^m$ and $\gamma \in SL(2, \mathbb{Z})$ with $\chi(\gamma) = \begin{pmatrix} a_\chi & b_\chi \\ c_\chi & d_\chi \end{pmatrix}$. The group $SL(2, \mathbb{Z})$ also acts on the space $S_{2,m}(\Gamma, \omega, \chi)$ of mixed cusp forms of type $(2, m)$ associated to Γ, ω and χ by

$$(f \mid \gamma)(z) = j(\gamma, z)^{-2} j(\chi(\gamma), \omega(z))^{-m} f(\gamma z)$$

for $f \in S_{2,m}(\Gamma, \omega, \chi)$, $z \in \mathcal{H}$ and $\gamma \in SL(2, \mathbb{Z})$. We denote by

$$\mathcal{E}_\Gamma = SL(2, \mathbb{Z})/\Gamma$$

the set of Γ-orbits in $SL(2, \mathbb{Z})$.

Definition 2.17 *Let* $e = \Gamma\gamma \in \mathcal{E}_\Gamma$ *be the* Γ-*orbit determined by* $\gamma \in SL(2, \mathbb{Z})$, *and let* $0 \leq k \leq m$. *Given an element*

$$f = (f_1, \overline{f}_2) \in S_{2,m}(\Gamma, \omega, \chi) \oplus \overline{S_{2,m}(\Gamma, \omega, \chi)},$$

the the complex number $r(e, k, f)$ *given by*

$$r(e, k, f) = \int_0^{i\infty} (f_1 \mid_{2,m} \gamma)\omega(z)^k dz + \int_0^{i\infty} \overline{(f_2 \mid_{2,m} \gamma)}\overline{\omega(z)}^k d\overline{z}, \tag{2.28}$$

is the period *of* f *associated to* e *and* k.

Note that, since $f_j \mid_{2,m} \gamma = f_j$ for $j = 1, 2$, the value of $r(e, k, f)$ given by (2.28) is independent of the choice of the representative γ of e. If $e \in \mathcal{E}_\Gamma$

is the Γ-orbit determined by $\gamma \in SL(2, \mathbb{Z})$ and if $0 \leq k \leq m$, we denote by $\xi(e, k)$ the modular symbol defined by

$$\xi(e, k) = \gamma \cdot \{0, \infty, 1_k, 1_m - 1_k\}_{\omega, \chi}, \tag{2.29}$$

where the vectors 1_k are as in (2.15) and the action of γ on the right hand side is given by (2.27).

Proposition 2.18 *Let $\langle \,, \rangle$ be the canonical pairing given in (2.20), and consider the element*

$$f = (f_1, \overline{f}_2) \in S_{2,m}(\Gamma, \omega, \chi) \oplus \overline{S_{2,m}(\Gamma, \omega, \chi)}.$$

Then we have

$$\langle \xi(e, k), f \rangle = r(e, k, f) \tag{2.30}$$

for each $e \in \mathcal{E}_\Gamma$ and $0 \leq k \leq m$.

Proof. Let $e = \Gamma\gamma \in \mathcal{E}_\Gamma$ with $\chi(\gamma) = \begin{pmatrix} a_\chi & b_\chi \\ c_\chi & d_\chi \end{pmatrix} \in SL(2, \mathbb{Z})$. Then by (2.29) we have

$$\xi(e, k) = \{\gamma\alpha, \gamma\beta, d_\chi q - c_\chi p, -b_\chi q + a_\chi p\}_{\omega, \chi},$$

where $\alpha = 0$, $\beta = i\infty$, $p = 1_k$ and $q = 1_m - 1_k$. Hence, using (2.24), we see that

$$\langle \xi(e, f), f \rangle = \int_{\gamma\alpha}^{\gamma\beta} f_1(z) \prod_{i=1}^{m} ((d_\chi p_i - c_\chi q_i)\omega(z) + (-b_\chi p_i + a_\chi q_i))dz$$

$$+ \int_{\gamma\alpha}^{\gamma\beta} \overline{f_2(z)} \prod_{i=1}^{m} ((d_\chi p_i - c_\chi q_i)\overline{\omega(z)} + (-b_\chi p_i + a_\chi q_i))d\overline{z}$$

$$= \int_{\alpha}^{\beta} f_1(\gamma z) \prod_{i=1}^{m} ((d_\chi p_i - c_\chi q_i)\chi(\gamma)\omega(z)$$

$$+ (-b_\chi p_i + a_\chi q_i))(cz + d)^{-2}dz$$

$$+ \int_{\alpha}^{\beta} \overline{f_2(\gamma z)} \prod_{i=1}^{m} ((d_\chi p_i - c_\chi q_i)\overline{\chi(\gamma)\omega(z)}$$

$$+ (-b_\chi p_i + a_\chi q_i))\overline{(cz + d)}^{-2}d\overline{z}.$$

Using the relation $\chi(\gamma)\omega(z) = (a_\chi\omega(z) + b_\chi)(c_\chi\omega(z) + d_\chi)^{-1}$, we obtain

$$\langle \xi(e, k), f \rangle = \int_0^{i\infty} (f_1 \mid_{2,m} \gamma)\omega(z)^k dz + \int_0^{i\infty} \overline{(f_2 \mid_{2,m} \gamma)} \overline{\omega(z)}^k d\overline{z}$$

$$= r(e, k, f);$$

hence the proposition follows. $\qquad\qquad\square$

By combining (2.30) with the fact that the pairing $\langle\,,\,\rangle$ is nondegenerate on the right, we see that the value of $\xi(e, k)$ given by (2.29) is also independent of the choice of representatives of $e \in \mathcal{E}_\Gamma$.

Proposition 2.19 *Let* $\alpha, \beta \in \widetilde{\mathbb{Q}}$ *and* $p, q \in \mathbb{Z}^m$. *Then the corresponding modular symbol can be written in the form*

$$\{\alpha, \beta, p, q\}_{\omega, \chi} = \sum_{k=0}^{m} \sum_{e \in \mathcal{E}_\Gamma} c(e, k)\xi(e, k)$$

for some integers $c(e, k)$.

Proof. As in the proof of Theorem 2.12, consider a finite sequence of points $\eta_1, \ldots, \eta_l \in \widetilde{\mathbb{Q}}$ such that

$$(\alpha, \eta_1) = (\gamma_1(0), \gamma_1(i\infty)), \ldots, (\eta_i, \eta_{i+1}) = (\gamma_{i+1}(0), \gamma_{i+1}(i\infty)), \ldots$$

$$\ldots, (\eta_l, \beta) = (\gamma_{l+1}(0), \gamma_{l+1}(i\infty))$$

for $\gamma_1, \ldots, \gamma_{l+1} \in SL(2, \mathbb{Z})$. Then by (2.26) we see that

$$\{\alpha, \beta, p, q\}_{\omega, \chi} = \{\alpha, \eta_1, p, q\}_{\omega, \chi} + \sum_{i=1}^{l-1}\{\eta_i, \eta_{i+1}, p, q\}_{\omega, \chi} + \{\eta_l, \beta, p, q\}_{\omega, \chi}.$$

Thus in order to prove the proposition it suffices to consider the case where $(\alpha, \beta) = (\gamma(0), \gamma(i\infty))$ for some element γ in $SL(2, \mathbb{Z})$. In this case, by using (2.27) we have

$$\{\alpha, \beta, p, q\}_{\omega, \chi} = \gamma \cdot \{0, i\infty, p', q'\}_{\omega, \chi} \tag{2.31}$$

with $(p', q') = (p, q)\chi(\gamma)^{-1}$ or $(p, q) = (p', q')\chi(\gamma)$; hence we have

$$p' = a_1 p + c_1 q, \quad q' = b_1 p + d_1 q$$

for $\chi(\gamma) = \begin{pmatrix} a_\chi & b_\chi \\ c_\chi & d_\chi \end{pmatrix} \in SL(2, \mathbb{Z})$. Let $p' = (p'_1, \ldots, p'_m)$, $q' = (q'_1, \ldots, q'_m)$, and define $c_k(p', q')$ by

$$\prod_{i=1}^{m}(p_i X + q_i Y) = \sum_{k=0}^{m} c_k(p, q)X^k Y^{m-k}. \tag{2.32}$$

Using this together with Lemma 2.15 and Lemma 2.16, it can be shown that

$$\{0, i\infty, p', q'\}_{\omega, \chi} = \sum_{k=0}^{m} c_k(p', q')\{0, i\infty, 1_k, 1_m - 1_k\}_{\omega, \chi}.$$

From this and (2.31) it follows that

$$\{\alpha, \beta, p, q\}_{\omega,\chi} = \sum_{k=0}^{m} c_k(p', q') \, \gamma \cdot \{0, i\infty, 1_k, 1_m - 1_k\}_{\omega,\chi}$$

$$= \sum_{k=0}^{m} c_k(p', q') \xi(e, k),$$

where $e = \Gamma\gamma \in \mathcal{E}_\Gamma$, which prove the proposition. $\square$

Now to discuss Eichler-Shimura relations for the periods of mixed cusp forms, we consider the elements

$$s = \begin{pmatrix} 0 & -1 \\ 1 & 0 \end{pmatrix}, \quad t = \begin{pmatrix} 1 & -1 \\ 1 & 0 \end{pmatrix}$$

of $SL(2, \mathbb{Z})$. Using the relations

$$(s(0), s(i\infty)) = (i\infty, 0), \quad (t(0), t(i\infty)) = (i\infty, 1), \quad (t^2(0), t^2(i\infty)) = (1, 0),$$

we see that

$$\gamma \cdot \{i\infty, 0, 1_k, 1_m - 1_k\}_{\omega,\chi} = \gamma s \cdot \{0, i\infty, v(s, k), w(s, k)\}_{\omega,\chi},$$

$$\gamma \cdot \{i\infty, 1, 1_k, 1_m - 1_k\}_{\omega,\chi} = \gamma t \cdot \{0, i\infty, v(t, k), w(t, k)\}_{\omega,\chi},$$

$$\gamma \cdot \{1, 0, 1_k, 1_m - 1_k\}_{\omega,\chi} = \gamma t^2 \cdot \{0, i\infty, v(t^2, k), w(t^2, k)\}_{\omega,\chi}$$

for all $\gamma \in SL(2, \mathbb{Z})$, where by (2.27)

$$v(\alpha, k) = a(1_k) + c(1_m - 1_k), \quad w(\alpha, k) = b(1_k) + d(1_m - 1_k)$$

if $\chi(\gamma) = \begin{pmatrix} a_\chi & b_\chi \\ c_\chi & d_\chi \end{pmatrix} \in SL(2, \mathbb{Z})$. Hence, if the integers $c_j(\cdot, \cdot)$ for $1 \leq j \leq m$ are as in (2.32), we have

$$\gamma \cdot \{i\infty, 0, 1_k, 1_m - 1_k\}_{\omega,\chi} = \sum_{j=1}^{m} c_j(v(s, k), w(s, k))\xi(es, j), \qquad (2.33)$$

$$\gamma \cdot \{i\infty, 1, 1_k, 1_m - 1_k\}_{\omega,\chi} = \sum_{j=1}^{m} c_j(v(t, k), w(t, k))\xi(et, j), \qquad (2.34)$$

$$\gamma \cdot \{1, 0, 1_k, 1_m - 1_k\}_{\omega,\chi} = \sum_{j=1}^{m} c_j(v(t^2, k), w(t^2, k))\xi(et^2, j), \qquad (2.35)$$

where $e = \Gamma\gamma \in \mathcal{E}_\Gamma$.

Proposition 2.20 *If $e \in \mathcal{E}_\Gamma$, then we have*

$$\xi(e,k) + \sum_{j=1}^{m} c_j(v(s,k), w(s,k))\xi(es,j) = 0,$$

$$\xi(e,k) + \sum_{j=1}^{m} c_j(v(t,k), w(t,k))\xi(et,j)$$

$$+ \sum_{j=1}^{m} c_j(v(t^2,k), w(t^2,k))\xi(et^2,j) = 0$$

for $0 \le k \le m$, where the $c_j(\cdot, \cdot)$ are integers given by (2.32).

Proof. Using Lemma 2.14, we see that

$$\gamma \cdot \{0, i\infty, 1_k, 1_m - 1_k\}_{\omega,\chi} + \gamma \cdot \{i\infty, 0, 1_k, 1_m - 1_k\}_{\omega,\chi} = 0,$$

$$\gamma \cdot \{0, i\infty, 1_k, 1_m - 1_k\}_{\omega,\chi} + \gamma \cdot \{i\infty, 1, 1_k, 1_m - 1_k\}_{\omega,\chi}$$

$$+ \gamma \cdot \{1, 0, 1_k, 1_m - 1_k\}_{\omega,\chi} = 0$$

for all $\gamma \in SL(2, \mathbb{Z})$. Now the proposition follows combining these relations with the identities (2.33), (2.34) and (2.35). □

Theorem 2.21 *Let $f = (f_1, \overline{f}_2)$ with $f_1, f_2 \in S_{2,m}(\Gamma, \omega, \chi)$. Then we have*

$$r(e,k,f) + \sum_{j=1}^{m} c_j(v(s,k), w(s,k))r(es,j,f) = 0, \tag{2.36}$$

$$r(e,k,f) + \sum_{j=1}^{m} c_j(v(t,k), w(t,k))r(et,j,f) \tag{2.37}$$

$$+ \sum_{j=1}^{m} c_j(v(t^2,k), w(t^2,k))r(et^2,j,f) = 0$$

for all $e \in \mathcal{E}_\Gamma$ and $0 \le k \le m$.

Proof. The theorem follows immediately from Proposition 2.18 and Proposition 2.20. □

The relations (2.36) and (2.37) may be regarded as the generalized Eichler-Shimura relations for the periods $r(e,k,f)$ of mixed cusp forms.

We now want to discuss the period map for mixed cusp forms. Let $V = \bigoplus \mathbb{Q}(e,k)$ be the $\mathbb{Q}$-vector space generated by the pairs (e,k) with $e \in \mathcal{E}_\Gamma$ and $0 \le k \le m$. Let V^* denote the dual space of V, and let $\{(e,f)^*\}$ be the dual basis of $\{(e,f)\}$. If K is a subfield of $\mathbb{C}$, we denote by $R_{2,m}(\Gamma, \omega, \chi, K)$ the subspace of $V^*(K) = V^* \otimes_{\mathbb{Q}} K$ consisting of all the elements of the form

$$\sum_{k=0}^{m} \sum_{e \in E} \rho(e,k)(e,k)^*$$

such that the coefficients $\rho(e,k) \in K$ satisfy the following relations:

$$\rho(e,k) + \sum_{j=1}^{m} c_j(v(s,k), w(s,k))\rho(es,j) = 0, \tag{2.38}$$

$$\rho(e,k) + \sum_{j=1}^{m} c_j(v(t,k), w(t,k))\rho(et,j) \tag{2.39}$$

$$+ \sum_{j=1}^{m} c_j(v(t^2,k), w(t^2,k))\rho(et^2,j) = 0.$$

Now we define the *period map*

$$r : S_{2,m}(\Gamma,\omega,\chi) \oplus \overline{S_{2,m}(\Gamma,\omega,\chi)} \longrightarrow R_{2,m}(\Gamma,\omega,\chi,\mathbb{C})$$

for mixed cusp forms by

$$r(f) = \sum_{k=0}^{m} \sum_{e \in E} r(e,k,f)(e,k)^* \tag{2.40}$$

for all $f = (f_1, \overline{f}_2)$ with $f_1, f_2 \in S_{2,m}(\Gamma,\omega,\chi)$.

Lemma 2.22 *If K, K' are subfields of $\mathbb{C}$ with $K \subset K' \subset \mathbb{C}$, then there is a canonical isomorphism of K'-vector spaces*

$$R_{2,m}(\Gamma,\omega,\chi,K) \otimes_K K' \cong R_{2,m}(\Gamma,\omega,\chi,K'). \tag{2.41}$$

Proof. This follows from the fact that the equations (2.38) and (2.39) are defined over $\mathbb{Q}$. $\square$

Proposition 2.23 *The period map r is injective. Furthermore, there is a canonical isomorphism*

$$R_{2,m}(\Gamma,\omega,\chi,K)^* \cong H_1(X, \Sigma, (R_1\pi_*K)^m). \tag{2.42}$$

for each subfield K of $\mathbb{C}$.

Proof. Suppose $r(f) = 0$ for $f = (f_1, \overline{f}_2)$ with $f_1, f_2 \in S_{2,m}(\Gamma,\omega,\chi)$. Then from (2.30) and (2.40) it follows that

$$\langle \xi(e,k), \Phi \rangle = 0$$

for all $e \in E$ and $0 \le k \le m$. Using the nondegeneracy of the paring $\langle , \rangle$ and the fact that the modular symbols $\xi(e,k)$ generate the homology group

$$H_1(X, (R_1\pi_*\mathbb{Q})^m) \subset H_1(X, \Sigma, (R_1\pi_*\mathbb{Q})^m),$$

we see that $f = 0$; hence r is injective. Now note that the assignment $(e, k) \mapsto \xi(e, k)$ defines a linear map from V into the space $H_1(X, \Sigma, (R_1\pi_*\mathbb{Q})^m)$ and that there is a canonical isomorphism

$$H_1(X, \Sigma, (R_1\pi_*K)^m) \cong H_1(X, \Sigma, (R_1\pi_*\mathbb{Q})^m) \otimes_\mathbb{Q} K.$$

Thus the isomorphism in (2.42) is obtained by combining these results with (2.30), (2.41), and the definition of $R_{2,m}(\Gamma, \omega, \chi, K)^*$. $\square$

3

Mixed Automorphic Forms and Cohomology

Given a positive integer m, let $\mathrm{Sym}^m(\mathbb{C}^2)$ denote the m-th symmetric power of $\mathbb{C}^2$, and let $H_P^1(\Gamma, \mathrm{Sym}^m(\mathbb{C}^2))$ be the associated parabolic cohomology of Γ, where the Γ-module structure of $\mathrm{Sym}^m(\mathbb{C}^2)$ is induced by the standard representation of $\Gamma \subset SL(2, \mathbb{R})$ on $\mathbb{C}^2$. Then the Eichler-Shimura isomorphism can be written in the form

$$H_P^1(\Gamma, \mathrm{Sym}^m(\mathbb{C}^2)) = S_{m+2}(\Gamma) \oplus \overline{S_{m+2}(\Gamma)},$$

where $S_{m+2}(\Gamma)$ is the space of cusp forms of weight $m+2$ for Γ (cf. [22, 112]). In particular, there is a canonical embedding of the space of cusp forms into the parabolic cohomology space. The Eichler-Shimura isomorphism can also be viewed as a Hodge structure on the parabolic cohomology (cf. [6]). If (ω, χ) is an equivariant pair considered in Chapter 1, we may consider another action of Γ on $\mathrm{Sym}^m(\mathbb{C}^2)$ which is induced by the homomorphism $\chi : \Gamma \to SL(2, \mathbb{R})$. If we denote the resulting Γ-module by $\mathrm{Sym}_\chi^m(\mathbb{C}^2)$, the associated parabolic cohomology $H_P^1(\Gamma, \mathrm{Sym}_\chi^m(\mathbb{C}^2))$ is linked to mixed automorphic forms associated to the equivariant pair (ω, χ). Indeed, mixed cusp forms can be embedded into such parabolic cohomology space, and they can also be used to determine a Hodge structure on $H_P^1(\Gamma, \mathrm{Sym}_\chi^m(\mathbb{C}^2))$, which provides an extension of the Eichler-Shimura isomorphism to mixed automorphic forms.

If $(\mathrm{Sym}^m(\mathbb{C}^2))^*$ denotes the dual of the complex vector space $\mathrm{Sym}^m(\mathbb{C}^2)$, there is a canonical paring

$$H^1(\Gamma, \mathrm{Sym}^m(\mathbb{C}^2)) \times H_1(\Gamma, (\mathrm{Sym}^m(\mathbb{C}^2))^*) \to \mathbb{C}$$

known as the Kronecker pairing (cf. [48]). In [48], Katok and Millson determined the value of the Kronecker pairing between the image of a cusp form for Γ of weight $2m+2$ in $H^1(\Gamma, S^{2m}V)$ and a certain 1-cycle in $H_1(\Gamma, S^{2m}V)$ associated to each element of Γ. They used this to determine a necessary and sufficient condition for the space of cusp forms to be spanned by a certain set of relative Poincaré series (see [47, 48]). Similar results can be obtained by using the Γ-module $\mathrm{Sym}_\chi^m(\mathbb{C}^2)$ and mixed cusp forms associated to (ω, χ).

In Section 3.1 and Section 3.2 we discuss relations between mixed cusp forms and parabolic cohomology of the corresponding discrete subgroup of $SL(2, \mathbb{R})$. We construct a map carrying a mixed cusp form to a parabolic cohomology class in Section 3.1 and show that the resulting map is injective in Section 3.2 by using a pairing on the space of mixed cusp forms. A Hodge structure of the parabolic cohomology in terms of mixed cusp forms is studied in Section 3.3. This generalizes the usual Eichler-Shimura isomorphism for cusp forms. In Section 3.4 we describe the value of the Kronecker pairing between the canonical image of a mixed cusp form f of type $(2, 2r)$ in $H^1(\Gamma, \operatorname{Sym}_\chi^{2r}(\mathbb{C}^2))$ and a cycle $\gamma \otimes Q_\gamma^r$ in $H_1(\Gamma, (\operatorname{Sym}_\chi^{2r}(\mathbb{C}^2))^*)$ for each $\gamma \in \Gamma$, where Q_γ^r is a certain element of $(\operatorname{Sym}^{2r}(\mathbb{C}^2))^*$ associated to γ, r, and the homomorphism χ.

3.1 Mixed Cusp Forms and Parabolic Cohomology

Connections between the cohomology of a discrete subgroup Γ of $SL(2, \mathbb{R})$ and automorphic forms for Γ were made by Eichler [22] and Shimura [112] decades ago. Indeed, they established an isomorphism between the space of cusp forms of weight $m + 2$ for Γ and the parabolic cohomology space of Γ with coefficients in the space of homogeneous polynomials of degree m in two variables over $\mathbb{R}$. A similar isomorphism for mixed cusp forms may not hold in general as can be seen in [20, Section 3] where mixed cusp forms of type $(0, 3)$ were studied in connection with elliptic surfaces (see Section 3.3). In this section we construct a map from the space of mixed cusp forms of type (k, ℓ) associated to Γ, ω and χ with $k \geq 2$ to the parabolic cohomology space of Γ with coefficients in some Γ-module.

If R is a commutative ring R we denote by $\mathcal{P}_{X,Y}^n(R)$ the R-algebra of homogeneous polynomials of degree n in two variables X and Y. Then the semigroup $M(2, R)$ of 2×2 matrices with entries in R acts on $\mathcal{P}_{X,Y}^n(R)$ on the left by

$$M^n(\gamma)\phi(X, Y) = \phi((X, Y)(\gamma^\iota)^t), \tag{3.1}$$

where $(\cdot)^t$ denotes the transpose of the matrix $(\cdot)$ and

$$\gamma^\iota = \operatorname{tr}(\gamma) \cdot I - \gamma = \det(\gamma)\gamma^{-1} \tag{3.2}$$

with I being the identity matrix.

Let Γ be a discrete subgroup of $SL(2, \mathbb{R})$, and let $\omega : \mathcal{H} \to \mathcal{H}$ and $\chi : \Gamma \to SL(2, \mathbb{R})$ be as in Section 1.1. For fixed nonnegative integers k and m we set

$$\mathcal{P}^{k,m}(\mathbb{C}) = \mathcal{P}_{X_1,Y_1}^k(\mathbb{C}) \otimes \mathcal{P}_{X_2,Y_2}^m(\mathbb{C}).$$

Then Γ acts on $\mathcal{P}^{k,m}(\mathbb{C})$ by $M_\chi^{k,m}(\gamma) = M^k(\gamma) \otimes M^m(\chi(\gamma))$, that is,

$$M_\chi^{k,m}(\gamma)\big(\phi(X_1, Y_1) \otimes \psi(X_2, Y_2)\big)$$
$$= \big(M^k(\gamma)\phi(X_1, Y_1)\big) \otimes \big(M^m(\chi(\gamma))\psi(X_2, Y_2)\big)$$

for all $\gamma \in \Gamma$, $\phi(X_1, Y_1) \in \mathcal{P}^k_{X_1, Y_1}(\mathbb{C})$ and $\psi(X_2, Y_2) \in \mathcal{P}^m_{X_2, Y_2}(\mathbb{C})$, where $M^k(\gamma)$ and $M^m(\chi(\gamma))$ are as in (3.1). Thus we can consider the parabolic cohomology $H^1_P(\Gamma, \mathcal{P}^{k,m}(\mathbb{C}))$ of Γ with coefficients in $\mathcal{P}^{k,m}(\mathbb{C})$, which can be described as follows.

Let $Z^1(\Gamma, \mathcal{P}^{k,m}(\mathbb{C}))$ be the set of 1-cocycles for the action of Γ on $\mathcal{P}^{k,m}(\mathbb{C})$. Thus it consists of maps $u : \Gamma \to \mathcal{P}^{k,m}(\mathbb{C})$ such that

$$u(\gamma\delta) = u(\gamma) + M^{k,m}_\chi(\gamma)u(\delta) \tag{3.3}$$

for all $\gamma, \delta \in \Gamma$. We denote by $Z^1_P(\Gamma, \mathcal{P}^{k,m}(\mathbb{C}))$ the subspace of $Z^1(\Gamma, \mathcal{P}^{k,m}(\mathbb{C}))$ consisting of the maps $u : \Gamma \to \mathcal{P}^{k,m}(\mathbb{C})$ satisfying

$$u(\pi) \in (M^{k,m}_\chi(\pi) - 1)\mathcal{P}^{k,m}(\mathbb{C}) \tag{3.4}$$

for $\pi \in P$, where P is the set of parabolic elements of Γ. We also denote by $B^1(\Gamma, \mathcal{P}^{k,m}(\mathbb{C}))$ the set of maps $u : \Gamma \to \mathcal{P}^{k,m}(\mathbb{C})$ satisfying

$$u(\gamma) = (M^{k,m}_\chi(\gamma) - 1)x \tag{3.5}$$

for all $\gamma \in \Gamma$, where x is an element of $\mathcal{P}^{k,m}(\mathbb{C})$ independent of γ. Then the parabolic cohomology of Γ with coefficients in $\mathcal{P}^{k,m}(\mathbb{C})$ is given by

$$H^1_P(\Gamma, \mathcal{P}^{k,m}(\mathbb{C})) = Z^1_P(\Gamma, \mathcal{P}^{k,m}(\mathbb{C}))/B^1(\Gamma, \mathcal{P}^{k,m}(\mathbb{C}))$$

(see e.g. [41, Appendix], [114, Chapter 8] for details).

Now we denote by $\Delta_{k,m}(z)$ the differential form on $\mathcal{H}$ with values in the space $\mathcal{P}^{k,m}(\mathbb{C})$ given by

$$\Delta_{k,m}(z) = (X_1 - zY_1)^k \otimes (X_2 - \omega(z)Y_2)^m dz \tag{3.6}$$

for all $z \in \mathcal{H}$.

Lemma 3.1 *Given $\gamma \in \Gamma$, we have*

$$\gamma^* \Delta_{k,m}(z) = j(\gamma, z)^{-k-2} j(\chi(\gamma), \omega(z))^{-m} (M^{k,m}_\chi(\gamma)\Delta_{k,m}(z))$$

for all $z \in \mathcal{H}$, where $\gamma^* \Delta_{k,m}(z) = \Delta_{k,m}(\gamma z)$.

Proof. Let $\gamma = \begin{pmatrix} a & b \\ c & d \end{pmatrix} \in \Gamma \subset SL(2, \mathbb{R})$. Then from (3.6) we obtain

$$\gamma^* \Delta_{k,m}(z) = (X_1 - (\gamma z)Y_1)^k \otimes (X_2 - \omega(\gamma z)Y_2)^m d(\gamma z) \tag{3.7}$$

for all $z \in \mathcal{H}$. The first factor on the right hand side of (3.7) can be written as

$$(X_1 - (\gamma z)Y_1)^k = \left((X_1, Y_1) \begin{pmatrix} 0 & 1 \\ -1 & 0 \end{pmatrix} \begin{pmatrix} \gamma z \\ 1 \end{pmatrix} \right)^k$$

$$= \left((X_1, Y_1) \begin{pmatrix} 0 & 1 \\ -1 & 0 \end{pmatrix} \gamma \begin{pmatrix} z \\ 1 \end{pmatrix} \right)^k (cz + d)^{-k}.$$

However, using (3.2), we have

$$(\gamma^\iota)^t \begin{pmatrix} 0 & 1 \\ -1 & 0 \end{pmatrix} = (\gamma^t)^{-1} \begin{pmatrix} 0 & 1 \\ -1 & 0 \end{pmatrix} = \begin{pmatrix} c & d \\ -a & -b \end{pmatrix} = \begin{pmatrix} 0 & 1 \\ -1 & 0 \end{pmatrix} \gamma;$$

hence we obtain

$$(X_1 - (\gamma z)Y_1)^k = \left((X_1, Y_1)(\gamma^\iota)^t \begin{pmatrix} 0 & 1 \\ -1 & 0 \end{pmatrix} \begin{pmatrix} z \\ 1 \end{pmatrix} \right)^k (cz + d)^{-k} \qquad (3.8)$$
$$= M^k(\gamma)(X_1 - zY_1)^k j(\gamma, z)^{-k},$$

where we used (3.1). Similarly, we have

$$(X_2 - \omega(\gamma z)Y_2)^m = \left((X_2, Y_2) \begin{pmatrix} 0 & 1 \\ -1 & 0 \end{pmatrix} \begin{pmatrix} \omega(\gamma z) \\ 1 \end{pmatrix} \right)^m$$
$$= \left((X_2, Y_2) \begin{pmatrix} 0 & 1 \\ -1 & 0 \end{pmatrix} \begin{pmatrix} \chi(\gamma)\omega(z) \\ 1 \end{pmatrix} \right)^m$$
$$= M^m(\chi(\gamma))(X_2 - \omega(z)Y_2)^m j(\chi(\gamma), \omega(z))^{-m}.$$

Therefore the lemma follows by combining this with (3.2), (3.7), (3.8), and the relation $d(\gamma z) = j(\gamma, z)^{-2}dz$. □

Given a mixed cusp form $f \in S_{k+2,m}(\Gamma, \omega, \chi)$ of type $(k+2, m)$, we also define the differential form $\Omega(f)$ on $\mathcal{H}$ by

$$\Omega(f) = 2\pi i f(z)\Delta_{k,m}(z). \qquad (3.9)$$

Corollary 3.2 *Given a mixed cusp form f in $S_{k+2,m}(\Gamma, \omega, \chi)$, we have*

$$\gamma^* \Omega(f) = M_\chi^{k,m}(\gamma)\Omega(f) \qquad (3.10)$$

for all $\gamma \in \Gamma$.

Proof. This follows immediately from Lemma 3.1 and the transformation formula in Definition 1.3(i) for mixed automorphic forms of type $(k+2, m)$ associated to Γ, ω and χ. □

We fix a point z in $\mathcal{H}^* = \mathcal{H} \cup \mathbb{Q} \cup \{\infty\}$ and for each $f \in S_{k+2,m}(\Gamma, \omega, \chi)$ we define the map $\mathcal{E}_z(f) : \Gamma \to \mathcal{P}^{k,m}(\mathbb{R})$ by

$$\mathcal{E}_z(f)(\gamma) = \int_z^{\gamma z} \mathrm{Re}(\Omega(f)) \in \mathcal{P}^{k,m}(\mathbb{R}) \qquad (3.11)$$

for each $\gamma \in \Gamma$, where $\mathrm{Re}(\cdot)$ denotes the real part of $(\cdot)$. Note that the integral is independent of the choice of the path joining z and γz, since $\Omega(f)$ is holomorphic. The integral is convergent even if z is a cusp because of the cusp condition for the mixed cusp form f given in Definition 1.3(ii)'.

Theorem 3.3 *For each mixed cusp form $f \in S_{k+2,m}(\Gamma, \omega, \chi)$ the associated map $\mathcal{E}_z(f) : \Gamma \to \mathcal{P}^{k,m}(\mathbb{C})$ is a 1-cocycle in $H^1_P(\Gamma, \mathcal{P}^{k,m}(\mathbb{C}))$ whose cohomology class is independent of the choice of the base point z.*

Proof. If $f \in S_{k+2,m}(\Gamma, \omega, \chi)$ and $\gamma, \gamma' \in \Gamma$, then by (3.11) we have

$$\mathcal{E}_z(f)(\gamma\gamma') = \int_z^{\gamma\gamma' z} \operatorname{Re}(\Omega(f)) \tag{3.12}$$

$$= \int_z^{\gamma z} \operatorname{Re}(\Omega(f)) + \int_{\gamma z}^{\gamma\gamma' z} \operatorname{Re}(\Omega(f))$$

$$= \int_z^{\gamma z} \operatorname{Re}(\Omega(f)) + \int_z^{\gamma' z} \operatorname{Re}(\gamma^* \Omega(f)).$$

However, using (3.10), we see that

$$\operatorname{Re}(\gamma^* \Omega(f)) = \operatorname{Re}(M^{k,m}_\chi(\gamma) \Omega(f)) = M^{k,m}_\chi(\gamma) \operatorname{Re}(\Omega(f)).$$

By substituting this and (3.11) into (3.12), we obtain

$$\mathcal{E}_z(f)(\gamma\gamma') = \mathcal{E}_z(f)(\gamma) + M^{k,m}_\chi(\gamma)\mathcal{E}_z(f)(\gamma'),$$

which implies by (3.3) that $\mathcal{E}_z(f)$ is a 1-cocycle for the Γ-module $\mathcal{P}^{k,m}(\mathbb{C})$. Now in order to show that it is a cocycle in the parabolic cohomology $H^1_P(\Gamma, \mathcal{P}^{k,m}(\mathbb{C}))$, let z, z' be elements of $\mathcal{H}^*$. Then we have

$$\mathcal{E}_{z'}(f)(\gamma) - \mathcal{E}_z(f)(\gamma) = \int_{z'}^{\gamma z'} \operatorname{Re}(\Omega(f)) - \int_z^{\gamma z} \operatorname{Re}(\Omega(f))$$

$$= \int_{\gamma z}^{\gamma z'} \operatorname{Re}(\Omega(f)) - \int_z^{z'} \operatorname{Re}(\Omega(f))$$

$$= M^{k,m}_\chi(\gamma) \int_z^{z'} \operatorname{Re}(\Omega(f)) - \int_z^{z'} \operatorname{Re}(\Omega(f))$$

$$= (M^{k,m}_\chi(\gamma) - 1) \int_z^{z'} \operatorname{Re}(\Omega(f));$$

hence by (3.5) it follows that $\mathcal{E}_{z'}(f)(\gamma)$ and $\mathcal{E}_z(f)(\gamma)$ determine the same cohomology class. On the other hand, if z' is a cusp $s \in \mathbb{Q} \cup \{\infty\}$ and if $\pi \in P$ is a parabolic element of Γ fixing s, then $\mathcal{E}_s(f)(\pi) = 0$, and therefore we have

$$\mathcal{E}_z(f)(\pi) = -(M^{k,m}_\chi(\pi) - 1) \int_z^s \operatorname{Re}(\Omega(f)) \in (M^{k,m}_\chi(\pi) - 1)\mathcal{P}^{k,m}(\mathbb{C}).$$

Thus by (3.4) $\mathcal{E}_z(f)(\gamma)$ is a 1-cocycle in $H^1_P(\Gamma, \mathcal{P}^{k,m}(\mathbb{C}))$, and the proof of the proposition is complete. $\qquad\square$

3.2 Pairings on Mixed Cusp Forms

In this section we continue our discussion of relations between mixed cusp forms and parabolic cohomology. We introduce a pairing on the space $S_{k+2,m}(\Gamma, \omega, \chi)$ of mixed cusp forms of type $(k+2, m)$ associated to Γ, ω and χ, and use this to show that the map $f \mapsto \mathcal{E}_z(f)$ described in Theorem 3.3 is injective.

If n is a positive integer, we denote by $S^n(\mathbb{C}^2)$ the n-th symmetric tensor power of $\mathbb{C}^2$, which can be identified with $\mathbb{C}^{n+1}$. Then each vector $(x, y)^t \in \mathbb{C}^2$ determines an element $(x, y)_n^t$ of $S^n(\mathbb{C}^2)$ given by

$$(x, y)_n^t = (x^n, x^{n-1}y, \ldots, xy^{n-1}, y^n)^t \in \mathbb{C}^{n+1}. \tag{3.13}$$

Note that in this section we consider elements of $\mathbb{C}^2$ or $\mathbb{C}^{n+1}$ as column vectors and denote the transpose of a matrix $(\cdot)$ by $(\cdot)^t$. As in [41, Section 6.2]), we now consider the $(n + 1) \times (n + 1)$ integral matrix Θ given by

$$\Theta = \left((-1)^i \binom{n}{j} \delta_{n-i,j} \right)_{0 \leq i,j \leq n}, \tag{3.14}$$

where $\delta_{n-i,j}$ is the Kronecker delta and $\binom{n}{j} = n!/(j!(n-j)!)$.

Lemma 3.4 *If Θ is as in (3.13), we have*

$$\Theta^{-1} = \left((-1)^{n-i} \binom{n}{j}^{-1} \delta_{i,n-j} \right)_{0 \leq i,j \leq n}, \tag{3.15}$$

$$(x, y)_n^t \Theta (x', y')_n^t = \det \begin{pmatrix} x & x' \\ y & y' \end{pmatrix}^n \tag{3.16}$$

for all $(x, y), (x', y') \in \mathbb{C}^2$.

Proof. For $0 \leq i, j \leq n$ we set

$$\theta_{ij} = (-1)^i \binom{n}{j} \delta_{n-i,j}, \quad \phi_{ij} = (-1)^{n-i} \binom{n}{j}^{-1} \delta_{i,n-j}.$$

Then we have

$$\sum_{j=0}^{n} \theta_{ij} \phi_{j\ell} = (-1)^i (-1)^{n-j} \binom{n}{j} \binom{n}{\ell}^{-1} \delta_{n-i,j} \delta_{j,n-\ell}$$

$$= (-1)^{2i} \binom{n}{n-i} \binom{n}{\ell}^{-1} \delta_{i,\ell} = \delta_{i,\ell};$$

hence we obtain (3.15). On the other hand, using (3.13) and (3.14), we see that

$$(x,y)_n \Theta(x',y')_n^t = \sum_{0 \le i,j \le n} (-1)^i \binom{n}{j} \delta_{n-i,j} x^{n-i} y^i x'^{n-j} y'^j$$

$$= \sum_{i=1}^{n} (-1)^i \binom{n}{n-i} (xy')^{n-i} (x'y)^i$$

$$= (xy' - x'y)^n,$$

which implies (3.16). $\qquad\square$

Using (3.15), we obtain a pairing on the space $S^n(\mathbb{C}^2)$ given by

$$\langle (x,y)_n^t, (x',y')_n^t \rangle = (x,y)_n \Theta(x',y')_n^t = \det \begin{pmatrix} x & x' \\ y & y' \end{pmatrix}^n. \tag{3.17}$$

We now identify the space $S^n(\mathbb{C}^2)$ with the dual of the space $\mathcal{P}^n_{X,Y}(\mathbb{C})$ of homogeneous polynomials of degree n in X and Y by allowing the dual of the basis vector $X^{n-i}Y^i$ with the i-th standard basis vector e_i in $\mathbb{C}^{n+1} = S^n(\mathbb{C}^2)$. Thus by (3.15) the pairing $\langle\,,\,\rangle$ in (3.17) induces the pairing $\langle\,,\,\rangle^n_{X,Y}$ on $\mathcal{P}^n_{X,Y}(\mathbb{C})$ given by

$$\left\langle \sum_{i=0}^{n} a_i X^{n-i} Y^i, \sum_{j=0}^{n} b_j X^{n-j} Y^j \right\rangle^n_{X,Y} \tag{3.18}$$

$$= (a_0, \ldots, a_n) \Theta^{-1} (b_0, \ldots, b_n)^t$$

$$= \sum_{0 \le j,\ell \le n} (-1)^{n-j} \binom{n}{\ell}^{-1} \delta_{j,n-\ell} a_j b_\ell$$

$$= \sum_{\ell=0}^{n} (-1)^\ell \binom{n}{\ell}^{-1} a_{n-\ell} b_\ell.$$

Hence we obtain a pairing of the form

$$\langle\!\langle\,,\,\rangle\!\rangle = \langle\,,\,\rangle^k_{X_1,Y_1} \cdot \langle\,,\,\rangle^m_{X_2,Y_2} \tag{3.19}$$

on the space

$$\mathcal{P}^{k,m}(\mathbb{C}) = \mathcal{P}^k_{X_1,Y_1}(\mathbb{C}) \otimes \mathcal{P}^m_{X_2,Y_2}(\mathbb{C}).$$

Lemma 3.5 *If $\langle\!\langle\,,\,\rangle\!\rangle$ is the pairing on $\mathcal{P}^{k,m}(\mathbb{C})$ in (3.19), we have*

$$\langle\!\langle (X_1 - zY_1)^k (X_2 - \omega(z)Y_2)^m, (X_1 - \bar{z}Y_1)^k (X_2 - \overline{\omega(z)}Y_2)^m \rangle\!\rangle$$

$$= (\bar{z} - z)^k (\overline{\omega(z)} - \omega(z))^m$$

for all $z \in \mathcal{H}$.

Proof. For each $z \in \mathcal{H}$, by (3.19), we have

$$\left\langle\!\!\left\langle (X_1 - zY_1)^k (X_2 - \omega(z)Y_2)^m, (X_1 - \bar{z}Y_1)^k (X_2 - \overline{\omega(z)}Y_2)^m \right\rangle\!\!\right\rangle$$
$$= \left\langle X_1 - zY_1, X_1 - \bar{z}Y_1 \right\rangle_{X_1,Y_1}^k \cdot \left\langle X_2 - \omega(z)Y_2, X_2 - \overline{\omega(z)}Y_2 \right\rangle_{X_2,Y_2}^m.$$

However, using (3.18), we obtain

$$\left\langle X_1 - zY_1, X_1 - \bar{z}Y_1 \right\rangle_{X_1,Y_1}^k$$
$$= \left\langle \sum_{i=0}^k (-z)^i \binom{k}{i} X_1^{k-i} Y_1^i, \sum_{j=0}^k (-\bar{z})^j \binom{k}{j} X_1^{k-j} Y_1^j \right\rangle$$
$$= \sum_{\ell=0}^k (-1)^\ell \binom{k}{\ell}^{-1} (-z)^{k-\ell} \binom{k}{k-\ell} (-\bar{z})^\ell \binom{k}{\ell}$$
$$= \sum_{\ell=0}^k (-1)^{k-\ell} \binom{k}{k-\ell} z^{k-\ell} \bar{z}^\ell = (\bar{z} - z)^k.$$

Similarly, we have

$$\left\langle X_2 - \omega(z)Y_2, X_2 - \overline{\omega(z)}Y_2 \right\rangle_{X_2,Y_2}^m = (\overline{\omega(z)} - \omega(z))^m.$$

Hence the lemma follows. □

Let Θ_1 (resp. Θ_2) be the matrix that determines the pairing on $S^k(\mathbb{C}^2)$ (resp. $S^m(\mathbb{C}^2)$) dual to $\langle , \rangle_{X_1,Y_1}^k$ (resp. $\langle , \rangle_{X_2,Y_2}^m$). Now let f and g be mixed cusp forms belonging to $S_{k+2,m}(\Gamma, \omega, \chi)$ so that $\operatorname{Re} \Omega(f)$, $\operatorname{Re} \Omega(g)$ are elements of $\mathcal{P}^{k,m}(\mathbb{R}) \subset \mathcal{P}^{k,m}(\mathbb{C})$. By identifying the element

$$\left(\sum_{i=0}^k a_i X_1^{k-i} Y_1^i \right) \otimes \left(\sum_{j=0}^m b_j X_2^{m-j} Y_2^j \right)$$

with the vector $(a_0, \ldots, a_k)^t \otimes (b_0, \ldots, b_n)^t$, we obtain

$$\langle\!\langle \operatorname{Re} \Omega(f), \operatorname{Re} \Omega(g) \rangle\!\rangle = \operatorname{Re}(\Omega(f))^t \wedge (\Theta_1^{-1} \otimes \Theta_2^{-1}) \operatorname{Re}(\Omega(g)). \tag{3.20}$$

We denote the form on the right hand side of (3.20) by $\Phi(f, g)$ and define the pairing $I(\cdot, \cdot) : S_{k+2,m}(\Gamma, \omega, \chi) \times S_{k+2,m}(\Gamma, \omega, \chi) \to \mathbb{C}$ on the space of mixed cusp forms $S_{k+2,m}(\Gamma, \omega, \chi)$ by

$$I(f, g) = \int_{\Gamma \backslash \mathcal{H}} \Phi(f, g) \tag{3.21}$$
$$= \int_{\Gamma \backslash \mathcal{H}} \operatorname{Re}(\Omega(f))^t \wedge (\Theta_1^{-1} \otimes \Theta_2^{-1}) \operatorname{Re}(\Omega(g))$$

for $f, g \in S_{k+2,m}(\Gamma, \omega, \chi)$.

Proposition 3.6 *The pairing $I(\cdot,\cdot)$ on $S_{k+2,m}(\Gamma,\omega,\chi)$ given by (3.21) is nondegenerate.*

Proof. Let $f, g \in S_{k+2,m}(\Gamma,\omega,\chi)$. Using the relations

$$\operatorname{Re}\Omega(f) = \frac{1}{2}(\Omega(f) + \overline{\Omega(f)}), \quad \operatorname{Re}\Omega(g) = \frac{1}{2}(\Omega(g) + \overline{\Omega(g)}),$$

we obtain

$$\Phi(f,g) = \frac{1}{4}\Big(\Omega(f)^t \wedge (\Theta_1^{-1} \otimes \Theta_2^{-1})\overline{\Omega(g)} \tag{3.22}$$
$$+ \overline{\Omega(f)}^t \wedge (\Theta_1^{-1} \otimes \Theta_2^{-1})\Omega(g)\Big).$$

However, by using (3.6), (3.9)and Lemma 3.5 we have

$$\Omega(f)^t \wedge (\Theta_1^{-1} \otimes \Theta_2^{-1})\overline{\Omega(g)} \tag{3.23}$$
$$= (2\pi i)^2 f(z)\overline{g(z)}\langle (X_1 - zY_1)^k, (X_1 - \overline{z}Y_1)^k \rangle_{X_1,Y_1}^k$$
$$\times \langle (X_2 - \omega(z)Y_2)^m, (X_2 - \overline{\omega(z)}Y_2)^m \rangle_{X_2,Y_2}^m dz \wedge d\overline{z}$$
$$= -4\pi^2 f(z)\overline{g(z)}(\overline{z} - z)^k(\overline{\omega(z)} - \omega(z))^m dz \wedge d\overline{z}$$
$$= -4\pi^2(-2i)f(z)\overline{g(z)}(\overline{z} - z)^k(\overline{\omega(z)} - \omega(z))^m dx \wedge dy$$
$$= -4\pi^2(-2i)^{k+m+1}f(z)\overline{g(z)}(\operatorname{Im} z)^k(\operatorname{Im}\omega(z))^m dx \wedge dy.$$

Similarly, we see that

$$\overline{\Omega(f)}^t \wedge (\Theta_1^{-1} \otimes \Theta_2^{-1})\Omega(g) \tag{3.24}$$
$$= -4\pi^2\overline{f(z)}g(z)(z - \overline{z})^k(\omega(z) - \overline{\omega(z)})^m d\overline{z} \wedge dz$$
$$= -4\pi^2(2i)^{k+m+1}\overline{f(z)}g(z)(\operatorname{Im} z)^k(\operatorname{Im}\omega(z))^m dx \wedge dy.$$

Thus from (3.22), (3.23) and (3.24) we obtain

$$\Phi(f,g) = -\pi^2(-2i)^{k+m+1}\Big(f(z)\overline{g(z)} + (-1)^{k+m+1}\overline{f(z)}g(z)\Big) \tag{3.25}$$
$$\times (\operatorname{Im} z)^k(\operatorname{Im}\omega(z))^m dx \wedge dy.$$

We now consider the Petersson inner product $\langle\,,\,\rangle_P$ on $S_{k+2,m}(\Gamma,\omega,\chi)$ given by

$$\langle f,g \rangle_P = \int_{\Gamma\backslash\mathcal{H}} f(z)\overline{g(z)}(\operatorname{Im} z)^k(\operatorname{Im}\omega(z))^m dxdy$$

(see [7, Proposition 2.1]). Then from this and (3.25) we obtain

$$I(f,g) = \int_{\Gamma\backslash\mathcal{H}} \Phi(f,g) = 4\pi^2(-2i)^{k+m-1}\Big(\langle f,g \rangle_P + (-1)^{k+m+1}\langle g,f \rangle_P\Big).$$

In particular, we have

$$I(f, i^{k+m-1}g) = 4\pi^2(-2i)^{k+m-1}\Big((-i)^{k+m-1}\langle f, g\rangle_P$$
$$+ (-1)^{k+m+1}i^{k+m-1}\langle g, f\rangle_P\Big)$$
$$= (-2)^{k+m+1}\pi^2 \operatorname{Re}\langle f, g\rangle_P,$$

$$I(f, i^{k+m}g) = 4\pi^2(-2i)^{k+m-1}\Big((-i)^{k+m}\langle f, g\rangle_P + (-1)^{k+m+1}i^{k+m}\langle g, f\rangle_P\Big)$$
$$= -(-2)^{k+m+2}\pi^2 \operatorname{Im}\langle f, g\rangle_P,$$

where we used the fact that $\langle g, f\rangle_P = \overline{\langle f, g\rangle_P}$. Hence the nondegeneracy of the pairing $I(\cdot, \cdot)$ follows from the nondegeneracy of the Petersson inner product $\langle\,,\,\rangle_P$.   □

In order to discuss the injectivity of the map $f \mapsto \mathcal{E}_z(f)$ described in Theorem 3.3, let $s \in \mathbb{Q} \cup \{\infty\}$ be a cusp of Γ such that $\sigma\infty = s$ with $\sigma \in SL(2,\mathbb{R})$. Given $\varepsilon > 0$, we set

$$V_{s,\varepsilon} = \{z \in \Gamma_s\backslash\mathcal{H} \mid \operatorname{Im}(\sigma^{-1}(z))^{-1} < \varepsilon\},$$

where Γ_s is the stabilizer of s in Γ as usual. We choose ε such that the members of $\{V_{s,\varepsilon} \mid s \in \Sigma\}$ are mutually disjoint, where Σ is the set of Γ-cusps. Let $X_0 = \Gamma\backslash\mathcal{H}$, $X = \Gamma\backslash\mathcal{H} \cup \Sigma$, and let

$$X_1 = X_0 - \bigcup_{s\in\Sigma} V_{s,\varepsilon}.$$

As is described in [41, Section 6.1], there is a triangulation $\mathcal{K}$ of X_1 satisfying the following conditions:

(i) Each element of Γ induces a simplicial map of $\mathcal{K}$ onto itself.
(ii) For each $s \in \Sigma$ the boundary of $V_{s,\varepsilon}$ is the image of a 1-chain of $\mathcal{K}$.
(iii) There is a fundamental domain D_1 in $\mathcal{H}_1$ whose closure consists of finitely many simplices in $\mathcal{K}$, where $\mathcal{H}_1$ is the inverse image of X_1 in $\mathcal{H}$.

If g denotes the genus of X and if ν is the number of cusps of Γ, then the Fuchsian group Γ is generated by $2g + \nu$ elements

$$\alpha_1, \ldots, \alpha_g, \beta_1, \ldots, \beta_g, \pi_1, \ldots, \pi_\nu$$

with the relation

$$\Big(\prod_{s\in\Sigma} \pi_s\Big)\alpha_1\beta_1\alpha_1^{-1}\beta_1^{-1} \cdots \alpha_g\beta_g\alpha_g^{-1}\beta_g^{-1} = 1.$$

Then the boundary ∂D_1 of the fundamental domain D_1 of X_1 is given by

$$\partial D_1 = \sum_{s \in \Sigma} t_s + \sum_{i=1}^{g} \Big((\alpha_i - 1)s_{\alpha_i} + (\beta_i - 1)s_{\beta_i} \Big),$$

where s_{α_i}, s_{β_i}, t_s denote the faces of D_1 corresponding to α_i, β_i, π_s, respectively.

Theorem 3.7 *Given $z \in \mathcal{H}$ and $f \in S_{k+2,m}(\Gamma, \omega, \chi)$, let $\mathcal{E}_z(f) : \Gamma \to \mathcal{P}^{k,m}(\mathbb{R})$ be as in (3.11). Then the associated map $\mathcal{E}_z : S_{k+2,m}(\Gamma, \omega, \chi) \to H^1_P(\Gamma, \mathcal{P}^{k,m}(\mathbb{C}))$ is injective.*

Proof. Since the pairing $I(\cdot, \cdot)$ on $S_{k+2,m}(\Gamma, \omega, \chi)$ in (3.21) is nondegenerate by Proposition 3.21, in order to establish the injectivity of $\mathcal{E}_z$ it suffices to show that, if

$$\mathcal{E}_z(f) = 0 \in H^1_P(\Gamma, \mathcal{P}^{k,m}(\mathbb{C})),$$

$I(f, g) = 0$ for all $g \in S_{k+2,m}(\Gamma, \omega, \chi)$. Thus suppose that $\mathcal{E}_z(f)$ determines the zero cohomology class in $H^1_P(\Gamma, \mathcal{P}^{k,m}(\mathbb{C}))$. Then there is an element $C \in \mathcal{P}^{k,m}(\mathbb{R})$ such that

$$\mathcal{E}_z(f)(\gamma) = (M^{k,m}_\chi(\gamma) - 1)C \tag{3.26}$$

for all $\gamma \in \Gamma$. We define a map $F : \mathcal{H} \to \mathcal{P}^{k,m}(\mathbb{R})$ by

$$F(w) = \int_z^w \operatorname{Re}(\Omega(f)) + C \tag{3.27}$$

for all $w \in \mathcal{H}$. Using this, (3.10), (3.11) and (3.26), we have

$$\begin{aligned} F(\gamma w) &= \int_z^{\gamma w} \operatorname{Re}(\Omega(f)) + C \tag{3.28} \\ &= \int_{\gamma z}^{\gamma w} \operatorname{Re}(\Omega(f)) + \int_z^{\gamma z} \operatorname{Re}(\Omega(f)) + C \\ &= \int_z^{w} \gamma^* \operatorname{Re}(\Omega(f)) + \mathcal{E}_z(f)(\gamma) + C \\ &= M^{k,m}_\chi(\gamma)(F(w) - C) + \mathcal{E}_z(f)(\gamma) + C \\ &= M^{k,m}_\chi(\gamma)F(w) + \mathcal{E}_z(f)(\gamma) - (M^{k,m}_\chi(\gamma) - 1)C \\ &= M^{k,m}_\chi(\gamma)F(w) \end{aligned}$$

for all $\gamma \in \Gamma$ and $w \in \mathcal{H}$. Let $g \in S_{k+2,m}(\Gamma, \omega, \chi)$, and set

$$G(w) = \int_z^w \operatorname{Re}(\Omega(g))$$

for all $w \in \mathcal{H}$, so that $dG = \operatorname{Re} \Omega(g)$. Since $dF = \operatorname{Re} \Omega(f)$ by (3.27), we see that

$$\Phi(f,g) = dF^t \wedge (\Theta_1^{-1} \otimes \Theta_2^{-1})dG$$
$$= d\left(F^t \cdot (\Theta_1^{-1} \otimes \Theta_2^{-1})dG\right)$$
$$= d\left(F^t \cdot (\Theta_1^{-1} \otimes \Theta_2^{-1}) \operatorname{Re}(\Omega(g))\right).$$

If $X_0 = \Gamma\backslash\mathcal{H}$, $X = \Gamma\backslash\mathcal{H}^*$ and $X_1 = X_0 - \bigcup V_{s,\varepsilon}$ as before, then we obtain

$$I(f,g) = \lim_{X_1 \to X} \int_{X_0} d\left(F^t \cdot (\Theta_1^{-1} \otimes \Theta_2^{-1}) \operatorname{Re}(\Omega(g))\right)$$
$$= \lim_{X_1 \to X} \int_{\partial D_1} F^t \cdot (\Theta_1^{-1} \otimes \Theta_2^{-1}) \operatorname{Re}(\Omega(g)).$$

However, using (3.10) and (3.28), for each simplex Ξ and $\gamma \in \Gamma$ we see that

$$\int_{\gamma\Xi} F^t \cdot (\Theta_1^{-1} \otimes \Theta_2^{-1}) \operatorname{Re}(\Omega(g))$$
$$= \int_{\Xi} (\gamma^*F)^t \cdot (\Theta_1^{-1} \otimes \Theta_2^{-1})\gamma^* \operatorname{Re}(\Omega(g))$$
$$= \int_{\Xi} F^t M_\chi^{k,m}(\gamma)^t \cdot (\Theta_1^{-1} \otimes \Theta_2^{-1}) M_\chi^{k,m}(\gamma) \operatorname{Re}(\Omega(g))$$
$$= \int_{\Xi} F^t \cdot (\Theta_1^{-1} \otimes \Theta_2^{-1}) \operatorname{Re}(\Omega(g)),$$

where we used (3.10) and (3.28). Hence the integral of the differential form

$$F^t \cdot (\Theta_1^{-1} \otimes \Theta_2^{-1}) \operatorname{Re}(\Omega(g))$$

over $(\alpha_i - 1)s_{\alpha_i} + (\beta_i - 1)s_{\beta_i}$ is zero for $1 \le i \le g$, and therefore we have

$$I(f,g) = \sum_{s \in \Sigma} \lim_{X_1 \to X} \int_{t_s} F^t \cdot (\Theta_1^{-1} \otimes \Theta_2^{-1}) \operatorname{Re}(\Omega(g)).$$

Since $F(w)$ is bounded near the cusps and $\operatorname{Re}(\Omega(g))$ is rapidly decreasing at each cusp of Γ, it follows that $I(f,g) = 0$. Hence the injectivity of the map $\mathcal{E}_z$ follows, and the proof of the theorem is complete. □

Remark 3.8 *For classical cusp forms the surjectivity of the map $\mathcal{E}_z$ in Theorem 3.7 also follows from the Eichler-Shimura isomorphism. However, for mixed cusp forms $\mathcal{E}_z$ may not be surjective in general. Although in this section we only considered mixed cusp forms of type (ℓ, m) with $\ell \ge 2$, it is known that $\mathcal{E}_z$ is not necessarily surjective for mixed cusp forms of type $(0,3)$ (see [20, Section 3]).*

3.3 Hodge Structures

The Eichler-Shimura isomorphism for cusp forms determines a Hodge structure on parabolic cohomology (cf. [6]). In this section we generalize the usual Eichler-Shimura isomorphism by determining a Hodge structure on parabolic cohomology by using mixed cusp forms.

Let $\mathcal{O}_{\mathcal{H}^*}$ be the sheaf on $\mathcal{H}^* = \mathcal{H} \cup \Sigma$ characterized by (2.3). In this section, if W is one of the spaces $\mathcal{H}$, $\mathcal{H}^*$, $\mathcal{H}^\#$, X_0, X, and $X_{\Gamma'}$, we denote by V_W the constant sheaf $\mathbb{C}^2$ on W and by V_W^m the m-th symmetric power of V_W, where m is a positive integer. Thus $V_{\mathcal{H}^*}$ and $V_{\mathcal{H}^*}^m$ are the sheaves on $\mathcal{H}^*$ given by

$$V_{\mathcal{H}^*} = \mathbb{C}^2 = \mathbb{C}\,e_1 \oplus \mathbb{C}\,e_2, \qquad V_{\mathcal{H}^*}^m = \operatorname{Sym}^m(\mathbb{C}^2) = \bigoplus_{j=0}^{m} \mathbb{C}\,e_1^j \cdot e_2^{m-j},$$

where $e_1 = (1,0)$ and $e_2 = (0,1)$ are the standard basis vectors of $\mathbb{C}^2$; here we are considering elements of $\mathbb{C}^2$ as row vectors as in Section 2.1. We set

$$\mathcal{O}_{\mathcal{H}^*}(V) = \mathcal{O}_{\mathcal{H}^*} \otimes V_{\mathcal{H}^*}, \qquad \mathcal{O}_{\mathcal{H}^*}(V^m) = \mathcal{O}_{\mathcal{H}^*} \otimes V_{\mathcal{H}^*}^m.$$

The Γ-action on $\mathcal{O}_{\mathcal{H}^*}(V)$ given by (2.4) induces a Γ-action on $\mathcal{O}_{\mathcal{H}^*}(V^m)$ on the right. Given a positive integer m, let $\mathcal{F}_\Gamma^m$ be the sheaf on $X = \Gamma\backslash\mathcal{H}^*$ given by (2.7), and let $\mathcal{F}_{\Gamma'}^m$ be the sheaf on $X_{\Gamma'} = \Gamma'\backslash\mathcal{H}^\#$ given by (2.8).

Proposition 3.9 *Let $\omega_X : X \to X_{\Gamma'}$ be the period map, and let Ω^1 be the sheaf of holomorphic 1-forms on X. Then there is a canonical isomorphism*

$$S_{2,m}(\Gamma, \omega, \chi) \cong H^0(X, (\omega_X^* \mathcal{F}_{\Gamma'}^m) \otimes \Omega^1), \tag{3.29}$$

where $S_{2,m}(\Gamma, \omega, \chi)$ is the space of mixed cusp forms of type $(2, m)$ associated to Γ, ω and χ.

Proof. If $f(e_1 - ze_2)^2 \in \mathcal{F}_{\mathcal{H}^*}^2$, then by (2.6) we have

$$f(z)(e_1 - ze_2)^2 \cdot \gamma = j(\gamma, z)^{-2} f(\gamma z)(e_1 - ze_2)^2$$

for all $z \in \mathcal{H}$ and $\gamma \in \Gamma$. Since $dz \in \Omega^1$ satisfies

$$d(\gamma z) = j(\gamma, z)^{-2} dz,$$

we see that the map $(e_1 - ze_2)^2 \mapsto dz$ induces a canonical isomorphism

$$\mathcal{F}_{\mathcal{H}}^2 \cong \Omega_{\mathcal{H}}.$$

Now the proposition follows from this and Proposition 2.6.   □

Let $\mathcal{O}_{\mathcal{H}}$ be the sheaf of holomorphic functions on $\mathcal{H}$, and set

$$\mathcal{O}_{\mathcal{H}}(V) = \mathcal{O}_{\mathcal{H}} \otimes V_{\mathcal{H}} = \mathcal{O}_{\mathcal{H}} \otimes \mathbb{C}^2.$$

Then the group Γ acts on $\mathcal{O}_{\mathcal{H}}(V)$ by (2.4). We now consider another right-action of Γ on $\mathcal{O}_{\mathcal{H}}(V)$ defined by

$$(f_1(z)e_1 + f_2(z)e_2) \cdot \gamma = (f_1(\gamma z), f_2(\gamma z)) \begin{pmatrix} a_\chi & b_\chi \\ c_\chi & d_\chi \end{pmatrix} \qquad (3.30)$$
$$= (a_\chi f_1(\gamma z) + c_\chi f_2(\gamma z))e_1$$
$$+ (b_\chi f_1(\gamma z) + d_\chi f_2(\gamma z))e_2$$

for $\gamma \in \Gamma$ with $\chi(\gamma) = \begin{pmatrix} a_\chi & b_\chi \\ c_\chi & d_\chi \end{pmatrix} \in \Gamma' = \chi(\Gamma)$. This action induces a Γ action on the sheaf

$$\mathcal{O}_{\mathcal{H}}(V^m) = \mathcal{O}_{\mathcal{H}} \otimes V_{\mathcal{H}}^m = \mathcal{O}_{\mathcal{H}} \otimes \mathrm{Sym}^m(\mathbb{C}^2).$$

We denote by the associated fixed sheaf on $X_0 = \Gamma \backslash \mathcal{H}$ by

$$\mathcal{O}_{X_0}(V^m) = (\mathcal{O}_{\mathcal{H}}(V^m))^{\Gamma, \chi},$$

where we used the symbol $(\cdot)^{\Gamma, \chi}$ to distinguish it from the Γ-fixed sheaf $(\cdot)^\Gamma$ with respect to the action given by (2.4). We also denote by $\mathcal{F}_{\mathcal{H}}$ the restriction of $\mathcal{F}_{\mathcal{H}^*}$ to $\mathcal{H}$, that is,

$$\mathcal{F}_{\mathcal{H}} = \{f(z)(e_1 - ze_2) \mid f \in \mathcal{O}_{\mathcal{H}}\}. \qquad (3.31)$$

Let $\omega^* \mathcal{F}_{\mathcal{H}}$ be the subsheaf of $\mathcal{O}_{\mathcal{H}}(V)$ on $\mathcal{H}$ obtained by pulling $\mathcal{F}_{\mathcal{H}}$ back via the equivariant holomorphic map $\omega : \mathcal{H} \to \mathcal{H}$.

Lemma 3.10 *The subsheaf $\omega^* \mathcal{F}_{\mathcal{H}}$ of $\mathcal{O}_{\mathcal{H}}(V)$ on $\mathcal{H}$ is invariant under the Γ-action on $\mathcal{O}_{\mathcal{H}}(V)$ given by* (3.30).

Proof. By (3.31) and element $\phi \in \omega^* \mathcal{F}_{\mathcal{H}}$ can be written in the form

$$\phi(z) = f(\omega(z))(e_1 - \omega(z)e_2)$$

for some $f \in \mathcal{O}_{\mathcal{H}}$. Given $\gamma \in \Gamma$ with $\chi(\gamma) = \begin{pmatrix} a_\chi & b_\chi \\ c_\chi & d_\chi \end{pmatrix}$, by (3.30) we have

$$\phi(z) \cdot \gamma = f(\omega(\gamma z))((a_\chi - c_\chi \omega(\gamma z))e_1 + (b_\chi - d_\chi \omega(\gamma z))e_2).$$

However, we have

$$a_\chi - c_\chi \omega(\gamma z) = a_\chi - c_\chi \chi(\gamma)\omega(z)$$
$$= a_\chi - c_\chi \left(\frac{a_\chi \omega(\gamma z) + b_\chi}{c_\chi \omega(\gamma z) + d_\chi}\right) = \frac{1}{c_\chi \omega(\gamma z) + d_\chi}$$
$$b_\chi - d_\chi \omega(\gamma z) = b_\chi - d_\chi \left(\frac{a_\chi \omega(\gamma z) + b_\chi}{c_\chi \omega(\gamma z) + d_\chi}\right) = \frac{-\omega(\gamma z)}{c_\chi \omega(\gamma z) + d_\chi},$$

where we used the relation $a_\chi d_\chi - b_\chi c_\chi = 1$. Hence we obtain

$$\phi(z) \cdot \gamma = f(\chi(\gamma)\omega(z))(c_\chi \omega(\gamma z) + d_\chi)^{-1}(e_1 - \omega(z)e_2) \in \omega^* \mathcal{F}_\mathcal{H},$$

which proves the lemma. $\square$

For each nonnegative integer k we set

$$F_\mathcal{H}^k = (\omega^* \mathcal{F}_\mathcal{H}^k) \cdot \mathcal{O}_\mathcal{H}(V^{m-k}) \tag{3.32}$$

with $\mathcal{F}_\mathcal{H}^k = (\mathcal{F}_\mathcal{H})^{\otimes k}$. Then by Lemma 3.10 the filtration $\mathcal{O}_\mathcal{H}(V) \supset \omega^* \mathcal{F}_\mathcal{H}$ induces the Γ-invariant filtration

$$\mathcal{O}_\mathcal{H}(V^m) = F_\mathcal{H}^0 \supset F_\mathcal{H}^1 \supset \cdots \supset F_\mathcal{H}^m \supset F_\mathcal{H}^{m+1} = 0.$$

By considering the associated Γ-fixed sheaves, we obtain the filtration

$$\mathcal{O}_{X_0}(V^m) = F^0 \supset F^1 \supset \cdots \supset F^m \supset F^{m+1} = 0 \tag{3.33}$$

of sheaves on $X_0 = \Gamma \backslash \mathcal{H}$, where $F^k = (F_\mathcal{H}^k)^{\Gamma,\chi}$ for each $k \geq 0$. We denote by $\mathcal{E}_{X_0}$ the sheaf of complex C^∞-functions on X_0, and consider the sheaf

$$\mathcal{E}_{X_0}(V^m) = \mathcal{E}_{X_0} \otimes_{\mathcal{O}_{X_0}} \mathcal{O}_{X_0}(V^m)$$

of V^m-valued C^∞ functions on X_0.

Proposition 3.11 Let $\omega_{X_0}^*(\mathcal{F}_{X_0}^k)_\mathcal{E}$ and $F_\mathcal{E}^k$ be the sheaves on X_0 defined by

$$\omega_{X_0}^*(\mathcal{F}_{X_0}^k)_\mathcal{E} = (\omega^* \mathcal{F}_\mathcal{H}^k)^\Gamma \otimes_{\mathcal{O}_{X_0}} \mathcal{E}_{X_0}, \quad F_\mathcal{E}^k = F^k \otimes_{\mathcal{O}_{X_0}} \mathcal{E}_{X_0}$$

for $k \geq 0$. Then we have

$$\mathcal{E}_{X_0}(V^m) = \bigoplus_{p+q=m} \omega_{X_0}^*(\mathcal{F}_{X_0}^p)_\mathcal{E} \cdot \overline{\omega_{X_0}^*(\mathcal{F}_{X_0}^q)}_\mathcal{E} \tag{3.34}$$

and

$$\omega_{X_0}^*(\mathcal{F}_{X_0}^p)_\mathcal{E} \cdot \overline{\omega_{X_0}^*(\mathcal{F}_{X_0}^q)}_\mathcal{E} = F_\mathcal{E}^p \cap \overline{F_\mathcal{E}^q}$$

for $0 \leq p, q \leq m$.

Proof. Since $\mathcal{E}_{X_0}(V^m)$ can be identified with $(\mathcal{E}_{X_0}(V))^m$, the m-th symmetric power of $\mathcal{E}_{X_0}(V)$, it suffices to prove the proposition for $m = 1$. The second identity in this case is trivial. To prove the first identity, let $v = f(z)e_1 + g(z)e_2$ be a section of

$$\mathcal{E}_\mathcal{H}(V) = \mathcal{E}_\mathcal{H} \, e_1 \oplus \mathcal{E}_\mathcal{H} \, e_2,$$

where $\mathcal{E}_\mathcal{H}$ is the sheaf of complex C^∞-functions on $\mathcal{H}$. Then v can be expressed uniquely in the form

$$v = f_1(z)(e_1 - \omega(z)e_2) + g_1(z)(e_1 - \overline{\omega(z)}e_2)$$

by using the functions

$$f_1(z) = (\overline{\omega(z)} - \omega(z))^{-1}(f(z)\overline{\omega(z)} + g(z)),$$
$$g_1(z) = (\omega(z) - \overline{\omega(z)})^{-1}(f(z)\omega(z) + g(z))$$

for $z \in \mathcal{H}$. Since $f_1, g_1 \in \mathcal{E}_\mathcal{H}$, using (3.31), we see that

$$f_1(z)(\omega(z)e_1 + e_2) \in (\omega^* \mathcal{F}_\mathcal{H}^1) \otimes \mathcal{E}_\mathcal{H},$$
$$g_1(z)(\overline{\omega(z)}e_1 + e_2) \in \overline{(\omega^*_{X_0} \mathcal{F}^1_{X_0})} \otimes \mathcal{E}_\mathcal{H}.$$

Hence we have

$$\mathcal{E}_{X_0}(V) = (\omega^*_{X_0} \mathcal{F}^1_{X_0})_\mathcal{E} \oplus \overline{(\omega^*_{X_0} \mathcal{F}^1_{X_0})}_\mathcal{E},$$

and therefore the lemma follows.   □

Given a nonnegative integer k, we set

$$\Omega^1_\mathcal{H}(V^k) = \Omega^1_\mathcal{H} \otimes V^k_\mathcal{H},$$

where $\Omega^1_\mathcal{H}$ is the sheaf of holomorphic 1-forms on $\mathcal{H}$, and consider the associated Γ-fixed sheaf

$$\Omega^1_{X_0}(V^k) = (\Omega^1_\mathcal{H}(V^k))^{\Gamma, \chi} = \Omega^1_{X_0} \otimes_{\mathcal{O}_0} \mathcal{O}_{X_0}(V^k)$$

on X_0. Then the connection on the constant sheaf V^m on X_0 is the $\mathbb{C}$-linear map

$$\triangledown : \mathcal{O}_{X_0}(V^k) \longrightarrow \Omega^1_{X_0}(V^k) \tag{3.35}$$

induced from the Γ-invariant map

$$d \otimes 1 : \mathcal{O}_\mathcal{H}(V^k) \longrightarrow \Omega^1_\mathcal{H}(V^k)$$

with

$$(d \otimes 1)(f(z)\zeta) = (df(z))\zeta$$

for all $f \in \mathcal{O}_\mathcal{H}$ and $\zeta \in V^m$.

Proposition 3.12 *The image $\triangledown F^k$ of F^k under the map $\triangledown$ in (3.35) is contained in the sheaf $F^{k-1} \otimes_{\mathcal{O}_{X_0}} \Omega^1_{X_0}$.*

Proof. Since $\omega^* \mathcal{F}^k_\mathcal{H}$ can be regarded as $\mathcal{O}_\mathcal{H}(e_1 - \omega(z)e_2)^k$, the sheaf $F^k_\mathcal{H}$ in (3.32) can be written in the form

$$F^k_\mathcal{H} = (\omega^* \mathcal{F}^k_\mathcal{H}) \cdot \mathcal{O}_\mathcal{H}(V^{m-k}) = \mathcal{O}_\mathcal{H}(e_1 - \omega(z)e_2)^k \cdot V^{m-k}_\mathcal{H}.$$

If $f(z)$ and ξ are sections of $\mathcal{O}_\mathcal{H}$ and $V^{m-k}_\mathcal{H}$, respectively, we have

$$(d \otimes 1)(f(z)(e_1 - \omega(z)e_2)^k \cdot \xi) = d(f(z)(e_1 - \omega(z)e_2)^k) \cdot \xi$$
$$= (e_1 - \omega(z)e_2)^k \cdot (df(z))\xi$$
$$- kf(z)(e_1 - \omega(z)e_2)^{k-1} \cdot (d\omega(z))e_1 \cdot \xi.$$

Since $e_1 \cdot \xi$ is a section of $V_{\mathcal{H}}^{m-(k-1)}$, we see that

$$f(z)(e_1 - \omega(z)e_2)^{k-1} \cdot (d\omega(z))e_1 \cdot \xi$$

is a section of the sheaf

$$\omega^* \mathcal{F}_{\mathcal{H}}^{k-1} \otimes \mathcal{O}_{\mathcal{H}}(V^{m-(k-1)}) \otimes_{\mathcal{O}_{\mathcal{H}}} \Omega_{\mathcal{H}}^1 = F^{k-1} \otimes_{\mathcal{O}_{\mathcal{H}}} \Omega_{\mathcal{H}}^1.$$

On the other hand, $(e_1 - \omega(z)e_2)^k \cdot (df(z))\xi$ is a section of

$$\omega^* \mathcal{F}_{\mathcal{H}}^k \otimes \mathcal{O}_{\mathcal{H}}(V^{m-k}) \otimes_{\mathcal{O}_{\mathcal{H}}} \Omega_{\mathcal{H}}^1 = F^k \otimes_{\mathcal{O}_{\mathcal{H}}} \Omega_{\mathcal{H}}^1.$$

Using these and the fact that

$$F^k \otimes_{\mathcal{O}_{\mathcal{H}}} \Omega_{\mathcal{H}}^1 \subset F^{k-1} \otimes_{\mathcal{O}_{\mathcal{H}}} \Omega_{\mathcal{H}}^1,$$

we see that

$$(d \otimes 1)(f(z)(e_1 - \omega(z)e_2)^k \cdot \xi) \in F_{\mathcal{H}}^{k-1} \otimes_{\mathcal{O}_{\mathcal{H}}} \Omega_{\mathcal{H}}^1.$$

Thus it follows that

$$\triangledown F^k \subset (F_{\mathcal{H}}^{k-1} \otimes \Omega_{\mathcal{H}})^{\Gamma} = F^{k-1} \otimes_{X_0} \Omega_{X_0}^1,$$

and therefore the proposition follows.   □

From Proposition 3.12 it follows that the filtration $\{F^k\}$ in (3.33) determines a variation of Hodge structure on $\mathcal{O}_{X_0}(V^m)$ over X_0. We shall now introduce a polarization on $\mathcal{O}_{X_0}(V^m)$. First, we consider the bilinear form B^1 on $\mathbb{C}^2$ given by

$$B^1(ae_1 + be_2, ce_1 + de_2) = \det \begin{pmatrix} a & b \\ c & d \end{pmatrix} = ad - bc. \tag{3.36}$$

We define the bilinear form B^m on $V_{\mathcal{H}}^m = \mathrm{Sym}^m(\mathbb{C}^2)$ by

$$B^m(u_1 \cdots u_m, v_1 \cdots v_m) = \frac{1}{(m!)^2} \sum_{\sigma, \tau \in \mathfrak{S}_m} \prod_{i=1}^m B^1(u_{\sigma(i)}, v_{\tau(i)}), \tag{3.37}$$

where $\mathfrak{S}_m$ is the group of permutations of the set $\{1, \ldots, m\}$. This can be extended to the Γ-invariant bilinear form

$$B^m : \mathcal{E}_{\mathcal{H}}(V^m) \times \mathcal{E}_{\mathcal{H}}(V^m) \longrightarrow \mathcal{E}_{\mathcal{H}}.$$

Taking the Γ-fixed sheaves, we obtain the nondegenerate $\mathcal{O}_{X_0}$-bilinear form

$$B^m : \mathcal{E}_{X_0}(V^m) \times \mathcal{E}_{X_0}(V^m) \longrightarrow \mathcal{E}_{X_0} \tag{3.38}$$

on the $\mathcal{E}_{X_0}$-module $\mathcal{E}_{X_0}(V^m)$. By (3.36) we have $B(v, u) = -B(u, v)$ for all $u, v \in V$; hence, using this and (3.38), we see that

$$B^m(y, x) = (-1)^m B^m(x, y)$$

for all $x, y \in \mathcal{E}_{X_0}(V^m)$. We now consider another bilinear form $\langle \, , \, \rangle$ on $\mathcal{E}_{X_0}(V^m)$ given by

$$\langle x, y \rangle = i^{q-p} B^m(x, \bar{y}) \tag{3.39}$$

for all $x, y \in \mathcal{E}_{X_0}(V^m)$.

Proposition 3.13 *The bilinear form* $\langle \, , \, \rangle$ *on* $\mathcal{E}_{X_0}(V^m)$ *in (3.39) is positive definite on the* $\mathcal{E}_{X_0}$-*module*

$$(\omega_{X_0}^* \mathcal{F}_{X_0}^p)_{\mathcal{E}} \cdot \overline{(\omega_{X_0}^* \mathcal{F}_{X_0}^q)}_{\mathcal{E}}$$

for each pair (p, q) *of nonnegative integers with* $p + q = m$, *and the decomposition of the sheaf* $\mathcal{E}_{X_0}(V^m)$ *given by (3.34) is orthogonal with respect to the bilinear form* $\langle \, , \, \rangle$.

Proof. Let $v = e_1 - \omega(z)e_2 \in \omega_{X_0}^* \mathcal{F}_{X_0}$, so that $\bar{v} = e_1 - \overline{\omega(z)}e_2 \in \overline{\omega_{X_0}^* \mathcal{F}_{X_0}}$. If $s' \in \mathcal{H}$ and $s = \pi(s') \in X_0$, then v is $\Gamma_{s'}$-invariant basis of $\omega^* \mathcal{F}_{\mathcal{H}}$, where $\Gamma_{s'} = \{\gamma \in \Gamma \mid \gamma s' = s'\}$. Hence the stalk of $\mathcal{E}_{X_0}(V^m)$ at s is given by

$$\mathcal{E}_{X_0}(V^m)_s = \bigoplus_{p+q=m} (\omega_{X_0}^* \mathcal{F}_{X_0}^p)_{\mathcal{E},s} \cdot \overline{(\omega_{X_0}^* \mathcal{F}_{X_0}^q)}_{\mathcal{E},s}$$

$$= \bigoplus_{p+q=m} \mathcal{E}_{X_0,s} v^p \cdot \bar{v}^q.$$

Consider pairs (p, q) and (p', q') of nonnegative integers satisfying $p + q = p' + q' = m$. If $p \neq p'$ and if B^m is the bilinear form in (3.38), then by using (3.37) we see that

$$B^m(v^p \cdot \bar{v}^q, v^{p'} \cdot \bar{v}^{q'}) = 0.$$

On the other hand, we have

$$B^m(v^p \cdot \bar{v}^q, v^p \cdot \bar{v}^q) = \frac{1}{(m!)^2} m! p! q! (2i \operatorname{Im}(\omega(z)))^p (-2i \operatorname{Im}(\omega(z)))^q$$

$$= \binom{m}{p}^{-1} i^{p-q} (2 \operatorname{Im}(\omega(z)))^m.$$

Thus, using this and (3.39), we obtain

$$\langle v^p \cdot \bar{v}^q, v^{p'} \cdot \bar{v}^{q'} \rangle = \begin{cases} 0 & \text{if } p \neq p' \\ \binom{m}{p}^{-1} (2 \operatorname{Im}(\omega(z))) & \text{if } p = p'. \end{cases}$$

The proposition follows from this result. $\qquad \square$

If $\chi : \Gamma \to SL(2, \mathbb{R})$ with $\Gamma' = \chi(\Gamma)$ is the monodromy representation as before, then Γ acts on $V^m = \text{Sym}^m(\mathbb{C}^2)$ via χ and the standard representation of $SL(2, \mathbb{R})$. Thus, if we set

$$(x, y)_m = (x^m, x^{m+1}y, \ldots, xy^{m-1}, y^m) \in V^m = \text{Sym}^m(\mathbb{C}^2)$$

for $(x, y) \in \mathbb{C}^2$, then we have

$$(x, y)_m \cdot \gamma = ((x, y)\chi(\gamma))_m \tag{3.40}$$

for all $\gamma \in \Gamma$. We denote by $H^1_P(\Gamma, V^m)$ the parabolic cohomology with respect to this representation. The following proposition shows that this bilinear form determines a polarization on the variation of Hodge structure over X_0 determined by the filtration $\{F^k\}$ in (3.33).

Proposition 3.14 *If $\eta : X_0 \to X$ is the inclusion map, then the parabolic cohomology $H^1_P(\Gamma, V^m)$ is naturally isomorphic to the cohomology $H^1(X, \eta_* V^m_{X_0})$ of $X = \Gamma \backslash \mathcal{H}^*$ with coefficients in the sheaf $\eta_* V^m_{X_0}$.*

Proof. See [125, Proposition 12.5]. $\qquad\qquad\qquad\qquad\qquad\qquad\qquad\square$

Let $\Omega^\bullet_{X_0}$ be the holomorphic de Rham complex on X_0, and set

$$\Omega^\bullet_{X_0}(V^m) = \Omega^\bullet_{X_0} \otimes_{\mathcal{O}_{X_0}} \mathcal{O}_{X_0}(V^m).$$

We denote by $\Omega^\bullet_{(2)}(V^m)$ the complex of sheaves on X whose sections are the sections of the complex $\eta_* \Omega^\bullet_{X_0}(V^m)$ that are square-summable near the Γ-cusps.

Proposition 3.15 *The cohomology $H^i(X, \eta_* V^m_{X_0})$ is isomorphic to the hypercohomology*

$$\mathbb{H}^i(X, \Omega^\bullet(V^m)_{(2)})$$

of the complex $\Omega^\bullet(V^m)_{(2)}$.

Proof. The proposition follows from Theorem 4.8, Corollary 6.15 and Corollary 7.13 in [125]. $\qquad\qquad\qquad\qquad\qquad\qquad\qquad\square$

Let $\mathcal{O}_{\mathcal{H}^*}$ and $V^m_{\mathcal{H}^*}$ be as in Section 2.1, and let

$$\mathcal{O}_X(V^m) = (\mathcal{O}_{\mathcal{H}^*} \otimes V^m_{\mathcal{H}^*})^{\Gamma, \chi}.$$

We consider the complex $K^\bullet$ given by $K^0 = \mathcal{O}_X(V^m)$ and

$$K^1 = \text{im}[\nabla : \mathcal{O}_X(V^m) \longrightarrow \Omega^1_X(\log \Sigma) \otimes \mathcal{O}_X(V^m)].$$

The complexes $\Omega^\bullet(V^m)_{(2)}$ and $K^\bullet$ can be filtered by the filtrations induced from the filtration $\{F^k\}$ of $\mathcal{O}_{\mathcal{H}}(V^m)$.

Proposition 3.16 *The inclusion map $\Omega^\bullet(V^m)_{(2)} \to K^\bullet$ of filtered complexes is a quasi-isomorphism.*

Proof. See [125, Proposition 9.1]. □

Now we state our main theorem in this section, which describes a Hodge structure on the parabolic cohomology $H^1_P(\Gamma, V^m)$ in terms of mixed cusp forms.

Theorem 3.17 *The parabolic cohomology $H^1_P(\Gamma, V^m)$ of the group Γ has a polarized Hodge decomposition of the form*

$$H^1_P(\Gamma, V^m) \cong S_{2,m}(\Gamma, \omega, \chi) \oplus W \oplus \overline{S_{2,m}(\Gamma, \omega, \chi)},$$

where $W = \oplus H^{p,q}$ is the direct sum of the Hodge components of $H^1_P(\Gamma, V^m)$ for $p, q \geq 1$ and $p + q = m + 1$.

Proof. By Propositions 3.15 and 3.16, the hypercohomology $\mathbb{H}^1(X, Gr^p K^\bullet)$ determines the subspace $H^{p,m+1-p}$ in the Hodge decomposition of the cohomology space $H^1(X, \eta_* V^m_{X_0})$. Thus we have

$$H^1(X, \eta_* V^m_{X_0}) = \bigoplus_{p=0}^{m+1} H^{p,m+1-p}$$

$$= H^{m+1,0} \oplus \left(\bigoplus_{p=1}^{m} H^{p,m+1-p} \right) \oplus \overline{H^{m+1,0}}.$$

By Proposition 3.13 the bilinear form B^m determines a polarization on this Hodge structure. Since $H^1_P(\Gamma, V^m)$ is isomorphic to $H^1(X, \eta_* V^m_{X_0})$ by Proposition 3.14, it remains to prove that $H^{m+1,0}$ is isomorphic to $S_{2,m}(\Gamma, \omega, \chi)$. However, we have

$$H^{m+1,0} = F^{m+1} H^1(X, \eta_* V^m_{X_0})$$
$$= \mathbb{H}(X, F^{m+1} \Omega^\bullet(V^m)_{(2)})$$
$$= H^0(X, (\omega^* \mathcal{F}^m \otimes \Omega^1)_{(2)}),$$

where $(\omega^* \mathcal{F}^m \otimes \Omega^1)_{(2)}$ is the extension of $\omega^*_{X_0} \mathcal{F}^m \otimes \Omega^1_{X_0}$ in $\Omega^\bullet(V^m)_{(2)}$. By [125, Proposition 4.4] the sheaf $(\omega^* \mathcal{F}^m \otimes \Omega^1)_{(2)}$ is isomorphic to $\omega^*_X(\mathcal{F}^m_0) \otimes \Omega^1$. Hence from Proposition 3.9 it follows that

$$H^{m+1,0} = H^0(X, \omega^*_X(\mathcal{F}^m_0) \otimes \Omega^1) \cong S_{2,m}(\Gamma, \omega, \chi);$$

hence the proof of the theorem is complete. □

Remark 3.18 *If the homomorphism $\chi : \Gamma \to SL(2, \mathbb{R})$ is the inclusion map and if ω is the identity map on $\mathcal{H}$, then we have $W = \{0\}$ and Theorem 3.17 reduces to the well-known Eichler-Shimura isomorphism (see [6, 22, 112]). The space W in Theorem 3.17, however, is not trivial in general as can be seen in [20, Section 3] for $m = 1$.*

3.4 The Kronecker Pairing

The Kronecker pairing is a bilinear map defined on the product of cohomology and homology of a discrete subgroup Γ of $SL(2, \mathbb{R})$. In this section we determine the value of the Kronecker pairing between the image of a mixed automorphic form and a special 1-cocycle associate to each element of Γ.

Let $\Gamma \subset SL(2, \mathbb{R})$ and the equivariant pair (ω, χ) be as in Section 3.1. Thus $\omega : \mathcal{H} \to \mathcal{H}$ is a holomorphic map that is equivariant with respect to the homomorphism $\chi : \Gamma \to SL(2, \mathbb{R})$. Let $V = \mathbb{C}^2$ be the space of the standard two-dimensional complex representation of $SL(2, \mathbb{R})$, and let V^* be its dual space. Given a positive integer r, we denote by V^{2r} and $(V^{2r})^*$ be the $2r$-th symmetric power of V and its dual space, respectively. If $\{u_1, u_2\}$ is the standard basis for V^*, then $(V^{2r})^*$ can be regarded as the space of all homogeneous polynomials $P_{2r}(u_1, u_2)$ of degree $2r$ in u_1 and u_2. As in (3.40), the discrete subgroup Γ of $SL(2, \mathbb{R})$ acts on V^{2r} on the right by

$$(x, y)_{2r} \cdot \gamma = ((x, y)\chi(\gamma))_{2r}$$

for all $\gamma \in \Gamma$, where elements of $V = \mathbb{C}^2$ were considered as row vectors and

$$(x, y)_{2r} = (x^{2r}, x^{2r+1}y, \ldots, xy^{2r-1}, y^{2r}) \in V^{2r}$$

for $(x, y) \in \mathbb{C}^2$. On the other hand, Γ also acts on $(V^{2r})^*$ on the right by

$$P_{2r}(u_1, u_2) \cdot \gamma = P_{2r}((u_1, u_2)\chi(\gamma)) \tag{3.41}$$
$$= P_{2r}(a_\chi u_1 + c_\chi u_2, b_\chi u_1 + d_\chi u_2)$$

for each $\gamma \in \Gamma$ with $\chi(\gamma) = \begin{pmatrix} a_\chi & b_\chi \\ c_\chi & d_\chi \end{pmatrix} \in SL(2, \mathbb{R})$. Thus we can consider the group cohomology $H^1(\Gamma, V^{2r})$ and the group homology $H_1(\Gamma, (V^{2r})^*)$ with respect to the Γ-module structures of V^{2r} and $(V^{2r})^*$ described above. For each $\gamma \in \Gamma$ we consider the element Q_γ^r of $(V^{2r})^*$ given by

$$Q_\gamma^r = (c_\chi u_2^2 + (a_\chi - d_\chi)u_1 u_2 - b_\chi u_1^2)^r \in (V^{2r})^*, \tag{3.42}$$

which is a modification of ξ_γ in [48, p. 744] (see also [47]).

Lemma 3.19 *Given $\gamma \in \Gamma$, the element $Q_\gamma^m \in (V^{2r})^*$ given by (3.42) is invariant under the action of $\gamma \in G$.*

Proof. For $\gamma \in G$ with $\chi(\gamma) = \begin{pmatrix} a_\chi & b_\chi \\ c_\chi & d_\chi \end{pmatrix}$, using (3.41), we have

$$\gamma \cdot Q_\gamma^r = (c_\chi(b_\chi u_1 + d_\chi u_2)^2 + (a_\chi - d_\chi)(a_\chi u_1 + c_\chi u_2)(b_\chi u_1 + d_\chi u_2)$$
$$- b_\chi(a_\chi u_1 + c_\chi u_2)^2)^r$$
$$= ((a_\chi c_\chi d_\chi - b_\chi c_\chi^2)u_2^2 + (a_\chi d_\chi - b_\chi c_\chi)(a_\chi - d_\chi)u_1 u_2$$
$$+ (c_\chi b_\chi^2 - a_\chi b_\chi d_\chi)u_1^2)^r$$
$$= (c_\chi u_2^2 + (a_\chi - d_\chi)u_1 u_2 - b_\chi u_2^2)^r = Q_\gamma^r,$$

and therefore the lemma follows. $\square$

Given a positive integer r, we consider the canonical isomorphism in (3.29) for $m = 2r$ and denote it by

$$\Phi : S_{2,2r}(\Gamma, \omega, \chi) \to H^0(X, (\omega_X^* \mathcal{F}_{\Gamma'}^r) \otimes \Omega^1), \tag{3.43}$$

where $\Gamma' = \chi(\Gamma)$, and $S_{2,2r}(\Gamma, \omega, \chi)$ is the space of mixed cusp forms of type $(2, 2r)$ associated to Γ, ω and χ. If $f \in S_{2,2r}(\Gamma, \omega, \chi)$, using the descriptions in Section 3.3, we see that $\Phi(f)$ can be represented by the 1-form

$$\Phi(f) = (e_1 - \omega(z)e_2)^{2r} f(z)dz \tag{3.44}$$

for $z \in \mathcal{H}$.

Lemma 3.20 If $\langle \, , \, \rangle$ denotes the natural pairing between V^{2r} and $(V^{2r})^*$, then we have

$$\langle \Phi(f), (au_1^2 + bu_1u_2 + cu_2^2)^r \rangle = f(z)(a - b\omega(z) + c\omega(z)^2)^r dz \tag{3.45}$$

for all $f \in S_{2,2r}(\Gamma, \omega, \chi)$ and $a, b, c \in \mathbb{C}$.

Proof. Given $f \in S_{2,2r}(\Gamma, \omega, \chi)$, by (3.44) we have

$$\Phi(f) = f(z)(e_1 - \omega(z)e_2)^{2r} dz. \tag{3.46}$$

We consider the factorization $au_1^2 + bu_1u_2 + cu_2^2 = a(u_1 + su_2)(u_1 + tu_2)$ with $s, t \in \mathbb{C}$. Using this, (3.46) and the fact that $\langle e_i, u_j \rangle = \delta_{ij}$ for $1 \leq i, j \leq 2$, we see that

$$\begin{aligned}
\langle \Phi(f), (au_1^2 + bu_1u_2 + cu_2^2)^r \rangle &= f(z)a^r((e_1 - \omega(z)e_2)(u_1 + su_2))^r \\
&\quad \times ((e_1 - \omega(z)e_2)(u_1 + tu_2))^r dz \\
&= f(z)(a(1 - s\omega(z))(1 - t\omega(z)))^r dz \\
&= f(z)(a - b\omega(z) + c\omega(z)^2)^r dz,
\end{aligned}$$

which prove the lemma. $\square$

Proposition 3.21 Let k be a positive integer, and let $\omega_X^* \mathcal{F}_{\Gamma'}^k$ be the sheaf on X in (3.29). Then there is a canonical isomorphism

$$H^0(X_0, (\omega_X^* \mathcal{F}_{\Gamma'}^k) \otimes \Omega^1 \,|_{X_0}) \cong H^1(X_0, (\omega_X^* \mathcal{F}_{\Gamma'}^k) \,|_{X_0}) \tag{3.47}$$

where $X_0 = \Gamma \backslash \mathcal{H}$ and Ω^1 is the sheaf of holomorphic 1-forms on X.

Proof. We shall first construct a pairing

$$\langle , \rangle : H^0(X_0, (\omega_X^* \mathcal{F}_{\Gamma'}^k) \otimes \Omega^1 \,|_{X_0}) \times H_1(X_0, (\omega_X^* \mathcal{F}_{\Gamma'}^k)^* \,|_{X_0}) \to \mathbb{C}, \tag{3.48}$$

where $(\omega_X^* \mathcal{F}_{\Gamma'}^k)^*$ denotes the dual of the sheaf $\omega_X^* \mathcal{F}_{\Gamma'}^k$. Consider an element

$$\zeta \in H^0(X_0, (\omega_X^* \mathcal{F}_{\Gamma'}^k) \otimes \Omega^1 \,|_{X_0}),$$

which is regarded as a closed 1-form on X_0 with values in $(\omega_X^* \mathcal{F}_{\Gamma'}^k) \mid_{X_0}$. We consider an oriented 1-simplex (α, β) in X_0 and a section v of the sheaf $(\omega_X^* \mathcal{F}_{\Gamma'}^k)^* \mid_{X_0}$ restricted to (α, β). Then $(\alpha, \beta) \otimes v$ is a simplicial 1-chain in X_0 with coefficients in $(\omega_X^* \mathcal{F}_{\Gamma'}^k)^* \mid_{X_0}$. If $\xi \in H_1(X_0, (\omega_X^* \mathcal{F}_{\Gamma'}^k)^* \mid_{X_0})$ denotes the homology class of $(\alpha, \beta) \otimes v$, then the pairing (3.48) is defined by

$$\langle \zeta, \xi \rangle = \int_{(\alpha, \beta)} \langle \zeta \mid_{(\alpha, \beta)}, v \rangle$$

(see [46, Section 4] for details). From the pairing (3.48) we obtain

$$H^0(X_0, (\omega_X^* \mathcal{F}_{\Gamma'}^k) \otimes \Omega^1 \mid_{X_0}) \cong (H_1(X_0, (\omega_X^* \mathcal{F}_{\Gamma'}^k)^* \mid_{X_0})^*$$
$$\cong H^1(X_0, (\omega_X^* \mathcal{F}_{\Gamma'}^k) \mid_{X_0});$$

hence the proposition follows. □

By combining the isomorphisms (3.43) and (3.47) with the restriction map

$$H^0(X_0, (\omega_X^* \mathcal{F}_{\Gamma'}^k) \otimes \Omega^1) \to H^0(X_0, (\omega_X^* \mathcal{F}_{\Gamma'}^k) \otimes \Omega^1 \mid_{X_0}),$$

we obtain a map

$$S_{2,k}(\Gamma, \omega, \chi) \to H^1(X_0, (\omega_X^* \mathcal{F}_{\Gamma'}^k) \mid_{X_0}).$$

On the other hand, it is well-known that $H^1(X_0, V^k)$, where V^k is regarded as the sheaf of locally constant functions on X_0 with values in V^k, is canonically isomorphic to $H^1(\Gamma, V^k)$. Since $(\omega_X^* \mathcal{F}_{\Gamma'}^k) \mid_{X_0}$ is a subsheaf of V^k, the inclusion induces a map

$$H^1(X_0, (\omega_X^* \mathcal{F}_{\Gamma'}^k) \mid_{X_0}) \to H^1(X_0, V^k) \cong H^1(\Gamma, V^k).$$

Thus, for $k = 2r$, we obtain the map

$$\Psi : S_{2,2r}(\Gamma, \omega, \chi) \to H^1(\Gamma, V^{2r}).$$

We denote by $\langle\langle \, , \, \rangle\rangle$ the canonical pairing

$$\langle\langle \, , \, \rangle\rangle : H^1(\Gamma, V^{2r}) \otimes H_1(\Gamma, (V^{2r})^*) \to \mathbb{C}, \qquad (3.49)$$

known as the *Kronecker pairing* (see [48, p. 738 and p. 745]). The Kronecker pairing can be interpreted in terms of the de Rham cohomology and the simplicial homology as in [48, Section 2] by identifying the group cohomology $H^1(\Gamma, V^{2r})$ with the de Rham cohomology $H^1(X_0, V^{2r})$ and the group cohomology $H_1(\Gamma, (V^{2r})^*)$ with the simplicial homology $H_1(X_0, (V^{2r})^*)$, where $(V^{2r})^*$ is the sheaf of locally constant functions on X_0 with values in $(V^{2r})^*$. Thus let ζ be a closed 1-form on X_0 with values in V^{2r} and let $\sigma \otimes \mu$, where σ is a simplicial 1-cycle in X_0 and μ is a parallel section of $(S^{2r}V)^*$ on σ. Then the Kronecker pairing (3.49) is determined by

$$\langle\langle \zeta, \sigma \otimes \mu \rangle\rangle = \int_\sigma \langle \zeta, \mu \rangle.$$

Lemma 3.22 *If $\gamma \in \Gamma$, then the 1-chain $\gamma \otimes Q_\gamma^r$ is a cycle in $H_1(\Gamma, (S^{2r}V)^*)$, where Q_γ^r is as in (3.42).*

Proof. Note that the boundary map ∂ for 1-chains is given by

$$\partial(\gamma \otimes v) = \gamma \cdot v - v$$

for all $\gamma \in \Gamma$ and $v \in (V^{2r})^*$. Thus from Lemma 3.19 it follows that

$$\partial(\gamma \otimes Q_\gamma^r) = 0,$$

and therefore $(\gamma \otimes Q_\gamma^r$ is a cycle.   □

Theorem 3.23 *For each $\gamma \in \Gamma$ and $f \in S_{2,2r}(\Gamma, \omega, \chi)$, we have*

$$\langle\!\langle \Psi(f), \gamma \otimes Q_\gamma^r \rangle\!\rangle = \int_{z_0}^{\gamma z_0} f(z)(c_\chi \omega(z)^2 + (d_\chi - a_\chi)\omega(z) - b_\chi)^r dz,$$

where z_0 is an arbitrary point in the Poincaré upper half plane $\mathcal{H}$ and the integral is taken along any piecewise continuous path joining z_0 and gz_0.

Proof. We shall use the interpretation of the Kronecker pairing in terms of de Rham cohomology and simplicial homology. For $z_0 \in \mathcal{H}$, let σ_γ be the image of a simplicial path joining z_0 to γz_0 under the natural projection $\mathcal{H} \to X_0 = \Gamma\backslash\mathcal{H}$, and let ζ_γ^r be the parallel section of $(S^{2r}V)^*$ restricted to σ_γ obtained by parallel translation of Q_γ^r around σ_γ. Then $\sigma_\gamma \otimes \zeta_\gamma^r$ is the cycle in $H_1(X_0, (V^{2r})^*)$ corresponding to $\gamma \otimes Q_\gamma^r$ (see [46, Section 4] and [48, Section 2]). If the map

$$\Phi : S_{2,2r}(\Gamma, \omega, \chi) \to H^0(X, \mathcal{F}_0^{2r} \otimes \Omega^1)$$

is as in (??), the closed 1-form on X_0 with values in V^{2r} corresponding to $\Psi(f)$ is simply $\Phi(f)$ restricted to X_0. Using (3.42) and (3.45), we see that

$$\langle \Phi(f), \zeta_\gamma^r \rangle = \langle \Phi(f), Q_\gamma^r \rangle = f(z)(c_\chi \omega(z)^2 + (d_\chi - a_\chi)\omega(z) - b_\chi)^r dz.$$

Hence the theorem follows.   □

Mixed Hilbert and Siegel Modular Forms

As was discussed in Section 2.2, a holomorphic form of the highest degree on an elliptic variety can be identified with a mixed automorphic form of one variable. An elliptic variety E is a fiber bundle over a quotient $X = \Gamma \backslash \mathcal{H}$ whose generic fiber is the product of a finite number of elliptic curves. Thus E is a family of abelian varieties parametrized by the complex curve $\Gamma \backslash \mathcal{H}$. In this chapter we discuss more general families of abelian varieties in connection with mixed automorphic forms of several variables.

If F is a totally real number field of degree n over $\mathbb{Q}$, then $SL(2, F)$ can be embedded in $SL(2, \mathbb{R})^n$. Given a subgroup Γ of $SL(2, F)$ whose embedded image in $SL(2, \mathbb{R})^n$ is a discrete subgroup, we can consider the associated Hilbert modular variety $\Gamma \backslash \mathcal{H}^n$ obtained by the quotient of the n-fold product $\mathcal{H}^n$ of the Poincaré upper half plane $\mathcal{H}$ by the action of Γ given by linear fractional transformations. Let $\omega : \mathcal{H}^n \to \mathcal{H}^n$ be a holomorphic map that is equivariant with respect to a homomorphism $\chi : \Gamma \to SL(2, F)$. Then the equivariant pair (ω, χ) can be used to define mixed Hilbert modular forms, which can be regarded as mixed automorphic forms of n variables. On the other hand, the same equivariant pair also determines a family of abelian varieties parametrized by $\Gamma \backslash \mathcal{H}^n$. As is expected, holomorphic forms of the highest degree on such a family of abelian varieties can be interpreted as mixed Hilbert modular forms of certain type.

Another type of mixed automorphic forms of several variables can be obtained by generalizing Siegel modular forms. Let $\mathcal{H}_m$ be the Siegel upper half space of degree m on which the symplectic group $Sp(m, \mathbb{R})$ acts, and let Γ_0 be a discrete subgroup of $Sp(m, \mathbb{R})$. If $\tau : \mathcal{H}_m \to \mathcal{H}_{m'}$ is a holomorphic map of $\mathcal{H}_m$ into another Siegel upper half space $\mathcal{H}_{m'}$ that is equivariant with respect to a homomorphism $\rho : \Gamma_0 \to Sp(m', \mathbb{R})$, then the equivariant pair (τ, ρ) can be used to define mixed Siegel modular forms. The same pair can also be used to construct a family of abelian varieties parametrized by the Siegel modular variety $\Gamma \backslash \mathcal{H}_m$ such that holomorphic forms of the highest degree on the family are mixed Siegel modular forms.

In Section 4.1 we introduce mixed Hilbert modular forms and describe some of their properties. The construction of families of abelian varieties parametrized by a Hilbert modular variety and the interpretation of some of their holomorphic forms in terms of mixed Hilbert modular forms are discussed in Section 4.2. Section 4.3 concerns mixed Siegel modular forms as well as their connections with families of abelian varieties parametrized by a Siegel modular variety. In Section 4.4 we extend some of the results in Section 1.3 to the Siegel modular case by considering Fourier expansions of Siegel modular forms associated to Mixed Siegel modular forms.

4.1 Mixed Hilbert Modular Forms

In this section we define mixed Hilbert modular forms associated to an equivariant pair and discuss some of their properties.

We fix a positive integer n, and let $\mathcal{H}^n = \mathcal{H} \times \cdots \times \mathcal{H}$ be the Cartesian product of n copies of the Poincaré upper half plane $\mathcal{H}$. Then the usual operation of $SL(2, \mathbb{R})$ on $\mathcal{H}$ by linear fractional transformations induces an action of the n-fold product $SL(2, \mathbb{R})^n$ of $SL(2, \mathbb{R})$ on $\mathcal{H}^n$. Let F be a totally real number field with $[F : \mathbb{Q}] = n$. Then there are n embeddings of F into $\mathbb{R}$, which we denote by

$$a \mapsto a^{(j)}, \quad F \hookrightarrow \mathbb{R} \tag{4.1}$$

for $1 \leq j \leq n$. These embeddings induce the embedding

$$\iota : SL(2, F) \to SL(2, \mathbb{R})^n \tag{4.2}$$

defined by

$$\iota \begin{pmatrix} a & b \\ c & d \end{pmatrix} = \left(\begin{pmatrix} a^{(1)} & b^{(1)} \\ c^{(1)} & d^{(1)} \end{pmatrix}, \ldots, \begin{pmatrix} a^{(n)} & b^{(n)} \\ c^{(n)} & d^{(n)} \end{pmatrix} \right)$$

for all $\left(\begin{smallmatrix} a & b \\ c & d \end{smallmatrix} \right) \in SL(2, F)$. Throughout this section we shall identify an element g of $SL(2, F)$ with its embedded image $\iota(g)$ of $SL(2, \mathbb{R})^n$ under the embedding ι in (4.2). Thus, in particular, an element $g = \left(\begin{smallmatrix} a & b \\ c & d \end{smallmatrix} \right) \in SL(2, F)$ acts on $\mathcal{H}^n$ by

$$gz = \left(\frac{a^{(1)} z_1 + b^{(1)}}{c^{(1)} z_1 + d^{(1)}}, \ldots, \frac{a^{(n)} z_n + b^{(n)}}{c^{(n)} z_n + d^{(n)}} \right)$$

for all $z = (z_1, \ldots, z_n) \in \mathcal{H}^n$. If $z = (z_1, \ldots, z_n) \in \mathcal{H}^n$, $g = \left(\begin{smallmatrix} a & b \\ c & d \end{smallmatrix} \right) \in SL(2, F)$ and $\boldsymbol{\ell} = (\ell_1, \ldots, \ell_n) \in \mathbb{Z}^n$, we set

$$N(g, z)^{\boldsymbol{\ell}} = \prod_{j=1}^{n} \left(c^{(j)} z_j + d^{(j)} \right)^{\ell_j}. \tag{4.3}$$

Then we see easily that the resulting map

$$(g, z) \mapsto N(g, z)^{\boldsymbol{\ell}} : SL(2, F) \times \mathcal{H}^n \to \mathbb{C}$$

satisfies the cocycle condition

$$N(g_1 g_2, z)^{\boldsymbol{\ell}} = N(g_1, g_2 z)^{\boldsymbol{\ell}} N(g_2, z)^{\boldsymbol{\ell}} \tag{4.4}$$

for all $z \in \mathcal{H}^n$ and $g_1, g_2 \in SL(2, F)$.

Let Γ be a subgroup of $SL(2, F)$ whose embedded image in $SL(2, \mathbb{R})^n$ is a discrete subgroup. Let $\omega : \mathcal{H}^n \to \mathcal{H}^n$ be a holomorphic map, and let $\chi : \Gamma \to SL(2, F)$ be a homomorphism satisfying the condition

$$\omega(gz) = \chi(g)\omega(z)$$

for all $g \in \Gamma$ and $z \in \mathcal{H}^n$, so that (ω, χ) becomes an equivariant pair. We assume that the the image $\chi(\Gamma)$ of Γ under χ is also a discrete subgroup of $SL(2, \mathbb{R})^n$ and that the parabolic elements of Γ and those of $\chi(\Gamma)$ correspond. Thus the inverse image of the set of parabolic elements of $\chi(\Gamma)$ coincides with the set of parabolic elements of Γ. Let $\mathbf{k} = (k_1, \dots, k_n)$ and $\mathbf{m} = (m_1, \dots, m_n) \in \mathbb{Z}^n$ be elements of $\mathbb{Z}^n$ with $k_i, m_i \geq 0$ for each $i \in \{1, \dots, n\}$. If $g \in \Gamma \subset SL(2, F)$ and $z \in \mathcal{H}^n$, we set

$$J_{\omega,\chi}^{2\mathbf{k},2\mathbf{m}}(g, z) = N(g, z)^{2\mathbf{k}} N(\chi(g), \omega(z))^{2\mathbf{m}}, \tag{4.5}$$

where $N(\cdot, \cdot)^{(\cdot)}$ is as in (4.4). Using the cocycle condition in (4.3), we see that the resulting map $J_{\omega,\chi}^{2\mathbf{k},2\mathbf{m}} : \Gamma \times \mathcal{H}^n \to \mathbb{C}$ is an automorphy factor, that is., it satisfies the relation

$$J_{\omega,\chi}^{2\mathbf{k},2\mathbf{m}}(gh, z) = J_{\omega,\chi}^{2\mathbf{k},2\mathbf{m}}(g, hz) \cdot J_{\omega,\chi}^{2\mathbf{k},2\mathbf{m}}(h, z) \tag{4.6}$$

for all $g, h \in \Gamma$ and $z \in \mathcal{H}^n$. If $\mathbf{k} = (k, \dots, k)$ and $\mathbf{m} = (m, \dots, m)$ for some nonnegative integers k and m, then $J_{\omega,\chi}^{2\mathbf{k},2\mathbf{m}}$ will also be denoted simply by $J_{\omega,\chi}^{2k,2m}$.

In order to discuss Fourier expansions we assume that $f : \mathcal{H}^n \to \mathbb{C}$ is a function that satisfies the functional equation

$$f(gz) = J_{\omega,\chi}^{2\mathbf{k},2\mathbf{m}}(g, z)f(z)$$

for all $g \in \Gamma$ and $z \in \mathcal{H}^n$. Then we can consider the Fourier expansion of f at the cusps of Γ as follows. Suppose first that ∞ is a cusp of Γ. We set

$$\Lambda = \Lambda(\Gamma) = \{\lambda \in F \mid (\begin{smallmatrix} 1 & \lambda \\ 0 & 1 \end{smallmatrix}) \in \Gamma\},$$

and identify it with a subgroup of $\mathbb{R}^n$ via the natural embedding $F \hookrightarrow \mathbb{R}^n :$ $\lambda \mapsto (\lambda^{(1)}, \dots, \lambda^{(n)})$ induced by (4.1). Since the homomorphism χ carries parabolic elements to parabolic elements, for each $\lambda \in \Lambda$ we see that

$$\chi \begin{pmatrix} 1 & \lambda \\ 0 & 1 \end{pmatrix} = \begin{pmatrix} 1 & \lambda_\chi \\ 0 & 1 \end{pmatrix}$$

for some $\lambda_\chi \in F$, and therefore we obtain

$$J_{\omega,\chi}^{2\mathbf{k},2\mathbf{m}} \left(\begin{pmatrix} 1 & \lambda \\ 0 & 1 \end{pmatrix}, z \right) = 1.$$

Thus f is periodic with $f(z + \lambda) = f(z)$ for all $z \in \mathcal{H}$ and $\lambda \in \Lambda$, and hence it has a Fourier expansion. Let Λ^* denote the dual lattice given by

$$\Lambda^* = \{\xi \in F \mid T(\xi\lambda) \in \mathbb{Z} \quad \text{for all} \quad \lambda \in \Lambda\},$$

where $T(\xi\lambda) = \sum_{j=1}^n \xi_j \lambda_j$. Then the *Fourier expansion of f at ∞* is given by

$$f(z) = \sum_{\xi \in \Lambda^*} a_\xi e^{2\pi i T(\xi z)},$$

where $T(\xi z) = \sum_{j=1}^n \xi_j z_j$.

Now we consider an arbitrary cusp s of Γ. Let σ be an element of $SL(2, F) \subset SL(2, \mathbb{R})^n$ such that $\sigma(\infty) = s$. We assume that the homomorphism $\chi : \Gamma \to SL(2, F)$ can be extended to a map $\chi : \Gamma' \to SL(2, F)$, where

$$\Gamma' = \Gamma \cup \{\alpha \in SL(2, F) \mid \alpha(\infty) = s, \ s \text{ a cusp of } \Gamma\}.$$

Given elements $k, m \in \mathbb{Z}^n$, we set

$$\Gamma^\sigma = \sigma^{-1}\Gamma\sigma,$$

$$(f \mid \sigma)(z) = J_{\omega,\chi}^{2k,2m}(\sigma, z)^{-1} f(\sigma z). \tag{4.7}$$

for all $z \in \mathcal{H}^n$.

Lemma 4.1 *If $f : \mathcal{H}^n \to \mathbb{C}$ is a function satisfying*

$$f(gz) = J_{\omega,\chi}^{2k,2m}(g, z) f(z)$$

for all $g \in \Gamma$ and $z \in \mathcal{H}^n$, then the function $f \mid \sigma : \mathcal{H}^n \to \mathbb{C}$ given by (4.7) satisfies the functional equation

$$(f \mid \sigma)(gz) = J_{\omega,\chi}^{2k,2m}(g, z)(f \mid \sigma)(z)$$

for all $g \in \Gamma^\sigma$ and $z \in \mathcal{H}^n$.

Proof. Let $g = \sigma^{-1}\gamma\sigma \in \Gamma^\sigma$ with $\gamma \in \Gamma$. Then we have

$$\begin{aligned}
(f \mid \sigma)(gz) &= J_{\omega,\chi}^{2k,2m}(\sigma, \sigma^{-1}\gamma\sigma z)^{-1} f(\sigma\sigma^{-1}\gamma\sigma z) \\
&= J_{\omega,\chi}^{2k,2m}(\sigma, \sigma^{-1}\gamma\sigma z)^{-1} J_{\omega,\chi}^{2k,2m}(\gamma, \sigma z) f(\gamma\sigma z) \\
&= J_{\omega,\chi}^{2k,2m}(\sigma^{-1}\gamma, \sigma z) f(\sigma z),
\end{aligned}$$

where we used the relation

$$\begin{aligned}
J_{\omega,\chi}^{2k,2m}(\gamma, \sigma z) &= J_{\omega,\chi}^{2k,2m}(\sigma\sigma^{-1}\gamma, \sigma z) \\
&= J_{\omega,\chi}^{2k,2m}(\sigma, \sigma^{-1}\gamma\sigma z) J_{\omega,\chi}^{2k,2m}(\sigma^{-1}\gamma, \sigma z)
\end{aligned}$$

that follows from (4.6). However, using (4.6) again, we have

$$J_{\omega,\chi}^{2k,2m}(\sigma^{-1}\gamma\sigma, z) = J_{\omega,\chi}^{2k,2m}(\sigma^{-1}\gamma, \sigma z)J_{\omega,\chi}^{2k,2m}(\sigma, z).$$

Thus we obtain

$$(f \mid \sigma)(gz) = J_{\omega,\chi}^{2k,2m}(\sigma^{-1}\gamma\sigma, z)J_{\omega,\chi}^{2k,2m}(\sigma, z)^{-1}f(\sigma z)$$
$$= J_{\omega,\chi}^{2k,2m}(\sigma^{-1}\gamma\sigma, z)(f \mid \sigma)(\sigma z);$$

hence the lemma follows. $\qquad\qquad\qquad\qquad\qquad\qquad\qquad\qquad$ □

Since ∞ is a cusp of Γ^σ, the function $f \mid \sigma$ has a Fourier expansion at ∞ of the form

$$(f \mid \sigma)(z) = \sum_{\xi \in \Lambda^*} a_\xi e^{2\pi i T(\xi z)}.$$

This series is called a *Fourier expansion of f at the cusp s*, and the coefficients a_ξ are called the *Fourier coefficients of f at s*.

Definition 4.2 *Let $\Gamma \in SL(2, \mathbb{R})^n$ be a discrete subgroup with cusp s, and let $f : \mathcal{H}^n \to \mathbb{C}$ be a holomorphic function satisfying the relation*

$$f(gz) = J_{\omega,\chi}^{2k,2m}(g, z)f(z).$$

(i) The function f is regular *at s if the Fourier coefficients of f at s satisfy the condition that $\xi \geq 0$ whenever $a_\xi \neq 0$.*
(ii) The function f vanishes *at s if the Fourier coefficients of f at s satisfy the condition that $\xi > 0$ whenever $a_\xi \neq 0$.*

Remark 4.3 *Given a cusp s of Γ there may be more than one element $\sigma \in SL(2, F)$ such that $\sigma(\infty) = s$. However the above definition makes sense because of the next lemma.*

Lemma 4.4 *Let s be a cusp of Γ and assume that $\sigma(\infty) = \sigma'(\infty) = s$ for $\sigma, \sigma' \in SL(2, F)$. Then $f \mid \sigma$ is regular (resp. vanishes) at ∞ if and only if $f \mid \sigma'$ is regular (resp. vanishes) at ∞.*

Proof. It is sufficient to prove the lemma for the case when σ' is the identity element in $SL(2, F)$ and $s = \infty$. Then we have $\sigma(\infty) = \infty$, and hence

$$\sigma = \begin{pmatrix} \delta & 0 \\ 0 & \delta^{-1} \end{pmatrix} \begin{pmatrix} 1 & b \\ 0 & 1 \end{pmatrix}$$

for some $b, \delta \in F$. Let $\Lambda_\sigma = \Lambda(\Gamma^\sigma) = \Lambda(\sigma^{-1}\Gamma\sigma)$. Then $\lambda \in \Lambda_\sigma$ if and only if

$$\sigma \begin{pmatrix} 1 & \lambda \\ 0 & 1 \end{pmatrix} \sigma^{-1} = \begin{pmatrix} 1 & \delta^2\lambda \\ 0 & 1 \end{pmatrix} \in \Gamma;$$

hence we have $\Lambda_\sigma = \delta^{-2}\Lambda$. Therefore $\Lambda_\sigma^* = \delta^2\Lambda^*$, and we have the Fourier expansions

$$f(z) = \sum_{\xi \in \Lambda^*} a_\xi e^{2\pi i T(\xi z)}, \qquad (f \mid \sigma)(z) = \sum_{\xi \in \Lambda_\sigma^*} a_\xi^\sigma e^{2\pi i T(\xi z)}.$$

On the other hand, we have

$$
\begin{aligned}
(f \mid \sigma)(z) &= J_{\omega,\chi}^{2k,2m}(\sigma,z)^{-1} f(\sigma z) = J_{\omega,\chi}^{2k,2m}(\sigma,z)^{-1} f(\delta^2(z+b)) \\
&= J_{\omega,\chi}^{2k,2m}(\sigma,z)^{-1} \sum_{\xi \in \Lambda^*} a_\xi e^{2\pi i T(\delta^2 b \xi)} e^{2\pi i T(\xi \delta^2 z)} \\
&= J_{\omega,\chi}^{2k,2m}(\sigma,z)^{-1} \sum_{\xi \in \Lambda_\sigma^*} a_{\xi \delta^{-2}} e^{2\pi i T(b\xi)} e^{2\pi i T(\xi z)}
\end{aligned}
$$

Thus we obtain

$$a_\xi^\sigma = J_{\omega,\chi}^{2k,2m}(\sigma,z)^{-1} e^{2\pi i T(b\xi)} a_{\xi \delta^{-2}}$$

for all $\xi \in \Lambda_\sigma^*$. The lemma follows from this relation.   □

Definition 4.5 *Let Γ, χ, and ω be as above, and assume that the quotient space $\Gamma \backslash \mathcal{H}^n \cup \{cusps\}$ is compact. A* mixed Hilbert modular form *of type $(2k, 2m)$ associated to Γ, ω and χ is a holomorphic function $f : \mathcal{H}^n \to \mathbb{C}$ satisfying the following conditions:*
  (i) $f(\gamma z) = J_{\omega,\chi}^{2k,2m}(\gamma, z) f(z)$ for all $\gamma \in \Gamma$.
  (ii) f is regular at the cusps of Γ.
The holomorphic function f is a mixed Hilbert cusp form *of the same type if (ii) is replaced with the following condition:*
  (ii)′ f vanishes at the cusps of Γ.
  If $\mathbf{k} = (k, \dots, k)$ and $\mathbf{m} = (m, \dots, m)$ with nonnegative integers k and m, then a mixed Hilbert modular form of type $(2\mathbf{k}, 2\mathbf{m})$ will also be called a mixed Hilbert modular form *of type $(2k, 2m)$.*

As in the case of the usual Hilbert modular forms, Koecher's principle also holds true in the mixed case as is described in the next proposition. Thus the condition (ii) is not necessary for $n \geq 2$.

Proposition 4.6 *If $n \geq 2$, then any holomorphic function $f : \mathcal{H}^n \to \mathbb{C}$ satisfying the condition (i) in Definition 4.5 is a mixed Hilbert modular form of type $(2\mathbf{k}, 2\mathbf{m})$ associated to Γ, ω and χ.*

Proof. Let ε be an element in F such that the transformation $z \mapsto \varepsilon z + b$ is contained in Γ for some b. Then we have

$$
\begin{aligned}
f(\varepsilon z + b) &= J_{\omega,\chi}^{2k,2m}\left(\begin{pmatrix} \varepsilon^{1/2} & b\varepsilon^{-1/2} \\ 0 & \varepsilon^{-1/2} \end{pmatrix}, z \right) f(z) \\
&= \varepsilon_1^{-k_1} \cdots \varepsilon_n^{-k_n} \varepsilon_{\chi,1}^{-m_1} \cdots \varepsilon_{\chi,n}^{-m} f(z) \\
&= N(\varepsilon^{-\mathbf{k}}) N(\varepsilon_\chi^{-\mathbf{m}}) f(z)
\end{aligned}
$$

if

$$\chi \begin{pmatrix} \varepsilon^{1/2} & b\varepsilon^{-1/2} \\ 0 & \varepsilon^{-1/2} \end{pmatrix} = \begin{pmatrix} \varepsilon_\chi^{1/2} & d\varepsilon_\chi^{-1/2} \\ 0 & \varepsilon_\chi^{-1/2} \end{pmatrix}$$

for some elements $\varepsilon_\chi, d \in F$ (note that the image of a parabolic element under χ is a parabolic element). Hence, if $f(z) = \sum_{\xi \in \Lambda^*} a_\xi e^{2\pi i T(\xi z)}$ is the Fourier expansion of $f(z)$ at ∞, then we have

$$a_{\varepsilon\xi} = a_\xi e^{2\pi i T(\xi b)} N(\varepsilon^{-k}) N(\varepsilon_\chi^{-m}).$$

Now suppose $\xi = (\xi_1, \ldots, \xi_n) \in \mathbb{R}^n$ with $\xi_i < 0$ for some i, and choose a unit $\varepsilon \gg 0$ such that $\varepsilon_i > 1$ and $\varepsilon_j < 1$ for $j \neq i$. Let c be any positive real number, and consider the subseries

$$\sum_{m=1}^\infty a_{\varepsilon^{2m}\xi} e^{2\pi i T(\varepsilon^{2m}\xi ic)}$$

$$= a_\xi e^{2\pi i T(\xi b)} \sum_{m=1}^\infty N(\varepsilon^{-2mk}) N((\varepsilon^{2m})_\chi^{-m}) e^{-2\pi c T(\varepsilon^{2m}\xi)}$$

of the Fourier series of $f(ic)$. Since we have

$$T(\varepsilon^{2m}\xi) = \varepsilon_i^{2m}\xi_i + \sum_{j \neq i} \varepsilon_j^{2m}\xi_j,$$

the above subseries cannot converge unless $a_\xi = 0$. Therefore ξ is positive whenever $a_\xi \neq 0$. □

4.2 Families of Abelian Varieties over Hilbert Modular Varieties

In this section we discuss connections between mixed Hilbert modular forms and holomorphic forms on families of abelian varieties parametrized by a Hilbert modular variety.

Let $\mathcal{H}^n$, Γ, ω and χ be as in Section 4.1. Thus $\Gamma \subset SL(2, F)$ is a discrete subgroup of $SL(2, \mathbb{R})^n$, $\chi : \Gamma \to SL(2, F) \subset SL(2, \mathbb{R})^n$ is a homomorphism of groups, and $\omega : \mathcal{H}^n \to \mathcal{H}^n$ is a holomorphic map equivariant with respect to χ. Throughout the rest of this paper, we shall assume that

$$\gamma \cdot (\mathbb{Z} \times \mathbb{Z})^n \subset (\mathbb{Z} \times \mathbb{Z})^n$$

for all $\gamma \in \Gamma$.

Consider the semidirect product $\Gamma \ltimes (\mathbb{Z} \times \mathbb{Z})^{mn}$ consisting of the elements of the form

$$(g, (\mu, \nu)) = (g_1, \ldots, g_n; (\mu, \nu)_1, \ldots (\mu, \nu)_n)$$
$$= (g_1, \ldots, g_n; (\mu_1^1, \nu_1^1), \ldots (\mu_1^m, \nu_1^m); \ldots ; (\mu_n^1, \nu_n^1), \ldots (\mu_n^m, \nu_n^m))$$

with its multiplication operation given by

$$(g, (\mu, \nu)) \cdot (g', (\mu', \nu')) = (gg', (\mu, \nu)g' + (\mu', \nu')),$$

where

$$(\mu, \nu) = ((\mu, \nu)_1, \ldots, (\mu, \nu)_n)$$
$$= ((\mu_1^1, \nu_1^1), \ldots, (\mu_1^m, \nu_1^m); \ldots; (\mu_n^1, \nu_n^1), \ldots, (\mu_n^m, \nu_n^m))$$

with $\mu_j^k, \nu_j^k \in \mathbb{Z}$ for $1 \le j \le n$ and $1 \le k \le m$, and

$$(\mu, \nu)g' = ((\mu, \nu)_1 g_1', \ldots, (\mu, \nu)_n g_n')$$
$$= ((\mu_1^1, \nu_1^1)g_1', \ldots, (\mu_1^m, \nu_1^m)g_1'; \ldots; (\mu_n^1, \nu_n^1)g_n', \ldots, (\mu_n^m, \nu_n^m)g_n')$$

for $g' = (g_1', \ldots, g_n') \in \Gamma \subset SL(2, \mathbb{R})^n$. Then the discrete group $\Gamma \ltimes (\mathbb{Z} \times \mathbb{Z})^{mn}$ operates on $\mathcal{H}^n \times \mathbb{C}^{mn}$ by

$$(g, (\mu, \nu)) \cdot (z, \zeta) \tag{4.8}$$
$$= \left(\frac{a_1 z_1 + b_1}{c_1 z_1 + d_1}, \ldots, \frac{a_n z_n + b_n}{c_n z_n + d_n}; \right.$$
$$\frac{\mu_1^1 \omega(z)_1 + \nu_1^1 + \zeta_1^1}{c_{\chi,1} \omega(z)_1 + d_{\chi,1}}, \ldots, \frac{\mu_1^m \omega(z)_1 + \nu_1^m + \zeta_1^m}{c_{\chi,1} \omega(z)_1 + d_{\chi,1}}; \ldots$$
$$\left. \ldots; \frac{\mu_n^1 \omega(z)_n + \nu_n^1 + \zeta_n^1}{c_{\chi,n} \omega(z)_n + d_{\chi,n}}, \ldots, \frac{\mu_n^m \omega(z)_n + \nu_n^m + \zeta_n^m}{c_{\chi,n} \omega(z)_n + d_{\chi,n}} \right),$$

where

$$g = (g_1, \ldots, g_n) \in \Gamma \quad \text{with} \quad g_j = \begin{pmatrix} a_j & b_j \\ c_j & d_j \end{pmatrix} \in SL(2, \mathbb{R}) \quad \text{for} \quad 1 \le j \le n,$$

$$(\mu, \nu) = ((\mu_1^1, \nu_1^1), \ldots, (\mu_1^m, \nu_1^m); \ldots; (\mu_n^1, \nu_n^1), \ldots, (\mu_n^m, \nu_n^m)) \in (\mathbb{Z} \times \mathbb{Z})^{mn},$$

$$(z, \zeta) = (z_1, \ldots, z_n; \zeta_1^1, \ldots, \zeta_1^m; \ldots; \zeta_n^1, \ldots, \zeta_n^m) \in \mathcal{H}^n \times \mathbb{C}^{mn},$$

$$\chi(g) = (\chi(g)_1, \ldots, \chi(g)_n) \quad \text{with} \quad \chi(g)_j = \begin{pmatrix} a_{\chi,j} & b_{\chi,j} \\ c_{\chi,j} & d_{\chi,j} \end{pmatrix} \in SL(2, \mathbb{R})$$

for $1 \le j \le n$, and $\omega(z) = (\omega(z)_1, \ldots, \omega(z)_n) \in \mathcal{H}^n$.

Now we assume that Γ does not contain elements of finite order, so that the corresponding quotient $\Gamma \backslash \mathcal{H}^n$ has the structure of a complex manifold, and set

$$E_{\omega, \chi}^{m,n} = \Gamma \ltimes (\mathbb{Z} \times \mathbb{Z})^{mn} \backslash \mathcal{H}^n \times \mathbb{C}^{mn}, \tag{4.9}$$

where the quotient is taken with respect to the operation of $\Gamma \ltimes (\mathbb{Z} \times \mathbb{Z})^{mn}$ on $\mathcal{H}^n \times \mathbb{C}^{mn}$ given by (4.8). If X_Γ denotes the Hilbert modular variety $\Gamma \backslash \mathcal{H}^n$, then the canonical projection map $\mathcal{H}^n \times \mathbb{C}^{mn} \to \mathcal{H}^n$ induces the map $\pi_\Gamma^{m,n} : E_{\omega, \chi}^{m,n} \to X_\Gamma$, which is a fiber bundle over X_Γ and whose fiber is the complex torus $(\mathbb{C}/\mathbb{Z} \times \mathbb{Z})^{mn}$. Now let $\Gamma' = \chi(\Gamma)$ be the image of Γ under χ regarded as a subgroup of $SL(2,\mathbb{R})^n$. If $1 : \mathcal{H}^n \to \mathcal{H}^n$ is the identity map and $\iota : \Gamma' \to SL(2,\mathbb{R})^n$ is the inclusion map, the corresponding quotient similar to (4.9) with Γ replaced with Γ' determines the fiber bundle $E_{1,\iota}^{m,n}$ over the Hilbert modular variety $X_{\Gamma'} = \Gamma'\backslash\mathcal{H}^n$. The next proposition shows that $E_{\omega,\chi}^{m,n}$ can be regarded as a family of abelian varieties parametrized by X_Γ.

Proposition 4.7 *(i) If $m = 1$, the corresponding map $\pi_\Gamma^{1,n} : E_{\omega,\chi}^{1,n} \to X_\Gamma$ is a fiber bundle over the Hilbert modular variety $X_\Gamma = \Gamma\backslash\mathcal{H}^n$ whose fiber $\mathbb{C}^n/(\mathbb{Z} \times \mathbb{Z})^n$ has a canonical structure of an abelian variety.*

(ii) If $m > 1$, the space $E_{\omega,\chi}^{m,n}$ is an m-fold fiber power of the fiber bundle $E_{\omega,\chi}^{1,n}$ in (i) over X_Γ

Proof. These statements are proved in [92, Proposition 7.4] for the case of $E_{1,\iota}^{m,n}$. The proof for the general case follows from the observation that $E_{\omega,\chi}^{m,n}$ can be obtained by pulling back the fiber bundle $E_{1,\iota}^{m,n}$ over $X_{\Gamma'} = \Gamma'\backslash\mathcal{H}^n = \chi(\Gamma)\backslash\mathcal{H}^n$ via the natural map $X_\Gamma \to X_{\chi(\Gamma)}$ induced by $\omega : \mathcal{H}^n \to \mathcal{H}^n$ so that the diagram

$$
\begin{array}{ccc}
E_{\omega,\chi}^{m,n} & \longrightarrow & E_{\chi(\Gamma),\mathrm{id},\mathrm{id}}^{m,n} \\
\downarrow & & \downarrow \\
X_\Gamma & \longrightarrow & X_{\chi(\Gamma)}
\end{array}
$$

is commutative (see also [61, 69], [108, Chapter IV]). $\square$

Given a nonnegative integer ν, let $J_{\omega,\chi}^{2,2\nu} : \Gamma \times \mathcal{H}^n \to \mathbb{C}$ be the automorphy factor described in (4.5), that is, the automorphy factor $J_{\omega,\chi}^{2k,2l}$ for $\boldsymbol{k} = (1,\ldots,1)$ and $\boldsymbol{m} = (\nu,\ldots,\nu)$. Then the discrete subgroup $\Gamma \subset G$ operates on $\mathcal{H}^n \times \mathbb{C}$ by

$$
g \cdot (z,\zeta) = (gz, J_{\omega,\chi}^{2,2\nu}(g,z)\zeta) \tag{4.10}
$$

for all $g \in \Gamma$ and $(z,\zeta) \in \mathcal{H}^n \times \mathbb{C}$. We set

$$
\mathcal{L}_{\omega,\chi}^{2,2\nu} = \Gamma\backslash\mathcal{H}^n \times \mathbb{C},
$$

where the quotient is taken with respect to the operation given by (4.10). Then the natural projection $\mathcal{H}^n \times \mathbb{C} \to \mathcal{H}^n$ induces on $\mathcal{L}_{\omega,\chi}^{2,2\nu}$ the structure of a line bundle over the arithmetic variety $X_\Gamma = \Gamma\backslash\mathcal{H}^n$, and holomorphic sections of this bundle can be identified with holomorphic functions $f : \mathcal{H}^n \to \mathbb{C}$ satisfying

$$
f(gz) = J_{\omega,\chi}^{2,2\nu}(g,z)f(z)
$$

for all $g \in \Gamma$ and $z \in \mathcal{H}^n$.

Theorem 4.8 *Let $\Omega^{(2\nu+1)n}$ be the sheaf of holomorphic $(2\nu+1)n$-forms on $E_{\omega,\chi}^{2,2\nu}$. Then the space of holomorphic sections of the line bundle $\mathcal{L}_{\omega,\chi}^{2,2\nu}$ over X_Γ is canonically isomorphic to the space $H^0(E_{\omega,\chi}^{2,2\nu}, \Omega^{(2\nu+1)n})$ of holomorphic $(2\nu+1)n$-forms on $E_{\omega,\chi}^{2,2\nu}$.*

Proof. From the construction of $E_{\omega,\chi}^{m,n}$ in (4.9) it follows that a holomorphic $(2\nu+1)n$-form on $E_{\omega,\chi}^{m,n}$ can be regarded as a holomorphic $(2\nu+1)n$-form on $\mathcal{H}^n \times \mathbb{C}^{2\nu n}$ that is invariant under the operation of $\Gamma \ltimes (\mathbb{Z} \times \mathbb{Z})^{2\nu n}$ given by (4.8). Since $(2\nu+1)n$ is the complex dimension of the space $\mathcal{H}^n \times \mathbb{C}^{2\nu n}$, a holomorphic $(2\nu+1)n$-form on $\mathcal{H}^n \times \mathbb{C}^{2\nu n}$ is of the form

$$\Theta = \tilde{f}(z,\zeta)dz \wedge d\zeta,$$

where

$$(z,\zeta) = (z_1,\dots,z_n;\zeta_1^1,\dots,\zeta_1^{2\nu};\dots;\zeta_n^1,\dots,\zeta_n^{2\nu}) \in \mathcal{H}^n \times \mathbb{C}^{2\nu n},$$

$$dz = dz_1 \wedge \cdots \wedge dz_n,$$

$$d\zeta = d\zeta_1^1 \wedge \cdots \wedge d\zeta_1^{2\nu} \wedge \cdots \wedge d\zeta_n^1 \wedge \cdots \wedge d\zeta_n^{2\nu},$$

and $\tilde{f}$ is a holomorphic function on $\mathcal{H}^n \times \mathbb{C}^{2\nu n}$. Given a fixed point $z_0 \in \mathcal{H}^n$, the holomorphic form Θ descends to a holomorphic $2\nu n$-form on the corresponding fiber of the fiber bundle $E_{\omega,\chi}^{m,n} \to X_\Gamma = \Gamma \backslash \mathcal{H}^n$. Since the complex dimension of the fiber is $2\nu n$, the dimension of the space of holomorphic $2\nu n$-forms is one. Thus the mapping $\zeta \mapsto \tilde{f}(z,\zeta)$ is a holomorphic $2\nu n$-variable function with $2\nu n$ independent variables, and therefore must be constant. Hence the function $\tilde{f}(z,\zeta)$ depends only on z, and $\tilde{f}(z,\zeta) = f(z)$ where f is a holomorphic function on $\mathcal{H}^n$. Given $(g,(\mu,\nu)) \in \Gamma \ltimes (\mathbb{Z} \times \mathbb{Z})^{2\nu n}$ as above, we have the operations

$$dz_j \mid (g,(\mu,\nu)) = (c_j z_j + d_j)^{-2} dz_j, \quad 1 \le j \le n,$$

$$d\zeta_j^k \mid (g,(\mu,\nu)) = (c_{\chi,j}\omega(z)_j + d_{\chi,j})^{-1} d\zeta_j^k + \sum_{i=1}^{n} F_i(z,\zeta)\, dz_i$$

for $1 \le k \le 2\nu$, $1 \le j \le n$, and some functions $F_i(z,\zeta)$. Thus the operation of $(g,(\mu,\nu))$ on Θ is given by

$$\Theta \mid (g,(\mu,\nu)) = f(gz) \prod_{j=1}^{n} (c_j z_j + d_j)^{-2} (c_{\chi,j}\omega(z)_j + d_{\chi,j})^{-2\nu} dz \wedge d\zeta.$$

Hence it follows that

$$f(gz) = f(z) \prod_{j=1}^{n} (c_j z_j + d_j)^2 (c_{\chi,j}\omega(z)_j + d_{\chi,j})^{2\nu}$$

$$= f(z) J_{\omega,\chi}^{2,2\nu}(g,z),$$

and therefore f can be identified with a holomorphic section of $\mathcal{L}_{\omega,\chi}^{2,2\nu}$. $\quad\square$

Corollary 4.9 *Let $\mathcal{A}_{2,2\nu}(\Gamma,\omega,\chi)$ be the space of mixed Hilbert modular forms of type $(2,2\nu)$ associated to Γ, ω and χ. If $n \ge 2$, then there is a canonical isomorphism*

$$\mathcal{A}_{2,2\nu}(\Gamma,\omega,\chi) \cong H^0(E_{\omega,\chi}^{2,2\nu}, \Omega^{(2\nu+1)n}).$$

Proof. The corollary follows from Theorem 4.8 and Proposition 4.6.   □

Arithmetic varieties such as the Hilbert modular variety $X_\Gamma = \Gamma \backslash \mathcal{H}^n$ considered above can be regarded as connected components of Shimura varieties (cf. [21]). Mixed Shimura varieties generalize Shimura varieties, and they play an essential role in the theory of compactifications of Shimura varieties (cf. [3, 38, 39]). A typical mixed Shimura variety is essentially a torus bundle over a family of abelian varieties parametrized by a Shimura variety (see [94, 103]). A Shimura variety and a family of abelian varieties which it parametrizes can also be considered as special cases of mixed Shimura varieties. We now want to discuss extensions of the results obtained above to the compactifications of families of abelian varieties using the theory of toroidal compactifications of mixed Shimura varieties developed in [3] (see also [38]).

Let $\pi_\Gamma : E^{2,2\nu}_{\omega,\chi} \to X_\Gamma$ be the family of abelian varieties parametrized by an arithmetic variety given by (4.9). Using the language of Shimura varieties, $E^{2,2\nu}_{\omega,\chi}$ can be regarded as the mixed Shimura variety $M^{K_f}(P,\mathcal{X})(\mathbb{C})$ associated to the group

$$P = \mathrm{Res}_{F/\mathbb{Q}} SL(2,F) \ltimes V_{4\nu n}$$

and the subgroup $K_f \subset P(\mathbb{A}_f)$ with $K_f \cap P(\mathbb{Q}) = \Gamma \ltimes (\mathbb{Z} \times \mathbb{Z})^{2\nu n}$, where Res is Weil's restriction map and $V_{4\nu n}$ is a $\mathbb{Q}$-vector space of dimension $4\nu n$. Thus $\mathcal{X}$ is a left homogeneous space under the subgroup $P(\mathbb{R}) \cdot U(\mathbb{C}) \subset P(\mathbb{C})$, where U is a subgroup of the unipotent radical W of P, and $M^{K_f}(P,\mathcal{X})(\mathbb{C}) = P(\mathbb{Q}) \backslash \mathcal{X} \times (P(\mathbb{A}_f)/K_f)$, where the operation of $P(\mathbb{Q})$ on $\mathcal{X}$ is via χ and ω (see [94, 103] for details). The arithmetic variety X_Γ is the mixed Shimura variety $M^{K_f}((P,\mathcal{X})/W)(\mathbb{C})$, which is in fact a pure Shimura variety. Furthermore, the mapping π_Γ can be considered as the natural projection map

$$M^{K_f}(P,\mathcal{X})(\mathbb{C}) \to M^{K_f}((P,\mathcal{X})/W)(\mathbb{C}).$$

There are number of ways of compactifying Shimura varieties. Among those are Baily-Borel compactifications (cf. [5]) and toroidal compactifications. The toroidal compactifications of mixed Shimura varieties were constructed by Pink in [103]. Let $\overline{X}_\Gamma$ be the Baily-Borel compactification of X_Γ, and denote by

$$\widetilde{E}^{2,2\nu}_{\omega,\chi} = M^{K_f}(P,\mathcal{X},\mathcal{S})(\mathbb{C})$$

the toroidal compactification of $E^{2,2\nu}_{\omega,\chi} = M^{K_f}(P,\mathcal{X})(\mathbb{C})$ associated to a K_f-admissible partial cone decomposition $\mathcal{S}$ for $(P,\mathcal{X})$. Then π_Γ induces the mapping $\widetilde{\pi}_\Gamma : \widetilde{E}^{2,2\nu}_{\omega,\chi} \to \overline{X}_\Gamma$ of compactifications (see [103] for details).

Theorem 4.10 *Let $\Omega^{(2\nu+1)n}(\log \partial \widetilde{E})$ be the sheaf of holomorphic $(2\nu+1)n$-forms on $\widetilde{E}^{2,2\nu}_{\omega,\chi}$ with logarithmic poles along the boundary*

$$\partial \widetilde{E} = \widetilde{E}^{2,2\nu}_{\omega,\chi} - E^{2,2\nu}_{\omega,\chi}.$$

Then there exists an extension $\overline{\mathcal{L}}_{\omega,\chi}^{2,2\nu}$ of $\mathcal{L}_{\omega,\chi}^{2,2\nu}$ to the Baily-Borel compactification $\overline{X}_\Gamma$ of X_Γ, which depends only on $(P,\mathcal{X})/U$ up to isomorphism, such that there is a canonical isomorphism

$$\Omega^{(2\nu+1)n}(\log \partial \widetilde{E}) \cong \tilde{\pi}_\Gamma^* \overline{\mathcal{L}}_{\omega,\chi}^{2,2\nu}$$

of sheaves, where the line bundle $\overline{\mathcal{L}}_{\omega,\chi}^{2,2\nu}$ is regarded as an invertible sheaf.

Proof. By Proposition 8.1 in [103], there is an invertible sheaf $\mathcal{F}$ on the Baily-Borel compactification $\overline{X}_\Gamma$ of X such that there is a canonical isomorphism

$$\pi_\Gamma^* \mathcal{F} \cong \Omega^{(2\nu+1)n}(\log \partial \widetilde{E}).$$

On the other hand, using Theorem 4.8 we obtain a canonical isomorphism $\pi_\Gamma^* \mathcal{L}_{\omega,\chi}^{2,2\nu} \cong \Omega^{(2\nu+1)n}$. Thus it follows that $\overline{\mathcal{L}}_{\omega,\chi}^{2,2\nu} = \mathcal{F}$ is the desired extension of $\mathcal{L}$. $\square$

Remark 4.11 Let $\widetilde{X}_\Gamma$ be the toroidal compactification of X_Γ, and let $\widetilde{\mathcal{L}}_{\omega,\chi}^{2,2\nu}$ be the canonical extension of $\mathcal{L}_{\omega,\chi}^{2,2\nu}$ to $\widetilde{X}_\Gamma$. Let $\iota : X_\Gamma \to \overline{X}_\Gamma$ be the canonical embedding of X_Γ into its Baily-Borel compactification $\overline{X}_\Gamma$. Then the image of the restriction map

$$H^0(\widetilde{X}_\Gamma, \widetilde{\mathcal{L}}_{\omega,\chi}^{2,2\nu}) \to H^0(X_\Gamma, \mathcal{L}_{\omega,\chi}^{2,2\nu}) \cong H^0(\overline{X}_\Gamma, \iota_* \mathcal{L}_{\omega,\chi}^{2,2\nu})$$

is the subspace of sections regular at infinity, and hence it is the space of sections in $H^0(\overline{X}_\Gamma, \iota_* \mathcal{L}_{\omega,\chi}^{2,2\nu})$ which vanish on $\overline{X}_\Gamma - X_\Gamma$, i.e., the space of mixed Hilbert modular cusp forms (see [39, p. 40], [5, Section 10]).

4.3 Mixed Siegel Modular Forms

In this section we introduce mixed Siegel modular forms, construct a family of abelian varieties parametrized by a Siegel modular variety, and show that holomorphic forms of the highest degree on such a family are mixed Siegel modular forms.

Given a positive integer m, let $Sp(m,\mathbb{R})$ and $\mathcal{H}_m$ be the symplectic group and the Siegel upper half space, respectively, of degree m. Then $Sp(m,\mathbb{R})$ acts on $\mathcal{H}_m$ as usual by

$$\begin{pmatrix} a & b \\ c & d \end{pmatrix} \cdot z = (az+b)(cz+d)^{-1}$$

for all $z \in \mathcal{H}_m$ and $\begin{pmatrix} a & b \\ c & d \end{pmatrix} \in Sp(m,\mathbb{R})$. Let Γ be an arithmetic subgroup of $Sp(m,\mathbb{R})$, and let m' be another positive integer. Let $\tau : \mathcal{H}_m \to \mathcal{H}_{m'}$ be a holomorphic map, and assume that there is a homomorphism $\rho : \Gamma \to Sp(m',\mathbb{R})$ such that τ is equivariant with respect to ρ, that is,

$$\tau(\gamma z) = \rho(\gamma)\tau(z)$$

for all $\gamma \in \Gamma$ and $z \in \mathcal{H}_m$. If ν is a positive integer and $g = \begin{pmatrix} a & b \\ c & d \end{pmatrix} \in Sp(\nu, \mathbb{R})$, we shall write

$$j(g, z) = \det(cz + d) \qquad (4.11)$$

for $z \in \mathcal{H}_\nu$. Then the resulting map $j : Sp(m, \mathbb{R}) \times \mathcal{H}_m \to \mathbb{C}$ satisfies the cocycle condition

$$j(gg', z) = j(g, g'z)j(g', z)$$

for all $g, g' \in Sp(m, \mathbb{R})$ and $z \in \mathcal{H}_m$.

Definition 4.12 *Let k and ℓ be nonnegative integers. A holomorphic function $f : \mathcal{H}_m \to \mathbb{C}$ is a mixed Siegel modular form of weight (k, ℓ) associated to Γ, τ and ρ if*

$$f(\gamma z) = j(\gamma, z)^k j(\rho(\gamma), \tau(z))^\ell f(z)$$

for all $z \in \mathcal{H}_m$ and $\gamma \in \Gamma \subset Sp(m, \mathbb{R})$.

Note that, if $\ell = 0$ and $n \geq 2$ in Definition 4.12, the function f is a usual Siegel modular form for Γ of weight k. We shall denote by $\mathcal{M}_k(\Gamma)$ the space of Siegel modular forms for Γ of weight k and by $\mathcal{M}_{k,\ell}(\Gamma, \tau, \rho)$ the space of mixed Siegel modular forms of type (k, ℓ) associated to Γ, τ and ρ. Thus we see that $\mathcal{M}_{k,0}(\Gamma, \tau, \rho) = \mathcal{M}_k(\Gamma)$.

Example 4.13 *Let Γ' be an arithmetic subgroup of $Sp(m', \mathbb{R})$ such that $\rho(\Gamma) \subset \Gamma'$, and let $\phi : \mathcal{H}_{m'} \to \mathbb{C}$ be an element of $\mathcal{M}_\ell(\Gamma')$. We denote by $\tau^*\phi : \mathcal{H}_m \to \mathbb{C}$ the pullback of ϕ via τ, that is, the function defined by $(\tau^*\phi)(z) = \phi(\tau(z))$ for all $z \in \mathcal{H}_m$. If $f : \mathcal{H}_m \to \mathbb{C}$ is an element of $\mathcal{M}_k(\Gamma)$, then we have*

$$\begin{aligned}
(f \cdot (\tau^*\phi))(\gamma z) &= f(\gamma z)\phi(\tau(\gamma z)) \\
&= f(\gamma z)\phi(\rho(\gamma)\tau(z)) \\
&= j(\gamma, z)^k f(z) \cdot j(\rho(\gamma), \tau(z))^\ell \phi(\tau(z)), \\
&= j(\gamma, z)^k \cdot j(\rho(\gamma), \tau(z))^\ell (f \cdot (\tau^*\phi))(z)
\end{aligned}$$

for all $\gamma \in \Gamma$ and $z \in \mathcal{H}_m$, and hence $f \cdot (\tau^*\phi)$ is an element of $\mathcal{M}_{k,\ell}(\Gamma, \rho, \tau)$. Thus we obtain a linear map $\mathcal{L}_{\phi, \tau} : \mathcal{M}_k(\Gamma) \to \mathcal{M}_{k,\ell}(\Gamma, \rho, \tau)$ sending f to $f \cdot (\tau^*\phi)$.

We assume that the arithmetic subgroup $\Gamma \subset Sp(m, \mathbb{R})$ is torsion-free, so that the corresponding quotient space $X = \Gamma \backslash \mathcal{H}_m$ has the structure of a complex manifold. Then we can construct a family of abelian varieties parametrized by the Siegel modular variety $X = \Gamma \backslash \mathcal{H}_m$ as described below.

Let $L \subset \mathbb{C}^{m'}$ be a lattice in $\mathbb{C}^{m'}$, and let Γ_0 be a torsion-free arithmetic subgroup of $Sp(m', \mathbb{R})$ such that

$$\Gamma_0 \cdot L \subset L, \quad \rho(\Gamma) \subset \Gamma_0. \qquad (4.12)$$

We shall first describe the standard family Y_0 of abelian varieties over $X_0 = \Gamma_0 \backslash \mathcal{H}_{m'}$. Let $Sp(m', \mathbb{R}) \ltimes \mathbb{R}^{2m'}$ denote the semidirect product whose multiplication operation is given by

$$\left(\begin{pmatrix} a & b \\ c & d \end{pmatrix}, (u, v) \right) \left(\begin{pmatrix} a' & b' \\ c' & d' \end{pmatrix}, (u', v') \right)$$
$$= \left(\begin{pmatrix} a & b \\ c & d \end{pmatrix} \cdot \begin{pmatrix} a' & b' \\ c' & d' \end{pmatrix}, (u, v) \begin{pmatrix} a' & b' \\ c' & d' \end{pmatrix} + (u', v') \right)$$

for $u, v, u', v' \in \mathbb{R}^{m'}$ and

$$\begin{pmatrix} a & b \\ c & d \end{pmatrix}, \ \begin{pmatrix} a' & b' \\ c' & d' \end{pmatrix} \in Sp(m', \mathbb{R}).$$

Then $Sp(m', \mathbb{R}) \ltimes \mathbb{R}^{2m'}$ acts on $\mathcal{H}_{m'} \times \mathbb{C}^{m'}$ by

$$\left(\begin{pmatrix} a & b \\ c & d \end{pmatrix}, (u, v) \right) \cdot (z, \zeta) \tag{4.13}$$
$$= ((az + b)(cz + d)^{-1}, (\zeta + uz + v)(cz + d)^{-1})$$

for $(z, \zeta) \in \mathcal{H}_{m'} \times \mathbb{C}^{m'}$. By identifying $\mathbb{C}^{m'}$ with $\mathbb{R}^{2m'}$ we may regard the lattice L as a subgroup of $\mathbb{R}^{2m'}$. Then the condition $\Gamma_0 \cdot L \subset L$ in (4.12) implies that the action of $Sp(m', \mathbb{R}) \ltimes \mathbb{R}^{2m'}$ on $\mathcal{H}_{m'} \times \mathbb{C}^{m'}$ induces an action of $\Gamma_0 \ltimes L$. We denote the associated quotient space by

$$Y_0 = \Gamma_0 \ltimes L \backslash \mathcal{H}_{m'} \times \mathbb{C}^{m'}.$$

Then the natural projection $\mathcal{H}_{m'} \times \mathbb{C}^{m'} \to \mathcal{H}_m$ induces the map

$$\pi_0 : Y_0 \to X_0 = \Gamma_0 \backslash \mathcal{H}_{m'},$$

which has the structure of a fiber bundle over the Siegel modular variety X_0 with fiber $\mathbb{C}^{m'}/L$. Each fiber of π_0 isomorphic to the complex torus $\mathbb{C}^{m'}/L$ in fact has the structure of an abelian variety, so that Y_0 is a family of abelian varieties parametrized by X_0. Such a family of abelian varieties is called a *standard family* (see e.g. [108, Chapter 4]).

Using the condition $\rho(\Gamma) \subset \Gamma_0$ in (4.12), we see that the holomorphic map $\tau : \mathcal{H}_m \to \mathcal{H}_{m'}$ induces the morphism $\tau_X : X \to X_0$ of Siegel modular varieties. We denote by Y be the fiber bundle over X obtained by pulling the standard family Y_0 back via τ_X so that the following diagram is commutative:

$$\begin{array}{ccc} Y = \tau_X^* Y_0 & \longrightarrow & Y_0 \\ \pi \downarrow & & \downarrow \pi_0 \\ X & \xrightarrow{\ \tau_X\ } & X_0 \end{array} \tag{4.14}$$

Then the fiber of $\pi : Y \to X$ is the same as that of $\pi_0 : Y_0 \to X_0$, and hence Y is a family of abelian varieties, each of which is isomorphic to $\mathbb{C}^{m'}/L$, parametrized by the Siegel modular variety $X = \Gamma \backslash \mathcal{H}_m$. Given a positive integer ν, we denote by Y^ν the ν-fold fiber power of Y over X, that is, the fiber product over X of the ν copies of Y. Thus Y^ν can be regarded as the quotient space

$$Y^\nu = \Gamma \ltimes L^\nu \backslash \mathcal{H}_m \times \mathbb{C}^{m'\nu}, \tag{4.15}$$

where the quotient is taken with respect to the operation described as follows. Let $L = L_1 + L_2 \subset \mathbb{C}^{m'}$ with $L_1, L_2 \subset \mathbb{R}^{m'}$ be the natural decomposition of the lattice L in (4.12) corresponding to the identification of $\mathbb{C}^{m'}$ with $\mathbb{R}^{m'} \oplus \mathbb{R}^{m'} = \mathbb{R}^{2m'}$. Then, using (4.13) and the fact that Y is the pullback bundle in (4.14), we see that the operation of the discrete subgroup $\Gamma \ltimes L^\nu$ of $Sp(m, \mathbb{R}) \times \mathbb{R}^{2m'\nu}$ on $\mathcal{H}_m \times \mathbb{C}^{m'\nu}$ is given by

$$(\gamma, (u, v)) \cdot (z, \zeta) = (\gamma z, (\zeta^{(1)} + u_1 \tau(z) + v_1)(c_\rho \tau(z) + d_\rho)^{-1}, \ldots \tag{4.16}$$
$$\ldots, (\zeta^{(\nu)} + u_\nu \tau(z) + v_\nu)(c_\rho \tau(z) + d_\rho)^{-1})$$

for all $\gamma \in \Gamma$ with

$$\rho(\gamma) = \begin{pmatrix} a_\rho & b_\rho \\ c_\rho & d_\rho \end{pmatrix} \in Sp(m', \mathbb{R}),$$

$\zeta = (\zeta^{(1)}, \ldots, \zeta^{(\nu)}) \in (\mathbb{C}^{m'})^\nu$, and

$$u = (u_1, \ldots, u_\nu) \in (L_1)^\nu, \quad v = (v_1, \ldots, v_\nu) \in (L_2)^\nu.$$

Theorem 4.14 *Let $\pi^\nu : Y^\nu \to X$ with $X = \Gamma \backslash \mathcal{H}_m$ be the ν-fold fiber power of the family of abelian varieties $\pi : Y \to X$ constructed by (4.14), and let $\langle m \rangle = m(m+1)/2 = \dim_{\mathbb{C}} \mathcal{H}_m$. Then the space $H^0(Y^\nu, \Omega^{\langle m \rangle + m'\nu})$ of all holomorphic forms of degree $\langle m \rangle + m'\nu$ on Y^ν is canonically isomorphic to the space $\mathcal{M}_{m+1,\nu}(\Gamma, \tau, \rho)$ of all mixed Siegel modular forms on $\mathcal{H}_m$ of type $(m+1, \nu)$ associated to Γ, τ and ρ.*

Proof. Note that Y^ν is given by the quotient (4.15), and let $z = (z_1, \ldots, z_{\langle m \rangle})$ and $\zeta = (\zeta^{(1)}, \ldots, \zeta^{(\nu)})$ with $\zeta^{(j)} = (\zeta_1^{(j)}, \ldots, \zeta_{m'}^{(j)})$ for $1 \leq j \leq \nu$ be the canonical coordinate systems for $\mathcal{H}_m$ and $\mathbb{C}^{m'\nu}$, respectively. Then a holomorphic form Φ of degree $\langle m \rangle + m'\nu$ on Y can be regarded as a holomorphic form on $\mathcal{H}_m \times \mathbb{C}^{m'\nu}$ of the same degree that is invariant under the action of $\Gamma \ltimes L^\nu$ given by (4.16). Thus there is a holomorphic function $f_\Phi(z, \zeta)$ on $\mathcal{H}_m \times \mathbb{C}^{m'\nu}$ such that

$$\Phi(z, \zeta) = f_\Phi(z, \zeta) dz \wedge d\zeta^{(1)} \wedge \cdots \wedge d\zeta^{(\nu)}, \tag{4.17}$$

where $dz = dz_1 \wedge \cdots \wedge dz_{\langle m \rangle}$ and $d\zeta^{(j)} = d\zeta_1^{(j)} \wedge \cdots \wedge d\zeta_{m'}^{(j)}$ for $1 \leq j \leq \nu$. Given an element $z_0 \in \mathcal{H}_m$, the restriction of the form Φ to the fiber $Y_{z_0}^\nu$ over the corresponding point in $X = \Gamma \backslash \mathcal{H}_m$ is the holomorphic $m'\nu$-form

$$\Phi(z_0, \zeta) = f_\Phi(z_0, \zeta) d\zeta^{(1)} \wedge \cdots \wedge d\zeta^{(\nu)},$$

where $\zeta \mapsto f_\Phi(z_0, \zeta)$ is a holomorphic function on $Y_{z_0}^\nu$. However, $Y_{z_0}^\nu$ is isomorphic to the complex torus $\mathbb{C}^{m'}/L$, and therefore is compact. Since any holomorphic function on a compact complex manifold is constant, we see that f_Φ is a function of z only. Thus (4.17) can be written in the form

$$\Phi(z, \zeta) = \widetilde{f_\Phi}(z)dz \wedge d\zeta^{(1)} \wedge \cdots \wedge d\zeta^{(\nu)}, \qquad (4.18)$$

where $\widetilde{f_\Phi}$ is a holomorphic function on $\mathcal{H}_m$. Now we consider the action of an element $(\gamma, (u, v)) \in \Gamma \ltimes L^\nu$ on Φ given by $\Phi \mapsto \Phi \circ (\gamma, (u, v))$. By (4.16), for the differential form dz with $z \in \mathcal{H}_m$, we have

$$dz \circ (\gamma, (u, v)) = d(\gamma z) = \det(cz + d)^{-m-1}dz. \qquad (4.19)$$

On the other hand, using (4.16) again, we obtain

$$(d\zeta^{(1)} \wedge \ldots \wedge d\zeta^{(\nu)}) \circ (\gamma, (u, v)) \qquad (4.20)$$
$$= \bigwedge_{j=1}^{\nu} d((\zeta^{(j)} + u_j \tau(z) + v_j)(c_\rho \tau(z) + d_\rho)^{-1})$$
$$= \bigwedge_{j=1}^{\nu} \left(\det(c_\rho \tau(z) + d_\rho)^{-1} d\zeta^{(j)} \right)$$
$$= \det(c_\rho \tau(z) + d_\rho)^{-\nu} d\zeta.$$

Thus by substituting (4.19) and (4.20) into (4.18) we see that

$$(\Phi \circ (\gamma, u, v))(z, \zeta) = \widetilde{f_\Phi}(\gamma z) \det(cz + d)^{-m-1} \det(c_\rho \tau(z) + d_\rho)^{-\nu} \qquad (4.21)$$
$$\times dz \wedge d\zeta^{(1)} \wedge \ldots \wedge d\zeta^{(\nu)}.$$

Now using the fact that Φ is $(\Gamma \ltimes L^\nu)$-invariant, from (4.18) and (4.21) we obtain

$$\widetilde{f_\Phi}(\gamma z) = \det(cz + d)^{m+1} \det(c_\rho \tau(z) + d_\rho)^\nu \widetilde{f_\Phi}(z)$$
$$= j(\gamma, z)^{m+1} j(\rho(\gamma), \tau(z))^\nu \widetilde{f_\Phi}(z);$$

hence by Definition 4.12 we see that $\widetilde{f_\Phi} \in \mathcal{M}_{m+1,\nu}(\Gamma, \tau, \rho)$. On the other hand, given an element f of $\mathcal{M}_{m+1,\nu}(\Gamma, \tau, \rho)$, we define the holomorphic form Φ_f on $\mathcal{H}_m \times \mathbb{C}^{m'\nu}$ by

$$\Phi_f(z, \zeta) = f(z)dz \wedge d\zeta.$$

Then for each $(\gamma, (u, v)) \in \Gamma \ltimes L^\nu$ we have

$$(\Phi_f \circ (\gamma, (u, v)))(z, \zeta) = f(\gamma z) \det(cz + d)^{-m-1}$$
$$\times \det(c_\rho \tau(z) + d_\rho)^{-\nu} dz \wedge d\zeta^{(1)} \wedge \ldots \wedge d\zeta^{(\nu)}$$
$$= f(z)dz \wedge d\zeta^{(1)} \wedge \ldots \wedge d\zeta^{(\nu)} = \Phi_f(z, \zeta)$$

for all $(z, \zeta) \in \mathcal{H}_m \times \mathbb{C}^{m'\nu}$. Therefore the map $f \mapsto \Phi_f$ gives an isomorphism between the spaces $\mathcal{M}_{m+1,\nu}(\Gamma, \tau, \rho)$ and $H^0(Y_0, \Omega^{\langle m \rangle + m'\nu})$. $\qquad \square$

4.4 Fourier Coefficients of Siegel Modular Forms

In Example 4.13 we described a map $\mathcal{L}_{\phi,\tau} : \mathcal{M}_k(\Gamma) \to \mathcal{M}_{k,\ell}(\Gamma, \rho, \tau)$ that carries a Siegel modular form to a mixed Siegel modular form. Thus the adjoint of $\mathcal{L}_{\phi,\tau}$ associates a Siegel modular form $\mathcal{L}_{\phi,\tau}^* f$ to a mixed Siegel modular form f. In this section we discuss connections of the Fourier coefficients of a Siegel modular form of the form $\mathcal{L}_{\phi,\tau}^* f$ with special values of certain Dirichlet series, when τ is the Eichler embedding.

Siegel modular forms are holomorphic functions on Siegel upper half spaces. We can consider C^∞ Siegel modular forms by using C^∞ functions instead. We shall first describe below a method of obtaining Siegel cusp forms from C^∞ Siegel modular forms.

Let Ξ be the set of all symmetric integral matrices of order n with even diagonal entries, and set

$$\Xi_0^+ = \{\xi \in \Xi \mid \xi \geq 0\}, \quad \Xi^+ = \{\xi \in \Xi \mid \xi > 0\}.$$

Let Γ be a discrete subgroup of $Sp(n, \mathbb{R})$, and let b be the smallest positive integer such that the set

$$\left\{ \begin{pmatrix} 1 & bu \\ 0 & 1 \end{pmatrix} \ \middle| \ u \in M_n(\mathbb{Z}), \quad u^t = u \right\} \tag{4.22}$$

is contained in Γ, where $M_n(\mathbb{Z})$ is the set of $n \times n$ matrices with entries in $\mathbb{Z}$. We set

$$GSp^+(n, \mathbb{Q}) \tag{4.23}$$
$$= \left\{ g \in M_{2n}(\mathbb{Q}) \ \middle| \ g^t \begin{pmatrix} 0 & -1 \\ 1 & 0 \end{pmatrix} g = r(g) \begin{pmatrix} 0 & -1 \\ 1 & 0 \end{pmatrix}, \ r(g) > 0 \right\},$$

so that $Sp(n, \mathbb{Q}) \subset GSp^+(n, \mathbb{Q})$. Let $f : \mathcal{H}_n \to \mathbb{C}$ be an element of $\mathcal{M}_k(\Gamma)$, that is, a Siegel modular form of weight k for Γ. Then, since Γ contains the set in (4.22), f is invariant under the map $z \mapsto z + u$ on $\mathcal{H}_n$; hence f has a Fourier expansion of the form

$$f(z) = \sum_{\xi \in b^{-1}\Xi_0^+} c(\xi)\mathbf{e}(\xi z) \tag{4.24}$$

for all $z \in \mathcal{H}_n$, where $\mathbf{e}(*) = e^{2\pi i \operatorname{Tr}(*)}$. If $\delta \in Sp(n, \mathbb{Q})$, then $\Gamma(\delta) = \delta^{-1}\Gamma\delta$ is a congruence subgroup of $Sp(n, \mathbb{R})$ and $f \mid_k \delta$ is an element of $\mathcal{M}_k(\Gamma(\delta))$ that has a Fourier expansion of the form

$$(f \mid_k \delta)(z) = \sum_{\xi \in b_\delta^{-1}\Xi_0^+} c_\delta(\xi)\mathbf{e}(\xi z)$$

with $c_\delta(\xi) \in \mathbb{C}$ and $b_\delta \in \mathbb{N}$; here

$$(f \mid_k \delta)(z) = j(\delta, z)^{-k} f(z)$$

for $\delta = \begin{pmatrix} a & b \\ c & d \end{pmatrix} \in Sp(n, \mathbb{R})$ with $j(\delta, z)$ as in (4.5).

Definition 4.15 *An element f of $\mathcal{M}_k(\Gamma)$ is a Siegel cusp form of weight k for Γ if for each $\delta \in Sp(n, \mathbb{Q})$ its Fourier expansion is of the form*

$$(f \mid_k \delta)(z) = \sum_{\xi \in b_\delta^{-1} \Xi^+} c_\delta(\xi) \mathbf{e}(\xi z)$$

for all $z \in \mathcal{H}_n$. We denote by $\mathcal{S}_k(\Gamma)$ the complex vector space of all Siegel cusp forms of weight k for Γ.

Given a positive integer N, we now consider the congruence subgroup

$$\Gamma_0(N) = \left\{ \gamma = \begin{pmatrix} a & b \\ c & d \end{pmatrix} \in Sp(n, \mathbb{Z}) \;\middle|\; c \equiv 0 \pmod{N} \right\}, \qquad (4.25)$$

and set

$$\mathcal{M}_k(N) = \mathcal{M}_k(\Gamma_0(N)), \qquad \mathcal{S}_k(N) = \mathcal{S}_k(\Gamma_0(N)).$$

Then a Siegel modular form in $\mathcal{M}_k(N)$ has a Fourier expansion of the form (4.24) with $b = 1$. If $f \in \mathcal{S}_k(N)$, $h \in \mathcal{M}_k(N)$ and $z = x + iy$, then the Petersson inner product $\langle f, h \rangle_k$ is given by

$$\langle f, h \rangle_k = \int_{\mathcal{F}} \overline{f(z)} h(z) \det y^k d^\times z, \qquad (4.26)$$

where $\mathcal{F} \subset \mathcal{H}_n$ is a fundamental domain for $\Gamma_0(N)$ and $d^\times z = \det y^{-n-1} dz$.

Definition 4.16 *A C^∞ function $F : \mathcal{H}_n \to \mathbb{C}$ is called a C^∞ Siegel modular form of weight k for $\Gamma_0(N)$ if it satisfies*

$$F(\gamma z) = \det(cz + d)^k F(z)$$

for all $\gamma = \begin{pmatrix} a & b \\ c & d \end{pmatrix} \in \Gamma_0(N)$ and $z \in \mathcal{H}_n$. We denote by $\mathcal{M}_k^\infty(N)$ the space of all C^∞ Siegel modular forms of weight k for $\Gamma_0(N)$.

If $F \in \mathcal{M}_k^\infty(N)$, then as a function of x it has a Fourier expansion of the form

$$F(z) = \sum_{\xi \in \Xi} A_\xi(y) \mathbf{e}(\xi x), \qquad (4.27)$$

where the $A_\xi(y)$ are C^∞ functions of y. We can also extend the Petersson inner product (4.26) to C^∞ Siegel modular forms so that for $f \in \mathcal{S}_k(N)$ and $F \in \mathcal{M}_k^\infty(N)$ their inner product $\langle f, F \rangle_k^\infty$ is given by

$$\langle f, F \rangle_k^\infty = \int_{\mathcal{F}} \overline{f(z)} F(z) \det y^k d^\times z, \qquad (4.28)$$

where $\mathcal{F}$ and $d^\times z$ are as in (4.26). We denote by $\boldsymbol{\Gamma}_n$ the Gamma function of degree n given by

$$\boldsymbol{\Gamma}_n(s) = \int_Y e^{-\operatorname{Tr}(y)} \det y^s d^\times y = \int_Y e^{-\operatorname{Tr}(y)} \det y^{s-(n+1)/2} dy, \qquad (4.29)$$

where $Y = \{ y \in M_n(\mathbb{R}) \mid {}^t y = y > 0 \}$.

Theorem 4.17 *Let F be an element of $\mathcal{M}_k^\infty(N)$ of bounded growth with $k \geq 2n$ whose Fourier expansion is given by (4.27), and let $\widetilde{F} : \mathcal{H}_n \to \mathbb{C}$ be the function defined by the series*

$$\widetilde{F}(z) = \sum_{\xi \in \Xi^+} a(\xi)\mathbf{e}(\xi z)$$

for all $z \in \mathcal{H}_n$, where

$$a(\xi) = \frac{\pi^{n(k-(n+1)/2)} \det(4\xi)^{k-(n+1)/2}}{\boldsymbol{\Gamma}_n(k - (n+1)/2)} \tag{4.30}$$
$$\times \int_Y A_\xi(y)\mathbf{e}(i\xi y) \det y^{k-1-n} dy.$$

Then $\widetilde{F}$ is a Siegel cusp form in $\mathcal{S}_k(N)$ and

$$\langle h, F \rangle_k^\infty = \langle h, \widetilde{F} \rangle_k$$

for all $h \in \mathcal{S}_k(N)$, where $\langle\,,\,\rangle_k$ and $\langle\,,\,\rangle_k^\infty$ are the Petersson inner products given by (4.26) and (4.28), respectively.

Proof. See Theorem 4.2 in [101, Chapter 2]. □

We now want to consider a special type of equivariant holomorphic maps of Siegel upper half spaces known as Eichler embeddings. Let σ be a real symmetric positive definite $r \times r$ matrix with entries in $\mathbb{Q}$. We define the holomorphic map $\tau_\sigma : \mathcal{H}_n \to \mathcal{H}_{nr}$ and the homomorphism $\rho_\sigma : Sp(n, \mathbb{R}) \to Sp(nr, \mathbb{R})$ by

$$\tau_\sigma(z) = \sigma \otimes z, \qquad \rho_\sigma(g) = \begin{pmatrix} \varepsilon \otimes a & \sigma \otimes b \\ \sigma^{-1} \otimes c & \varepsilon \otimes d \end{pmatrix}$$

for all $z \in \mathcal{H}_n$ and

$$g = \begin{pmatrix} a & b \\ c & d \end{pmatrix} \in Sp(n, \mathbb{R}),$$

where ε is the $r \times r$ identity matrix. The map τ_σ is known as the Eichler embedding (cf. [25, §II.4]) associated to σ, and it is equivariant with respect to ρ_σ. Let $\Gamma_0(N) \subset Sp(n, \mathbb{Z})$ be the congruence subgroup given by (4.25) for the symplectic group $Sp(n, \mathbb{R})$, and let Γ' be an arithmetic subgroup of $Sp(nr, \mathbb{R})$ such that $\rho_\sigma(\Gamma_0(N)) \subset \Gamma'$. Let $\phi : \mathcal{H}_{nr} \to \mathbb{C}$ be a Siegel modular form of weight ℓ for $\Gamma' \subset Sp(nr, \mathbb{R})$, and let $\tau_\sigma^* \phi$ be its pullback via $\tau_\sigma : \mathcal{H}_n \to \mathcal{H}_{nr}$. Thus $\tau_\sigma^* \phi$ is the function on $\mathcal{H}_n$ given by $\tau_\sigma^* \phi(z) = \phi(\tau_\sigma(z))$ for $z \in \mathcal{H}_n$. As discussed in Example 4.13, we obtain a linear map

$$\mathcal{L}_{\phi, \tau_\sigma} : \mathcal{M}_k(N) \to \mathcal{M}_{k, \ell}(\Gamma_0(N), \tau_\sigma, \rho_\sigma)$$

given by

$$(\mathcal{L}_{\phi, \tau_\sigma} f)(z) = (f \cdot \tau_\sigma^* \phi)(z) = f(z)\phi(\tau_\sigma(z))$$

for all $z \in \mathcal{H}_n$.

Lemma 4.18 *The image $\mathcal{L}_{\phi,\tau_\sigma} f$ of each $f \in \mathcal{M}_k(N)$ under $\mathcal{L}_{\phi,\tau_\sigma}$ is a Siegel modular form for $\Gamma_0(N)$ of weight $k + \ell r$, that is, $\mathcal{L}_{\phi,\tau_\sigma} f \in \mathcal{M}_{k+\ell r}(N)$.*

Proof. For each $\gamma = \begin{pmatrix} a & b \\ c & d \end{pmatrix} \in \Gamma_0(N)$ and $z \in \mathcal{H}_n$ we have

$$
\begin{aligned}
\tau_\sigma^* \phi(\gamma z) &= \phi(\tau_\sigma(\gamma z)) = \phi(\rho_\sigma(\gamma)\tau_\sigma(z)) \qquad (4.31) \\
&= \det((\sigma^{-1} \otimes c)(\sigma \otimes z) + \varepsilon \otimes d)^\ell \phi(\tau_\sigma(z)) \\
&= \det(cz + d)^{\ell r} \tau_\sigma^* \phi(z) = j(\gamma, z)^{\ell r} \tau_\sigma^* \phi(z).
\end{aligned}
$$

Thus for each $f \in \mathcal{M}_k(\Gamma_0(N))$ we have

$$
\begin{aligned}
\mathcal{L}_{\phi,\tau_\sigma} f(\gamma z) &= f(\gamma z) \cdot (\tau_\sigma^* \phi)(\gamma z) \\
&= j(\gamma, z)^k f(z) \cdot j(\gamma, z)^{\ell r} \tau_\sigma^* \phi(z) \\
&= j(\gamma, z)^{k+\ell r} \mathcal{L}_{\phi,\tau_\sigma} f(z),
\end{aligned}
$$

and hence the lemma follows. $\square$

By Lemma 4.18, if we restrict the linear map $\mathcal{L}_{\phi,\tau_\sigma}$ to the space $\mathcal{S}_k(N)$ of Siegel cusp forms on $\mathcal{H}_n$ for $\Gamma_0(N)$ of weight k, then we obtain the linear map

$$
\mathcal{L}_{\phi,\tau_\sigma} : \mathcal{S}_k(N) \to \mathcal{S}_{k+\ell r}(N)
$$

given by $\mathcal{L}_{\phi,\tau_\sigma}(h) = h \cdot (\tau_\sigma^* \phi)$ for $h \in \mathcal{S}_k(N)$. By taking the adjoint with respect to the Petersson inner product, we have the linear map

$$
\mathcal{L}_{\phi,\tau_\sigma}^* : \mathcal{S}_{k+\ell r}(N) \to \mathcal{S}_k(N)
$$

such that

$$
\langle \mathcal{L}_{\phi,\tau_\sigma}(h), f \rangle_{k+\ell r} = \langle h, \mathcal{L}_{\phi,\tau_\sigma}^*(f) \rangle_k
$$

for $h \in \mathcal{S}_k(N)$ and $f \in \mathcal{S}_{k+\ell r}(N)$. The following lemma will be used in the next section.

Lemma 4.19 *Let $\Gamma' \subset Sp(nr, \mathbb{R})$ be an arithmetic subgroup containing $\rho_\sigma(\Gamma_0(N))$ as above, and let $\phi : \mathcal{H}_{nr} \to \mathbb{C}$ be an element of $\mathcal{M}_\ell(\Gamma')$. If $f : \mathcal{H}_n \to \mathbb{C}$ is an element of $\mathcal{S}_{k+\ell r}(N)$, then the function*

$$
F(z) = f(z)\overline{\phi(\tau_\sigma(z))}(\det(\operatorname{Im} z))^{\ell r} \qquad (4.32)
$$

for $z \in \mathcal{H}_n$ is a C^∞ Siegel modular form of weight k for $\Gamma_0(N)$, where $\operatorname{Im} z$ is the imaginary part of the matrix z.

Proof. Given $\gamma = \begin{pmatrix} a & b \\ c & d \end{pmatrix} \in \Gamma_0(N)$ and $z \in \mathcal{H}_n$ we have

$$
f(\gamma z) = \det(cz + d)^{k+\ell r} f(z), \quad \det(\operatorname{Im} \gamma z) = |\det(cz + d)|^{-2} \det(\operatorname{Im} z).
$$

On the other hand by (4.31) we get

$$\overline{\phi(\tau_\sigma(\gamma z))} = \overline{\det(cz+d)}^{\ell r} \cdot \overline{\phi(\tau_\sigma(z))}.$$

Since $|\det(cz+d)|^2 = \det(cz+d) \cdot \overline{\det(cz+d)}$, we see that

$$
\begin{aligned}
F(\gamma z) &= f(\gamma z)\overline{\phi(\tau_\sigma(\gamma z))}(\det(\operatorname{Im}\gamma z))^{\ell r} \\
&= \det(cz+d)^{k+\ell r} \cdot \overline{\det(cz+d)}^{\ell r} \cdot |\det(cz+d)|^{-2\ell r} \cdot F(z) \\
&= \det(cz+d)^k F(z),
\end{aligned}
$$

and hence the lemma follows.   □

For the families of abelian varieties determined by Eichler embeddings we have the following result.

Proposition 4.20 *Let $\pi_\sigma : Y_\sigma \to X_\sigma$ be the family of abelian varieties over $X_\sigma = \Gamma_0(N)\backslash \mathcal{H}_n$ associated to the equivariant pair $(\tau_\sigma, \rho_\sigma)$, and let $(Y_\sigma)^\nu$ be its ν-fold fiber power. If $\langle n \rangle = n(n+1)/2 = \dim_{\mathbb{C}}\mathcal{H}_n$, then the space $H^0((Y_\sigma)^\nu, \Omega^{\langle n \rangle + nr\nu})$ of holomorphic forms on $(Y_\sigma)^\nu$ of degree $\langle n \rangle + nr\nu$ is canonically embedded in the space $\mathcal{M}_{n+r\nu+1}(N)$ of Siegel modular forms on $\mathcal{H}_n$ for $\Gamma_0(N)$ of weight $n + r\nu + 1$.*

Proof. By Theorem 4.14 the space $H^0((Y_\sigma)^\nu, \Omega^{\langle n \rangle + nr\nu})$ of holomorphic forms on $(Y_\sigma)^\nu$ of degree $n(n+1)/2 + nr\nu$ is canonically isomorphic to the space $\mathcal{M}_{n+1,\nu}(\Gamma_0(N), \tau_\sigma, \rho_\sigma)$ of mixed Siegel modular forms of type $(n+1, \nu)$ associated to $\Gamma_0(N)$, τ_σ and ρ_σ. However, using the arguments in the proof of Lemma 4.18 we see that $\mathcal{M}_{n+1,\nu}(\Gamma_0(N), \tau_\sigma, \rho_\sigma)$ is embedded in the space $\mathcal{M}_{n+r\nu+1}(N)$, and therefore the proposition follows.   □

Let q be a prime with $q \nmid N$, and let

$$\Delta_q = \left\{ \begin{pmatrix} a & b \\ c & d \end{pmatrix} \in GSp^+(n,\mathbb{Q}) \cap GL(2n, \mathbb{Z}_{(q)}) \,\middle|\, c \equiv 0 \pmod{q} \right\},$$

where $\mathbb{Z}_{(q)}$ is the ring of rational numbers whose denominators are prime to q and $GSp^+(n, \mathbb{Q})$ is as in (4.23). Let χ_1 be a Dirichlet character modulo N, and set $\chi(g) = \chi_1(\det a)$ for $g = \left(\begin{smallmatrix} a & b \\ c & d \end{smallmatrix} \right) \in \Delta_q$. If $h : \mathcal{H}_n \to \mathbb{C}$ is a function and if $\alpha \in \Delta_q$, then we set

$$(h \mid_k \alpha) = \chi(\alpha)j(\alpha, z)^{-k}h(z)$$

for all $z \in \mathcal{H}_n$. Let $\alpha \in \Delta_q$, and assume that the double coset $\Gamma_0(N)\alpha\Gamma_0(N)$ has a decomposition of the form

$$\Gamma_0(N)\alpha\Gamma_0(N) = \sum_{i=1}^{d} a_i \cdot \Gamma_0(N)\alpha_i$$

for some $\alpha_1, \ldots, \alpha_d \in GSp^+(n, \mathbb{Q})$ and $a_1, \ldots, a_d \in \mathbb{C}$. The the associated *Hecke operator* on the space $\mathcal{S}_k(N)$ of Siegel modular forms for $\Gamma_0(N)$ of weight k is given by

$$T_N(\alpha)f = \sum_{i=1}^{d} a_i(f \mid_k \alpha_i)$$

for all $f \in \mathcal{S}_k(N)$. We now consider the *Poincaré series of two variables* $P^k(z, w, \alpha, s)$ defined by

$$P^k(z, w, \alpha, s) = \sum_{\gamma \in \Gamma_0(N)\alpha\Gamma_0(N)} (cz + d)^{-k}|j(\gamma, z)|^{-2}$$
$$\times \det(\gamma z + w)^k |\det(\gamma z + w)|^{-2s}$$

for $s \in \mathbb{C}$ and $\gamma = \begin{pmatrix} a & b \\ c & d \end{pmatrix}$. Then it is known that the function

$$w \mapsto P^k(z, w, \alpha, s)$$

is a Siegel cusp form for $\Gamma_0(N)$ of weight k (cf. [101]).

Lemma 4.21 *Let $f : \mathcal{H}_n \to \mathbb{C}$ be a Siegel cusp form for $\Gamma_0(N)$ of weight k. Then we have*

$$\langle P^k(-\bar{z}, w, \alpha, s), f(w)\rangle_k = \mu(s)(T_N(\alpha)f)(z)$$

for all $z \in \mathcal{H}_n$ with

$$\mu(s) = 2^{n+1+n(n+1)/2 - 2ns} i^{-nk} \pi^{n(n+1)/2} \qquad (4.33)$$
$$\times \Gamma_n(k + s - (n+1)/2)\Gamma_n(k+s)^{-1},$$

where $\Gamma_n(\cdot)$ denotes the Gamma function given by (4.29).

Proof. See [101, p. 73]. □

Proposition 4.22 *Let $\widetilde{F}(z)$ be the Siegel cusp form for $\Gamma_0(N)$ associated to the C^∞ Siegel modular form $F(z)$ given by (4.32). Then we have $\mathcal{L}^*_{\phi,\tau_\sigma} f = \widetilde{F}$.*

Proof. Let $\langle\,,\,\rangle_m$ denote the Petersson inner product for Siegel modular forms of weight m, and let $w = u + iv$. Then we have

$$\langle P^k(-\bar{z}, w, \alpha, s), \mathcal{L}^*_{\phi,\tau_\sigma} f(w)\rangle_k$$
$$= \langle \mathcal{L}_{\phi,\tau_\sigma} P^k(-\bar{z}, w, \alpha, s), f(w)\rangle_{k+\ell r}$$
$$= \langle \phi(\tau_\sigma(w)) P^k(-\bar{z}, w, \alpha, s), f(w)\rangle_{k+\ell r}$$
$$= \int_{\mathcal{F}} \overline{\phi(\tau_\sigma(w))} P^k(-\bar{z}, w, \alpha, s) f(w) \det v^{k+\ell r - n - 1} du\, dv$$
$$= \int_{\mathcal{F}} \overline{P^k(-\bar{z}, w, \alpha, s)} (f(w)\overline{\phi(\tau_\sigma(w))} \det v^{\ell r}) \det v^{k-n-1} du\, dv$$
$$= \langle P^k(-\bar{z}, w, \alpha, s), F(w)\rangle_k^\infty,$$

where $\mathcal{F} \subset \mathcal{H}_n$ is a fundamental domain for $\Gamma_0(N)$. Thus by Theorem 4.17 we obtain

$$\langle P^k(-\bar{z}, w, \alpha, s), \mathcal{L}^*_{\phi, \tau_\sigma} f(w) \rangle_k = \langle P^k(-\bar{z}, w, \alpha, s), \widetilde{F}(w) \rangle_k$$

for all $z, w \in \mathcal{H}_n$, and hence by Lemma 4.21 we have

$$\mu(s)(T_N(\alpha)\mathcal{L}^*_{\phi, \tau_\sigma} f)(z) = \mu(s)(T_N(\alpha)\widetilde{F})(z)$$

for all $z \in \mathcal{H}_n$, where $\mu(s)$ is a nonzero number given by (4.33). Therefore the proposition follows by taking α to be the identity matrix in Δ_q.   □

Let $\phi : \mathcal{H}_{nr} \to \mathbb{C}$ be an element of $\mathcal{M}_\ell(\Gamma')$ with $\Gamma' \supset \rho_\sigma(\Gamma_0(N))$ as above so that the function $\tau_\sigma^*(\phi) : \mathcal{H}_n \to \mathbb{C}$ is an element of $\mathcal{M}_{\ell r}(N)$. For the elements $\xi \in \Xi^+$, let $A(\xi)$ and $B(\xi)$ be the Fourier coefficients of $f \in \mathcal{S}_{k+\ell r}(N)$ and $\tau_\sigma^*(\phi) \in \mathcal{M}_{\ell r}(N)$, respectively, so that

$$f(z) = \sum_{\xi \in \Xi^+} A(\xi)\mathbf{e}(\xi z), \qquad \tau_\sigma^*(\phi)(z) = \sum_{\xi \in \Xi} B(\xi)\mathbf{e}(\xi z). \tag{4.34}$$

Given $\xi \in \Xi^+$, we define the *Dirichlet series* $L_{f,\phi}^{\sigma,\xi}(s)$ associated to f, ϕ, σ and ξ by

$$L_{f,\phi}^{\sigma,\xi}(s) = \sum_{\eta \in \Xi} \frac{A(\xi + \eta)\overline{B(\eta)}}{\det(\xi + \eta)^s}. \tag{4.35}$$

for $s \in \mathbb{C}$.

Theorem 4.23 *Let* $\phi : \mathcal{H}_{nr} \to \mathbb{C}$ *be an element of* $\mathcal{M}_\ell(\Gamma')$ *with* $\Gamma' \supset \rho_\sigma(\Gamma_0(N))$, *and let*

$$\mathcal{L}^*_{\phi, \tau_\sigma} : \mathcal{S}_{k+\ell r}(N) \to \mathcal{S}_k(N)$$

be the associated linear map described above. If $f \in \mathcal{S}_{k+\ell r}(N)$, *then the Fourier expansion of* $\mathcal{L}^*_{\phi, \tau_\sigma} f \in \mathcal{S}_k(N)$ *is given by*

$$(\mathcal{L}^*_{\phi, \tau_\sigma} f)(z) = \sum_{\xi \in \Xi^+} a(\xi)\mathbf{e}(\xi z)$$

for all $z \in \mathcal{H}_n$, *where*

$$a(\xi) = \frac{(\det \xi)^{k-(n+1)/2} \boldsymbol{\Gamma}_n(k + \ell r - (n+1)/2)}{(4\pi)^{n(\ell r - (n-1)/2)} \boldsymbol{\Gamma}_n(k - (n+1)/2)} L_{f,\phi}^{\sigma,\xi}(k + \ell r - n) \tag{4.36}$$

for all $\xi \in \Xi^+$.

Proof. By Proposition 4.22 we see that $\mathcal{L}^*_{\phi, \tau_\sigma} f$ coincides with the Siegel cusp form $\widetilde{F} \in \mathcal{S}_k(N)$ associated to the C^∞ Siegel modular form

$$F(z) = f(z)\overline{\phi(\tau_\sigma(z))} \det(\operatorname{Im} z)^{\ell r}.$$

Given $\xi \in \Xi^+$, we set

$$K(\xi) = \frac{\pi^{n(k-(n+1)/2)} \det(4\xi)^{k-(n+1)/2}}{\Gamma_n(k-(n+1)/2)}. \tag{4.37}$$

Then by (4.30) we have

$$a(\xi) = K(\xi) \int_Y A_\xi(y) e(i\xi y) \det y^{k-1-n} dy.$$

Using the Fourier expansions in (4.27) and (4.34) we see that

$$F(z) = \sum_{\xi \in \Xi} A_\xi(y) e(\xi x)$$

$$= \sum_{\mu \in \Xi^+} \sum_{\eta \in \Xi} A(\mu) \overline{B(\eta)} e((\mu - \eta)x) e(i(\mu + \eta)y) \det y^{\ell r}.$$

Thus, using $\xi = \mu - \eta$ or $\mu = \xi + \eta$, we have

$$A_\xi(y) = \sum_{\eta \in \Xi} A(\xi + \eta) \overline{B(\eta)} e(i(\xi + 2\eta)y) \det y^{\ell r}.$$

Hence it follows that

$$a(\xi) = K(\xi) \int_Y \sum_{\eta \in \Xi} A(\xi + \eta) \overline{B(\eta)} e(2i(\xi + \eta)y) \det y^{k+\ell r-1-n} dy$$

$$= K(\xi) \sum_{\eta \in \Xi} A(\xi + \eta) \overline{B(\eta)} \int_Y e^{-4\pi \operatorname{Tr}((\xi + \eta)y)} \det y^{k+\ell r-1-n} dy.$$

If $v = 4\pi(\xi + \eta)y$, we have

$$\det v = (4\pi)^n \det(\xi + \eta) \det y, \qquad dv = (4\pi)^n \det(\xi + \eta) dy,$$

and therefore, for each $\xi \in \Xi^+$, we see that

$$a(\xi) = K(\xi)(4\pi)^{-n(k+\ell r-n)} \int_Y e^{-\operatorname{Tr} v} \det v^{k+\ell r-1-n} dv$$

$$\times \sum_{\eta \in \Xi^+} \frac{A(\xi + \eta) \overline{B(\eta)}}{\det(\xi + \eta)^{k+\ell r-n}}.$$

However, using (4.29) and (4.35), we get

$$\Gamma_n(k + \ell r - (n+1)/2) = \int_Y e^{-\operatorname{Tr} v} \det v^{k+\ell r-1-n} dv,$$

$$L_{f,\phi}^{\sigma,\xi}(k + \ell r - n) = \sum_{\eta \in \Xi} \frac{A(\xi + \eta) \overline{B(\eta)}}{\det(\xi + \eta)^{k+\ell r-n}}.$$

From these relations and (4.37) we obtain the formula (4.36); hence the proof of the theorem is complete.   □

Remark 4.24 *The method used for the proof of Theorem 4.23 was developed by Kohnen [53], where he considered modular forms. This method was extended to the case of Hilbert modular forms in [78], and the case of Siegel modular forms was investigated in [19, 75].*

Mixed Automorphic Forms on Semisimple Lie Groups

In the previous chapters we considered mixed automorphic forms on the Poincaré upper half plane $\mathcal{H}$, its product space $\mathcal{H}^n$, and the Siegel upper half space $\mathcal{H}_n$, which are Hermitian symmetric domains associated to the semisimple Lie groups $SL(2, \mathbb{R})$, $SL(2, \mathbb{R})^n$, and $Sp(n, \mathbb{R})$, respectively. In this chapter we discuss mixed automorphic forms associated to more general semisimple Lie groups and real reductive groups (cf. [71, 82]). We include nonholomorphic automorphic forms by using the representation theoretic interpretation of automorphic forms initiated by Selberg and Langlands (see e.g. [13, 15]).

Let G be a semisimple Lie group, K a maximal compact subgroup, and Γ a discrete subgroup of finite covolume. Let $Z(\mathfrak{g})$ be the center of the universal enveloping algebra of the complexification $\mathfrak{g}_{\mathbb{C}}$ of the Lie algebra $\mathfrak{g}$ of G, and let V be a finite-dimensional complex vector space. A slowly increasing analytic function $f : G \to V$ is an automorphic form for Γ if it is left Γ-invariant, right K-finite, and $Z(\mathfrak{g})$-finite. In order to describe mixed automorphic forms, let G' be another semisimple Lie group with the corresponding objects K', Γ' and V', and let $\rho : G \to G'$ be a homomorphism such that $\rho(K) \subset K'$ and $\rho(\Gamma) \subset \Gamma'$. Then associated mixed automorphic forms occur as linear combinations of functions of the form $f \otimes (f' \circ \rho) : G \to V \otimes V'$, where $f : G \to V$ is an automorphic form for Γ and $f' : G' \to V'$ is an automorphic form for Γ'. If G and G' are a real reductive groups, the construction described above produces mixed automorphic forms on real reductive groups. If G and G' are semisimple Lie groups of Hermitian type, we can define holomorphic mixed automorphic forms as follows. In this case the Riemannian symmetric spaces $\mathcal{D} = G/K$ and $\mathcal{D}' = G'/K'$ are Hermitian symmetric domains, and from the condition $\rho(K) \subset K'$ we see that ρ induces a holomorphic map $\tau : \mathcal{D} \to \mathcal{D}'$ that is equivariant with respect to ρ. Given complex vector spaces V and V' and automorphy factors $J : G \times \mathcal{D} \to GL(V)$ and $J' : G' \times \mathcal{D}' \to GL(V')$, a mixed automorphic form on $\mathcal{D}$ for Γ is a holomorphic function $f : \mathcal{D} \to V \otimes V'$ satisfying

$$f(\gamma z) = J(\gamma, z) \otimes J'(\rho(\gamma), \tau(z)) f(z)$$

for all $z \in \mathcal{D}$ and $\gamma \in \Gamma$.

Whittaker vectors associated to representations of real reductive groups generalize Whittaker's confluent hypergeometric functions that occur in nonconstant terms of Fourier coefficients of automorphic forms (cf. [88]), and over the last few decades various aspects of Whittaker vectors have been investigated in numerous papers (see e.g. [30, 45, 56, 110, 111]). In [93] Miatello and Wallach constructed an analogue of non-holomorphic Poincaré series for automorphic forms on reductive groups of real rank one using Whittaker vectors in order to construct square-integrable automorphic forms.

In this chapter we construct Poincaré series and Eisenstein series for mixed automorphic forms on semisimple Lie groups and discuss Whittaker vectors and Jacquet integrals associated to mixed automorphic forms on real reductive groups. We also construct an analogue of Poincaré series of Miatello and Wallach and express the Fourier coefficients of Eisenstein series for mixed automorphic forms in terms of Jacquet integrals.

Section 5.1 contains the discussion of holomorphic mixed automorphic forms on Hermitian symmetric domains. In Section 5.2 we introduce a representation theoretic description of mixed automorphic forms on semisimple Lie groups and construct Poincaré series as examples of such automorphic forms. The construction of Eisenstein series for automorphic forms on semisimple Lie groups is contained in Section 5.3. In Section 5.4 we describe mixed automorphic forms on real reductive groups and construct Eisenstein series as well as an analogue of Poincaré series of Miatello and Wallach. Whittaker vectors for such mixed automorphic forms are also discussed in this section. In Section 5.5 we study Fourier coefficients of Eisenstein series for mixed automorphic forms on real reductive groups in connection with Jacquet integrals and Whittaker vectors.

5.1 Mixed Automorphic Forms on Symmetric Domains

In this section we describe holomorphic mixed automorphic forms on Hermitian symmetric domains. We also construct Poincaré series which provide examples of such automorphic forms.

Let G be a semisimple Lie group of Hermitian type. Thus, if K is a maximal compact subgroup of G, the corresponding Riemannian symmetric space $\mathcal{D} = G/K$ has a G-invariant complex structure and is called a Hermitian symmetric domain. Let V be a finite-dimensional complex vector space, and let $J : G \times \mathcal{D} \to GL(V)$ be an automorphy factor satisfying

$$J(\gamma_1 \gamma_2, z) = J(\gamma_1, \gamma_2 z) J(\gamma_2, z)$$

for all $z \in \mathcal{D}$ and $\gamma_1, \gamma_2 \in G$.

Let $\mathcal{D}' = G'/K'$ be another Hermitian symmetric domain associated to a semisimple Lie group of Hermitian type G' and a maximal compact subgroup K' of G'. We assume that there is a holomorphic map $\tau : \mathcal{D} \to \mathcal{D}'$ that is equivariant with respect to a Lie group homomorphism $\rho : G \to G'$. This means that ρ and τ satisfies

$$\tau(\gamma z) = \rho(\gamma)\tau(z)$$

for all $z \in \mathcal{D}$ and $\gamma \in G$. Such equivariant pairs (τ, ρ) will be studied in Chapter 6 in connection with families of abelian varieties. We now consider another automorphy factor $J' : G' \times \mathcal{D}' \to GL(V')$ for some finite-dimensional complex vector space V'. Let Γ be a discrete subgroup of G.

Definition 5.1 *A* mixed automorphic form *for* Γ *on* $\mathcal{D}$ *of type* (J, J', τ, ρ) *is a holomorphic function* $f : \mathcal{D} \to V \otimes V'$ *satisfying*

$$f(\gamma z) = J(\gamma, z) \otimes J'(\rho(\gamma), \tau(z))f(z)$$

for all $z \in \mathcal{D}$ *and* $\gamma, \gamma' \in \Gamma$.

Remark 5.2 *If* Γ *is torsion-free and cocompact and if* G' *is a symplectic group, then it was shown in [74] that for some specific automorphy factors* J *and* J' *mixed automorphic forms on* $\mathcal{D}$ *of type* (J, J', τ, ρ) *can be identified with holomorphic forms on certain families of abelian varieties parametrized by a locally symmetric space. In the elliptic modular case, that is, when* $G = SL(2, \mathbb{R})$, *various results were discussed about the corresponding mixed automorphic forms in Chapters 1, 2 and 3 (see also [43, 68, 70]). Similar geometric aspects for the Siegel and Hilbert modular cases were treated in Chapter 4 (see also [71, 76]).*

If $\gamma \in G$ and $z \in \mathcal{D}$, let $j_{\mathcal{D}}(\gamma, z)$ denote the determinant of the Jacobian matrix of γ at z, so that the resulting map $j_{\mathcal{D}} : G \times \mathcal{D} \to \mathbb{C}$ is an automorphy factor. We denote by $j_{\mathcal{D}'} : G' \times \mathcal{D}' \to \mathbb{C}$ the similar automorphy factor for $\mathcal{D}'$. Let f be a bounded holomorphic function on $\mathcal{D}$. Given positive integers ℓ and m and a discrete subgroup Γ of G, we set

$$\mathcal{P}_{\ell,m}^{f}(z) = \sum_{\gamma \in \Gamma} f(\gamma z) j_{\mathcal{D}}(\gamma, z)^{\ell} j_{\mathcal{D}'}(\rho(\gamma), \tau(z))^{m} \tag{5.1}$$

for all $z \in \mathcal{D}$.

Theorem 5.3 *The function* $\mathcal{P}_{\ell,m}^{f} : \mathcal{D} \to \mathbb{C}$ *defined by* (5.1) *is a mixed automorphic form on* $\mathcal{D}$ *for* Γ *of type* $(j_{\mathcal{D}}^{-\ell}, j_{\mathcal{D}'}^{-m}, \tau, \rho)$, *where* $j_{\mathcal{D}}^{-\ell}(\gamma, z) = j_{\mathcal{D}}(\gamma, z)^{-\ell}$ *for* $\gamma \in \Gamma$ *and* $z \in \mathcal{D}$ *and similarly for* $j_{\mathcal{D}'}^{-m}$.

Proof. Given $\gamma \in \Gamma$ and $z \in \mathcal{D}$, we have

$$\mathcal{P}_{\ell,m}^{f}(\gamma z) = \sum_{\gamma' \in \Gamma} f(\gamma' \gamma z) j_{\mathcal{D}}(\gamma', \gamma z)^{\ell} j_{\mathcal{D}'}(\rho(\gamma'), \tau(\gamma z))^{m}.$$

Thus, using the relations

$$j_{\mathcal{D}}(\gamma', \gamma z) = j_{\mathcal{D}}(\gamma' \gamma, z) j_{\mathcal{D}}(\gamma, z)^{-1},$$

$$j_{\mathcal{D}'}(\rho(\gamma'), \tau(\gamma z)) = j_{\mathcal{D}'}(\rho(\gamma'\gamma), \tau(z))j_{\mathcal{D}'}(\rho(\gamma), \tau(z))^{-1},$$

we see that

$$\begin{aligned}
\mathcal{P}^f_{\ell,m}(\gamma z) &= \sum_{\gamma' \in \Gamma} f(\gamma'\gamma z)j_{\mathcal{D}}(\gamma'\gamma, z)^\ell j_{\mathcal{D}}(\gamma, z)^{-\ell} \\
&\qquad \times j_{\mathcal{D}'}(\rho(\gamma'\gamma), \tau(z))^m j_{\mathcal{D}'}(\rho(\gamma), \tau(z))^{-m} \\
&= j_{\mathcal{D}}(\gamma, z)^{-\ell} j_{\mathcal{D}'}(\rho(\gamma), \tau(z))^{-m} \mathcal{P}^f_{\ell,m}(z)
\end{aligned}$$

Hence it suffices to show that the series in (5.1) converges uniformly and absolutely on any compact subset of $\mathcal{D}$ for $\ell \geq 2$, which can be carried out by modifying the proof of Proposition 1 in [4, p. 44]. Let E and C be compact subsets of $\mathcal{D}$ such that E is contained in the interior of C. We choose a positive real number λ such that for each $e \in E$ there is a polydisc B_e of volume λ centered at e with $\overline{B_e} \subset C$. Given $e \in E$, we have

$$|j_{\mathcal{D}}(\gamma, e)|^2 \leq \lambda^{-1} \int_{B_e} |j_{\mathcal{D}}(\gamma, z)|^2 dv_z = \lambda^{-1} \operatorname{vol}(\gamma B_e)$$

for all $\gamma \in \Gamma$. Thus, if $z \in E$ and if M is a bound of $|f|$ on $\mathcal{D}$, we see that

$$\sum_{\gamma \in \Gamma} |f(\gamma z)||j_{\mathcal{D}}(\gamma, z)|^2 \leq \sum_{\gamma \in \Gamma} M|j_{\mathcal{D}}(\gamma, z)|^2 \leq \lambda^{-1} M \sum_{\gamma \in \Gamma} \operatorname{vol}(\gamma B_z).$$

Let q be the number of elements in the set

$$\Gamma \cap \{\gamma \in G \mid \gamma C \cap C \neq \emptyset\}.$$

If $\gamma' B_z \cap \gamma B_z \neq \emptyset$ with $\gamma, \gamma' \in \Gamma$ and $z \in E$, then we have $\gamma^{-1}\gamma' C \cap C \neq \emptyset$; hence for each $\gamma \in \Gamma$ the number of sets of the form $\gamma' B_z$ having nonempty intersection with γB_z is at most q. Thus it follows that the collection $\{\gamma B_z\}_{\gamma \in \Gamma}$ covers each point of $\mathcal{D}$ at most q times, and therefore we obtain

$$\sum_{\gamma \in \Gamma} |f(\gamma z)||j_{\mathcal{D}}(\gamma, z)|^2 \leq \lambda^{-1} q M \operatorname{vol}(\mathcal{D}). \tag{5.2}$$

This proves the uniform convergence for $\ell = 2$ and $m = 0$. The proof for the general case follows from (5.2) and the fact that

$$|j_{\mathcal{D}}(\gamma, z)| < 1, \quad |j_{\mathcal{D}'}(\rho(\gamma), \tau(z))| < 1$$

for all but a finite number of $\gamma \in \Gamma$. □

Definition 5.4 *The series* $\mathcal{P}^f_{\ell,m}(z)$ *given by* (5.1) *is called a* Poincaré series *associated to* τ, ρ *and* f.

5.2 Mixed Automorphic Forms on Semisimple Lie Groups

In this section we introduce a representation theoretic definition of mixed automorphic forms on semisimple Lie groups. As an example of such automorphic forms, we also construct Poincaré series.

First, we shall review the definition of the usual automorphic forms on semisimple Lie groups (see e.g. [13, 15, 27, 28]). Let G be a semisimple Lie group, and let $\mathfrak{g}$ be its Lie algebra. If V is a finite-dimensional complex vector space, then $\mathfrak{g}$ operates on the space of smooth functions $f : G \to V$ by

$$(Y \cdot f)(g) = \frac{d}{dt} f\big((\exp tY)g\big)\Big|_{t=0} \qquad (5.3)$$

for $Y \in \mathfrak{g}$ and $g \in G$. Let $Z(\mathfrak{g})$ be the center of the universal enveloping algebra $U(\mathfrak{g})$ of the complexification $\mathfrak{g}_{\mathbb{C}}$ of $\mathfrak{g}$. Then a vector-valued function $f : G \to V$ is said to be $Z(\mathfrak{g})$-finite if $Z(\mathfrak{g}) \cdot f$ is a finite-dimensional vector space. This is equivalent to the condition that f is annihilated by an ideal $\mathfrak{A}$ of $Z(\mathfrak{g})$ of finite codimension. If the ideal $\mathfrak{A}$ has codimension one, then f is an eigenfunction of every operator in $Z(\mathfrak{g})$.

Let W be a finite-dimensional vector space over $\mathbb{C}$, and let $\alpha : G \to GL(W)$ be a representation of G in W whose kernel is finite and whose image is closed in $\operatorname{End}(W)$. Then we can define a norm $\| \cdot \|_\alpha$ on G by

$$\|g\|_\alpha = \Big(\operatorname{Tr}\big(\alpha(g)^* \cdot \alpha(g)\big)\Big)^{1/2},$$

where $*$ denotes the adjoint with respect to a Hilbert space structure on W invariant under a maximal compact subgroup K of G. If β is another such representation, then there is a constant $M > 0$ and a positive integer m such that

$$\|x\|_\alpha \leq M \|x\|_\beta^m \qquad (5.4)$$

for all $x \in G$. A vector valued function $f : G \to V$ is said to be *slowly increasing* if there is a norm $\| \cdot \|$ on G, a constant $C > 0$, and a positive integer m such that

$$|f(g)| \leq C \|g\|^m,$$

where $| \cdot |$ is a norm on V.

Definition 5.5 *Let K be a maximal compact subgroup of G, and let $\sigma : K \to GL(V)$ be a representation of K in a finite-dimensional complex vector space V. Given a discrete subgroup Γ of G, a smooth vector-valued function $f : G \to V$ is an* automorphic form *for Γ and σ if the following conditions are satisfied:*

(i) $f(kg\gamma) = \sigma(k)f(g)$ for all $k \in K$ and $\gamma \in \Gamma$.
(ii) f is $Z(\mathfrak{g})$-finite.
(iii) f is slowly increasing.

Let the subgroups K and Γ of G and the representation $\sigma : K \to GL(V)$ be as in Definition 5.5. We now consider another semisimple Lie group G'. Let K' be a maximal compact subgroup and Γ' a discrete subgroup of G'. We assume that there is a homomorphism $\rho : G \to G'$ of Lie groups such that $\rho(K) \subset K'$ and $\rho(\Gamma) \subset \Gamma'$, and let $\sigma' : K' \to GL(V')$ be a representation of K' in a finite-dimensional complex vector space V'. Then we obtain the representation $\sigma \otimes (\sigma' \circ \rho|_K) : K \to GL(V \otimes V')$ of K in $V \otimes V'$, where $\rho|_K$ denotes the restriction of ρ to K.

Definition 5.6 *A* mixed automorphic form *for Γ of type (ρ, σ, σ') is an automorphic form for Γ and the representation $\sigma \otimes (\sigma' \circ \rho|_K)$ in the sense of Definition 5.5.*

Let $f_\sigma : G \to V$ (resp. $f'_{\sigma'} : G' \to V'$) be an automorphic form for Γ (resp. Γ') and $\sigma : K \to GL(V)$ (resp. $\sigma' : K' \to GL(V')$), where V (resp. V') is a finite-dimensional complex vector space. We denote by $f_{\rho,\sigma,\sigma'}$ the function from G to $V \otimes V'$ given by

$$f_{\rho,\sigma,\sigma'}(g) = (f_\sigma \otimes (f'_{\sigma'} \circ \rho))(g) = f_\sigma(g) \otimes f'_{\sigma'}(\rho(g)) \tag{5.5}$$

for all $g \in G$.

Proposition 5.7 *The function $f_{\rho,\sigma,\sigma'} : G \to V \otimes V'$ given by (5.5) is a mixed automorphic form for Γ of type (ρ, σ, σ').*

Proof. We must show that $f_{\rho,\sigma,\sigma'}$ satisfies the conditions (i), (ii) and (iii) in Definition 5.5 for the discrete group Γ and the representation $\sigma \otimes (\sigma' \circ \rho|_K)$. Given $g \in G$, $k \in K$ and $\gamma \in \Gamma$, using (5.5), we see that

$$
\begin{aligned}
f_{\rho,\sigma,\sigma'}(kg\gamma) &= f(kg\gamma) \otimes f'(\rho(k)\rho(g)\rho(\gamma)) \\
&= \sigma(k)f(g) \otimes \sigma'(\rho(k))f'(\rho(g)) \\
&= (\sigma \otimes (\sigma' \circ \rho|_K))(k) \cdot (f \otimes f' \circ \rho)(g),
\end{aligned}
$$

which verifies the condition (i). Now, given $Y \in Z(\mathfrak{g})$, by (5.3) we have

$$
\begin{aligned}
Y \cdot f_{\rho,\sigma,\sigma'}(g) &= \frac{d}{dt} f_{\rho,\sigma,\sigma'}((\exp tY)g) \Big|_{t=0} \\
&= \left[\frac{d}{dt} f_\sigma((\exp tY)g) \right]_{t=0} \otimes f'_{\sigma'}(\rho(g)) \\
&\quad + f_\sigma(g) \otimes \left[\frac{d}{dt} f'_{\sigma'}(\rho((\exp tY)g)) \right]_{t=0} \\
&= \left[\frac{d}{dt} f_\sigma((\exp tY)g) \right]_{t=0} \otimes f'_{\sigma'}(\rho(g)) \\
&\quad + f_\sigma(g) \otimes \left[\frac{d}{dt} f'_{\sigma'}((\exp td\rho Y)\rho(g)) \right]_{t=0}
\end{aligned}
$$

for all $g \in G$. Thus the condition (ii) follows from this and the fact that f_σ is $Z(\mathfrak{g})$-finite and $f'_{\sigma'}$ is $Z(\mathfrak{g}')$-finite. As for the condition (iii), since f_σ and $f'_{\sigma'}$ are slowly increasing, we have

$$|f_{\rho,\sigma,\sigma'}(g)| = |f_\sigma(g) \otimes f'_{\sigma'}(\rho(g))| \leq C_1 \|g\|_\alpha^{m_1} \cdot C_2 \|\rho(g)\|_\beta^{m_2}$$

for some constants C_1, C_2, positive integers m_1, m_2, and representations $\alpha : G \to GL(W)$, $\beta : G \to GL(W')$. However, by (5.4) we have

$$\|\rho(g)\|_\beta = \|g\|_{\beta \circ \rho} \leq C_3 \|g\|_\alpha^{m_3}$$

for some constant C_3 and positive integer m_3. Thus we see that

$$|f_{\rho,\sigma,\sigma'}(g)| \leq C_1 C_2 C_3 \|g\|_\alpha^{m_1+m_3},$$

and $f_{\rho,\sigma,\sigma'}$ is slowly increasing, and therefore the proof of the proposition is complete. $\square$

Example 5.8 *Let $\rho : G \to G'$, K, K', V and V' be as above. Assume that the symmetric spaces $\mathcal{D} = G/K$ and $\mathcal{D}' = G'/K'$ have G-invariant and G'-invariant complex structures, respectively. This assumption implies that $\mathcal{D}$ and $\mathcal{D}'$ are equivalent to bounded symmetric domains (see e.g. [36]). Since $\rho(K) \subset K'$, the map ρ induces a holomorphic map $\tau : \mathcal{D} \to \mathcal{D}'$ satisfying $\tau(gz) = \rho(g)\tau(z)$ for all $g \in G$ and $z \in \mathcal{D}$. Let $J : G \times \mathcal{D} \to GL(V)$ and $J' : G' \times \mathcal{D}' \to GL(V')$ be automorphy factors, and set*

$$J_\rho(g, z) = J(g, z) \otimes J'(\rho(g), \tau(z))$$

for all $g \in G$ and $z \in \mathcal{D}$. Then the resulting map $J_\rho : G \times \mathcal{D} \to GL(V) \otimes GL(V')$ satisfies

$$J_\rho(g_1 g_2, z) = J_\rho(g_1, g_2 z) J_\rho(g_2, z) \tag{5.6}$$

for all $g_1, g_2 \in \mathcal{D}$ and $z \in \mathcal{D}$. We define the homomorphisms $\sigma : G \to GL(V)$ and $\sigma' : G' \to GL(V')$ by

$$\sigma(g) = J(g, 0), \quad \sigma'(g') = J'(g', 0'),$$

where $0 \in \mathcal{D}$ and $0' \in \mathcal{D}'$ are the fixed points of K and K', respectively. Note that $0' = \tau(0)$ because of the condition $\rho(K) \subset K'$. These homomorphisms induce the map $\sigma_\rho : G \to GL(V) \otimes GL(V')$ given by

$$\sigma_\rho(g) = J(g, 0) \otimes J'(\rho(g), \tau(0)) = \sigma(g) \otimes \sigma'(\rho(g)) = (\sigma \otimes (\sigma' \circ \rho))(g).$$

We now consider a mixed automorphic form $f : \mathcal{D} \to V \otimes V'$ for Γ on $\mathcal{D}$ of type (J, J', τ, ρ) in the sense of Definition 5.1, so that f satisfies

$$f(\gamma z) = J_\rho(\gamma, z) f(z)$$

for all $\gamma \in \Gamma$ and $z \in \mathcal{D}$. We define the function $\widetilde{f} : G \to V \otimes V'$ by

$$\widetilde{f}(g) = \sigma_\rho(g)f(g^{-1}0)$$

for all $g \in G$. Then for $g \in G$, $k \in K$ and $\gamma \in \Gamma$ we have

$$
\begin{aligned}
\widetilde{f}(kg\gamma) &= \sigma_\rho(kg\gamma)f(\gamma^{-1}g^{-1}k^{-1}0) \\
&= \sigma_\rho(k)\sigma_\rho(g)\sigma_\rho(\gamma)f(\gamma^{-1}g^{-1}0) \\
&= \sigma_\rho(k)\sigma_\rho(g)\sigma_\rho(\gamma)J_\rho(\gamma^{-1},g^{-1}0)f(g^{-1}0) \\
&= \sigma_\rho(k)\sigma_\rho(g)\sigma_\rho(\gamma)J_\rho(\gamma^{-1}g^{-1},0)J_\rho(g^{-1},0)^{-1}f(g^{-1}0),
\end{aligned}
$$

where we used (5.6). Using this and the relation

$$J_\rho(\gamma^{-1}g^{-1},0)J_\rho(g^{-1},0)^{-1} = \sigma_\rho(\gamma)^{-1}\sigma_\rho(g)^{-1}\sigma_\rho(g),$$

we obtain

$$\widetilde{f}(kg\gamma) = \sigma_\rho(k)\sigma_\rho(g)f(g^{-1}0) = \sigma_\rho(k)\widetilde{f}(g).$$

In fact, it can be shown that $\widetilde{f}$ is a mixed automorphic form for Γ of type (ρ, σ, σ') in the sense of Definition 5.6.

We shall construct below Poincaré series which are examples of mixed automorphic forms on semisimple Lie groups. Let Γ (resp. Γ') be a discrete subgroup of a semisimple Lie group G (resp. G'), and let K (resp. K') be a maximal compact subgroup of G (resp. G'). Let $\rho : G \to G'$ be a homomorphism such that $\rho(K) \subset K'$. If $f : G \to V$ is a vector-valued function and if h is an element of G, we denote by $l(h)$ and $r(h)$ the translation operators given by

$$l(h)f(g) = f(h^{-1}g), \quad r(h)f(g) = f(gh)$$

for all $g \in G$.

Definition 5.9 A vector-valued function $f : G \to V$ on G is said to be left (resp. right) K-finite if the set of left (resp. right) translations

$$\{l(k)f \mid k \in K\} \quad (resp. \quad \{r(k)f \mid k \in K\})$$

of f by elements of K spans a finite-dimensional vector space.

Proposition 5.10 Let V and V' be finite-dimensional complex vector spaces, and assume that the following conditions are satisfied:

(i) $f \otimes (f' \circ \rho) \in L^1(G) \otimes (V \otimes V')$, where $L^1(G)$ denotes the set of integrable functions on G.

(ii) f is $Z(\mathfrak{g})$-finite and f' is $Z(\mathfrak{g}')$-finite.

(iii) f is right K-finite and f' is right K'-finite.

Then the series $P_{\rho,f,f'}(g)$ defined by

$$P_{\rho,f,f'}(g) = \sum_{\gamma \in \Gamma}(f \otimes (f' \circ \rho))(g \cdot \gamma) \tag{5.7}$$

converges absolutely and uniformly on compact sets. Furthermore, the series

$$\sum_{\gamma \in \Gamma} |(f \otimes (f' \circ \rho))(g \cdot \gamma)|$$

is bounded on G where $|\cdot|$ denotes the norm on $V \otimes V'$.

Proof. As in the proof of Proposition 5.7, it can be shown that the function $f \otimes (f' \circ \rho) : G \to V \otimes V'$ is $Z(\mathfrak{g})$-finite by using the condition (ii). For $k \in K$ and $g \in G$, we have

$$
\begin{aligned}
r(k)(f \otimes (f' \circ \rho))(g) &= (f \otimes (f' \circ \rho))(gk) \\
&= f(gk) \otimes f'(\rho(g)\rho(k)) \\
&= r(k)f(g) \otimes r(\rho(k))f'(\rho(k)).
\end{aligned}
$$

Hence the condition (iii) implies that the function $f \otimes (f' \circ \rho)$ is right K-finite. Therefore the proposition follows from [4, Theorem 23], [5, Theorem 5.4], or [13, Theorem 9.1]. $\qquad\square$

Definition 5.11 *The series $P_{\rho,f,f'}(g)$ given by (5.7) is called a* Poincaré series *associated to ρ, f and f'.*

Corollary 5.12 *Let $\sigma : K \to GL(V)$ and $\sigma' : K' \to GL(V')$ be finite-dimensional representations of K and K', respectively, over $\mathbb{C}$. Assume that the vector-valued functions $f : G \to V$ and $f' : G' \to V'$ satisfy the conditions (i), (ii) and (iii) of Proposition 5.10 together with the condition that*

$$f(kg) = \sigma(k)f(g), \quad f'(k'g') = \sigma'(k')f'(g')$$

for $k \in K$, $g \in G$, $k' \in K'$ and $g' \in G'$. Then the associated Poincaré series $P_{\rho,f,f'}$ is a mixed automorphic form for Γ of type (ρ, σ, σ').

Proof. Since $f \otimes (f' \circ \rho)$ is $Z(\mathfrak{g})$-finite on the left, so is the Poincaré series $P_{\rho,f,f'}$. From the definition of $P_{\rho,f,f'}(g)$ the right Γ-invariance of $P_{\rho,f,f'}$ follows immediately, and $P_{\rho,f,f'}$ is slowly increasing by Proposition 5.10. As in the proof of Proposition 5.7, we have

$$(f \otimes (f' \circ \rho))(kg) = (\sigma \otimes (\sigma' \circ \rho|_K))(k)(f \otimes (f' \circ \rho))(g);$$

hence it follows that

$$P_{\rho,f,f'}(kg) = (\sigma \otimes (\sigma' \circ \rho|_K))(k)P_{\rho,f,f'}(g),$$

and the proof of the corollary is complete. $\qquad\square$

Let G, G', $\mathcal{D}$, $\mathcal{D}'$, ρ, τ, $j_{\mathcal{D}} : G \times \mathcal{D} \to \mathbb{C}$ and $j_{\mathcal{D}'} : G' \times \mathcal{D}' \to \mathbb{C}$ be as in Section 5.1.

Lemma 5.13 *If ℓ and m are nonnegative integers with $\ell \geq 2$, then the $\mathbb{C}$-valued function on G given by*

$$g \mapsto j(g,0)^\ell \otimes j_{\mathcal{D}'}(\rho(g),0')^m : G \to \mathbb{C}$$

is an element of $L^1(G)$.

Proof. Note first that the function $g \mapsto |j(g,0)|^\ell$ (resp. $g' \mapsto |j_{\mathcal{D}'}(g',0')|^m$) is left and right K-invariant (resp. K'-invariant) (see [5, Section 5.8]); hence it can be considered as a function on $\mathcal{D}$ (resp. $\mathcal{D}'$), and we have

$$\int_G |j(g,0)^\ell j_{\mathcal{D}'}(\rho(g),0')^m| \, dg = \int_{\mathcal{D}} |j(z,0)|^\ell |j_{\mathcal{D}'}(\tau(z),0')|^m d\mu(z),$$

where $d\mu(z)$ denotes the invariant Bergmann measure on $\mathcal{D}$ (cf. [4, Section 4.3]). If dz is the usual Euclidean measure on $\mathcal{D}$, then we have

$$d\mu(z) = |j(z,0)|^{-2} dz$$

up to a positive factor. Thus we have

$$\int_G |j(g,0)^\ell j_{\mathcal{D}'}(\rho(g),0')^m| \, dg = \int_{\mathcal{D}} |j(z,0)|^{\ell-2} |j_{\mathcal{D}'}(\tau(z),0')|^m dz.$$

However, both $|j(z,0)|$ and $|j(\tau(z),0')|$ are bounded by [5, Proposition 1.12]; hence the lemma follows.   □

Let $F : \mathcal{D} \to V$ and $F' : \mathcal{D}' \to V'$ be functions such that $F \otimes (F' \circ \tau) : \mathcal{D} \to V \otimes V'$ is a polynomial function. Given nonnegative integers ℓ, m with $\ell \geq 2$, we define the functions $f : G \to V$ and $f' : G' \to V'$ by

$$f(g) = j(g,0)^\ell F(\pi g), \quad f'(g') = j_{\mathcal{D}'}(g',0')^m F'(\pi' g') \tag{5.8}$$

for $g \in G$ and $g' \in G'$, where $0 \in \mathcal{D}$ and $0' \in \mathcal{D}'$ are the fixed points of K and K', respectively, and $\pi : G \to \mathcal{D}$ and $\pi' : G' \to \mathcal{D}'$ are canonical projection maps. We set

$$\sigma(k) = j(k,0)^\ell, \quad \sigma'(k') = j(k',0')^m \tag{5.9}$$

for $k \in K$ and $k' \in K'$. Then σ and σ' are representations of G and G' in V and V', respectively. We set

$$P_\rho^{\ell,m}(g) = \sum_{\gamma \in \Gamma} (f \otimes (f' \circ \rho))(g\gamma) \tag{5.10}$$

for $g \in G$, which is indeed the Poincaré series $P_{\rho,f,f'}$ associated to the functions f and f' in (5.8) in the sense of Definition 5.11.

Theorem 5.14 *The Poincaré series $P_\rho^{\ell,m}(g)$ given by (5.10) is a mixed automorphic form for Γ of type (ρ, σ, σ').*

Proof. For each $g \in G$, using (5.8), we have

$$(f \otimes (f' \circ \rho))(g) = \left(j(g,0)^\ell F(\pi g) \right) \otimes \left(j_{\mathcal{D}'}(\rho(g),0')^m F'(\pi'(\rho(g))) \right)$$
$$= \left(j(g,0)^\ell \cdot j_{\mathcal{D}'}(\rho(g),0')^m \right) \left(F(\pi g) \otimes F'(\omega(\pi g)) \right).$$

Thus from Lemma 5.13 and the fact that $F \otimes (F' \circ \omega)$ is a polynomial mapping on the bounded symmetric domain $\mathcal{D}$ it follows that

$$\int_G |(f \otimes (f' \circ \rho))(g)| \, dg < \infty;$$

hence $f \otimes (f' \circ \rho)$ is an element of $L^1(G) \otimes (V \otimes V')$. Using (5.9) and the relation $j(g_1 g_2, 0) = j(g_1, 0) j(g_2, 0)$ for $g_1, g_2 \in G$, we have

$$f(kg) = j(gk, 0)^\ell F(\pi(gk))$$
$$= j(g,0)^\ell j(k,0)^\ell F(\pi g) = \sigma(k) f(g)$$

for $k \in K$ and $g \in G$. Similarly, we have $f'(k'g') = \sigma'(k') f'(g')$ for $k' \in K'$ and $g' \in G'$. Thus we see that f is $Z_{\mathfrak{g}}$-finite and f' is $Z_{\mathfrak{g}'}$-finite. For $k \in K$ and $g \in G$ the set of right translates $r(k)F(\pi(gk))$ are polynomials of the same degree as F; hence f is right K-finite. Similarly, f' is right K'-finite. Thus it follows from Proposition 5.10 that the series

$$\sum_{\gamma \in \Gamma} |(f \otimes (f' \circ \rho))(g\gamma)|$$

is bounded on G. In particular, the Poincaré series $P_\rho^{\ell,m}$ in (5.10) is slowly increasing. As in the proof of Proposition 5.7, the function $f \otimes (f' \circ \rho)$ is $Z_{\mathfrak{g}}$-finite, and for $k \in K$ and $g \in G$ we have

$$(f \otimes (f' \circ \rho))(kg) = (\sigma \otimes (\sigma' \circ \rho|_K))(k)(f \otimes (f' \circ \rho))(g);$$

hence we see that

$$P_\rho^{\ell,m}(kg) = (\sigma \otimes (\sigma' \circ \rho|_K))(k) P_\rho^{\ell,m}(g).$$

Therefore $P_\rho^{\ell,m}(g)$ is an automorphic form for Γ of type (ρ, σ, σ') in the sense of Definition 5.5. $\square$

5.3 Eisenstein Series

In this section we construct Eisenstein series and show that they are mixed automorphic forms on semisimple Lie groups in the sense of Definition 5.5. For this purpose, instead of the usual semisimple Lie group G, we need to consider an algebraic group whose set of real points will coincide with G.

Let $\mathbb{G}$ be a connected, semisimple, linear algebraic group defined over a subfield k of $\mathbb{R}$, and let $\mathbb{P}$ be a k-parabolic subgroup of $\mathbb{G}$ containing a maximal k-split torus $\mathbb{S}$ of $\mathbb{G}$. Let $\mathbb{P}_0$ be a minimal k-parabolic subgroup of $\mathbb{G}$ such that $\mathbb{S} \subset \mathbb{P}_0 \subset \mathbb{P}$. We define an ordering on the set Σ_k of k-roots of $\mathbb{G}$ with respect to $\mathbb{S}$ as follows: A k-root $\alpha \in \Sigma_k$ is positive if and only if the subgroup of $\mathbb{G}$ generated by the α-eigenspace of the adjoint representation of $\mathbb{S}$ is contained in $\mathbb{P}_0$. We denote by Δ_k the set of simple positive k-roots in Σ_k.

If $\varXi$ is a subset of Δ_k, we set

$$\mathbb{S}_\varXi = \left(\bigcap_{\alpha \in \varXi} \operatorname{Ker} \alpha \right)^0,$$

where $(\cdot)^0$ denotes the connected component of the identity. Let Θ be the subset of Δ_k such that $\mathbb{P}$ is generated by the unipotent radical $\mathbb{U}_0$ of $\mathbb{P}_0$ and by the centralizer $Z(\mathbb{S}_\Theta)$ of $\mathbb{S}_\Theta$. Let $\mathbb{U}_\Theta$ be the unipotent radical of $\mathbb{P}$, and let $\mathbb{M}_\Theta$ be a subgroup of $\mathbb{G}$ such that $Z_\mathbb{G}(\mathbb{S}_\Theta) = \mathbb{S}_\Theta \cdot \mathbb{M}_\Theta$ with $\mathbb{S}_\Theta \cap \mathbb{M}_\Theta$ finite. We set

$$P = \mathbb{P}(\mathbb{R}), \quad A = \mathbb{S}_\Theta(\mathbb{R})^0, \quad M = \mathbb{M}_\Theta(\mathbb{R}), \quad U = \mathbb{U}_\Theta(\mathbb{R}).$$

Then we obtain the Langlands decomposition

$$P = MAU$$

of P and the corresponding decomposition

$$G = KP = KMAU$$

of $G = G(\mathbb{R})$. If $g = kmau \in G$ with $k \in K$, $m \in M$, $a \in A$ and $u \in U$, then $k \cdot m$, a and u are uniquely determined. We write $a = a(g)$.

Let $\{\Lambda_\alpha\}_{\alpha \in \Delta_k}$ be the set of fundamental dominant k-weights of $\mathbb{G}$ that satisfy

$$\langle \Lambda_\alpha, \beta \rangle = d_\alpha \delta_{\alpha\beta}$$

for all $\alpha, \beta \in \Delta_k$, where $\delta_{\alpha\beta}$ is the Kronecker delta and d_α is a positive real number (see [13, Section 11]). Let $\mathfrak{u}$ be the Lie algebra of U and let $\chi = \det \operatorname{Ad}_\mathfrak{u}$ be the character of P with $p^\chi = \det \operatorname{Ad}_\mathfrak{u} p$ for $p \in P$. We set $\widetilde{\Theta} = \Delta_k - \Theta$. Then χ is a positive linear combination of the Λ_α for $\alpha \in \widetilde{\Theta}$, that is,

$$\chi = \sum_{\alpha \in \widetilde{\Theta}} e_\alpha \Lambda_\alpha$$

with $e_\alpha > 0$. If $\{s_\alpha\}_{\alpha \in \widetilde{\Theta}}$ is a set of complex numbers $s_\alpha \in \mathbb{C}$ indexed by the set $\widetilde{\Theta}$, and if $p \in P$, then we set

$$p^{\Lambda_s} = \prod_{\alpha \in \widetilde{\Theta}} |p^{\Lambda_\alpha}|^{s_\alpha}.$$

Lemma 5.15 *Let Γ be a discrete subgroup of G and let Γ_∞ be a subgroup of $\Gamma \cap MU$. Suppose that there is a set $\{s_\alpha\}_{\alpha \in \widetilde{\Theta}}$ of complex numbers satisfying the following conditions:*

(i) $a(\gamma)^{\Lambda_\alpha} \geq \varepsilon > 0$ for all $\gamma \in \Gamma$ and $\alpha \in \widetilde{\Theta}$.

(ii) MU/Γ_∞ has finite measure.

(iii) $\operatorname{Re} s_\alpha > e_\alpha$ for all $\alpha \in \widetilde{\Theta}$.

Then the series

$$E(g, s) = \sum_{\gamma \in \Gamma/\Gamma_\infty} a(g\gamma)^{-\Lambda_s}$$

converges uniformly on any compact subset of G.

Proof. See Lemma 4 in [4] or Lemma 11.1 in [13] (see also [29]). □

Now we consider another semisimple Lie group $G' = \mathbb{G}'(\mathbb{R})$ associated to a connected algebraic group $\mathbb{G}'$ defined over a subfield k of $\mathbb{R}$. We consider the corresponding subgroups K', P', M', A', U', etc. defined in a way similar to the case of G above. Thus we have decompositions

$$P' = K'A'U', \quad G' = K'P' = K'M'A'U'.$$

Let $\rho : G \to G'$ be a Lie group homomorphism such that

$$\rho(K) \subset K', \quad \rho(P) \subset P', \quad \rho(A) \subset A'.$$

Theorem 5.16 *Let Γ and Γ_∞ be as in Lemma 5.15, and let $f : G \to V$ and $f' : G' \to V'$ be smooth vector-valued functions, where V and V' are finite-dimensional vector spaces. Suppose that there is a set $\{s_\alpha\}_{\alpha \in \widetilde{\Theta}}$ of complex numbers satisfying the following conditions:*

(i) $a(\gamma)^{\Lambda_\alpha} \geq \varepsilon > 0$ for all $\gamma \in \Gamma$ and $\alpha \in \widetilde{\Theta}$.

(ii) MU/Γ_∞ has finite measure.

(iii) $\operatorname{Re} s_\alpha > e_\alpha$ for all $\alpha \in \widetilde{\Theta}$.

(iv) $(f \otimes (f' \circ \rho))(g\gamma) = (f \otimes (f' \circ \rho))(g)$ for all $\gamma \in \Gamma_\infty$.

(v) $|f(gp)||f'(\rho(gp))|\,|p^{\Lambda_s}|$ is bounded for $p \in P$ and g belonging to a fixed compact set.

Then the series

$$E_{\rho,f,f'}(g) = \sum_{\gamma \in \Gamma/\Gamma_\infty} (f \otimes (f' \circ \rho))(g\gamma) \tag{5.11}$$

converges absolutely and uniformly on any compact set of G.

Proof. Since $G = KP$, we have

$$\begin{aligned}
(f \otimes (f' \circ \rho))(g) \cdot a(g)^{\Lambda_s} &= (f \otimes (f' \circ \rho))(kp) \cdot a(kp)^{\Lambda_s} \\
&= (f \otimes (f' \circ \rho))(kp) \cdot a(p)^{\Lambda_s} \\
&= (f \otimes (f' \circ \rho))(kp) \cdot p^{\Lambda_s}
\end{aligned}$$

for $g = kp$ with $k \in K$ and $p \in P$. Hence by (v) $|(f \otimes (f' \circ \rho))(g) \cdot a(g)^{\Lambda_s}|$ is bounded for $g \in G$. Therefore, the series defining $E_{f,f',\rho}$ is majorized by a constant times the series

$$\sum_{\gamma \in \Gamma/\Gamma_\infty} a(g\gamma)^{-\Lambda_s},$$

which converges uniformly on any compact set by Lemma 5.15. Hence the theorem follows. $\qquad\square$

Definition 5.17 *The series* $E_{\rho,f,f'}(g)$ *given by the series in* (5.11) *is called an* Eisenstein series *for* Γ *associated to* ρ, f *and* f'.

Let G, P, M, A, and U be as before. Thus we have decompositions

$$P = MAU, \quad G = KP = KMAU.$$

Let $\sigma : K \to GL(V)$ be a representation of K in a finite-dimensional complex vector space V. Let $\pi : MU \to M$ be the natural projection, and let $K_M = \pi(K \cap MU)$. Then K_M is a maximal compact subgroup of M, and σ induces the representation σ_M of K_M given by

$$\sigma_M(\pi(k)) = \sigma(k)$$

for all $k \in K \cap MU$. Let Γ be an arithmetic subgroup of G, and let $\Gamma_M = \pi(\Gamma \cap MU)$ be the corresponding arithmetic subgroup of M. We denote by $L^2(M/\Gamma_M, \sigma_M)$ the space of square-integrable functions $\varphi : M \to V$ satisfying

$$\varphi(km\gamma) = \sigma_M(k)\varphi(m)$$

for all $k \in K_M$, $m \in M$ and $\gamma \in \Gamma_M$. Any function $\varphi : M \to V$ satisfying $\varphi(km) = \sigma_M(k)\varphi(m)$ for $k \in K_M$ and $m \in M$ can be extended to a function $\varphi_G : G \to V$ on G by the formula

$$\varphi_G(kmau) = \sigma(k)\varphi(m)$$

for all $k \in K$, $m \in M$, $a \in A$ and $u \in U$. Then φ_G is σ-equivariant, that is,

$$\varphi_G(kg) = \sigma(k)\varphi_G(g)$$

for $k \in K$ and $g \in G$. Although a decomposition $g = kmau$ is not unique, the extension φ_G is uniquely determined. We shall identify a σ_M-equivariant function φ on M and the corresponding σ-equivariant function φ_G on G. Thus each element of $L^2(M/\Gamma_M, \sigma_M)$ will be regarded as a function of G into V.

Let $\mathfrak{a}_\mathbb{C}^*$ be the dual space of the complexification $\mathfrak{a}_\mathbb{C}$ of the Lie algebra $\mathfrak{a}$ of A, and let

$$(\mathfrak{a}_\mathbb{C}^*)^- = \{\Lambda \in \mathfrak{a}_\mathbb{C}^* \mid \langle \operatorname{Re} \Lambda + \rho, \alpha \rangle < 0 \quad \text{for all} \quad \alpha \in \Sigma^0\},$$

where Σ^0 is the set of simple roots of the Lie algebra $\mathfrak{g}$ of G. For $\varphi \in L^2(M/\Gamma_M, \sigma_M)$ and $\Lambda \in \mathfrak{a}^*_{\mathbb{C}}$, we set

$$\varphi_\Lambda(x) = \varphi(x)e^{(\Lambda-\rho)(H(x))}$$

for $x \in G$, where $H(x)$ denotes $\log a(x)$. Then we have

$$\varphi_\Lambda(kg\gamma) = \sigma(k)\varphi_\Lambda(g)$$

for all $g \in G$, $k \in K$ and $\gamma \in U(\Gamma \cap P)$. Given $\varphi \in L^2(M/\Gamma_M, \sigma_M)$ and $\Lambda \in (\mathfrak{a}^*_{\mathbb{C}})^-$, the series

$$E(\Lambda, \varphi, x) = \sum_{\Gamma/\Gamma\cap P} \varphi_\Lambda(x\gamma)$$

is called an *Eisenstein series* (see [37]; see also [60, 64, 65, 66, 100]).

Let χ be a representation of $Z(\mathfrak{g})$ in V. We denote by $L^2(M/\Gamma_M, \sigma_M, \chi)$ the subgroup of $L^2(M/\Gamma_M, \sigma_M)$ consisting of functions $f : G \to V$ satisfying the condition

$$(Y \cdot f)(g) = f(g)\chi(Y) \tag{5.12}$$

for all $Y \in Z(\mathfrak{g})$.

Lemma 5.18 *If $\varphi \in L^2(M/\Gamma_M, \sigma_M, \chi)$ and $\gamma \in U(\Gamma \cap P)$, then there are a positive real number C and a positive integer N such that*

$$\sum_{\Gamma/\Gamma\cap P} |\varphi_\Lambda(x\gamma)| \le C\|x\|^N$$

for all $x \in G$.

Proof. This follows from [37, Lemma 24]. $\qquad\square$

Let G' be another semisimple Lie group, and consider the corresponding objects K', P', M', A', U', $\mathfrak{a}'$, $\mathfrak{a}'^*_{\mathbb{C}}$, $(\mathfrak{a}'^*_{\mathbb{C}})^-$ and the representation $\sigma' : K' \to GL(V')$ of K' in a finite-dimensional complex vector space V'. Let $\rho : G \to G'$ be a Lie group homomorphism such that

$$\rho(K) \subset K', \quad \rho(P) \subset P', \quad \rho(A) \subset A'.$$

As in the case of G, for $\varphi' \in L^2(M'/\Gamma'_{M'}, \sigma'_{M'})$ and $\Lambda' \in (\mathfrak{a}'^*_{\mathbb{C}})^-$, we set

$$\varphi'_{\Lambda'}(y) = \varphi'(y)e^{(\Lambda'-\rho')(H'(y))}$$

for $y \in G'$, where $H'(y)$ denotes $\log a'(y)$.

Proposition 5.19 *Let χ and χ' be representations of $Z(\mathfrak{g})$ and $Z(\mathfrak{g}')$ in V and V' respectively, and let*

$$\varphi_\Lambda \in L^2(M/\Gamma_M, \sigma_M, \chi), \quad \varphi'_{\Lambda'} \in L^2(M'/\Gamma'_{M'}, \sigma'_{M'}, \chi')$$

with $\Lambda \in (\mathfrak{a}^*_{\mathbb{C}})^-$ and $\Lambda' \in (\mathfrak{a}'^*_{\mathbb{C}})^-$. Then there are a positive real number C_0 and a positive integer N_0 such that

$$\sum_{\gamma \in \Gamma/\Gamma \cap P} |(\varphi_\Lambda \otimes (\varphi'_{\Lambda'} \circ \rho))(x\gamma)| \le C_0 \|x\|^{N_0}$$

for all $x \in G$.

Proof. By Lemma 5.18 there exist positive real numbers C, C' and positive integers N, and N' such that

$$\sum_{\gamma \in \Gamma/\Gamma \cap P} |\varphi_\Lambda(x\gamma)| \le C\|x\|^N,$$

$$\sum_{\gamma' \in \Gamma'/\Gamma' \cap P'} |\varphi'_{\Lambda'}(x'\gamma')| \le C'\|x'\|^{N'}$$

for all $x \in G$ and $x' \in G'$. In particular, we have

$$\sum_{\gamma \in \Gamma/\Gamma \cap P} |\varphi'_{\Lambda'}(\rho(x\gamma))| = \sum_{\gamma \in \Gamma/\Gamma \cap P} |\varphi'_{\Lambda'}(\rho(x)\rho(\gamma))|$$

$$\le C'\|\rho(x)\|^{N'} \le C'\|\rho\|^{N'}\|x\|^{N'}$$

for all $x \in G$, since $\rho(\gamma) \in \Gamma'/\Gamma' \cap P'$ whenever $\gamma \in \Gamma/\Gamma \cap P$. Thus we obtain

$$\sum_{\gamma \in \Gamma/\Gamma \cap P} |(\varphi_\Lambda \otimes (\varphi'_{\Lambda'} \circ \rho))(x\gamma)| = \sum_{\gamma \in \Gamma/\Gamma \cap P} |\varphi_\Lambda(x\gamma)| \cdot |\varphi'_{\Lambda'}(\rho(x\gamma))|$$

$$\le \sum_{\gamma \in \Gamma/\Gamma \cap P} |\varphi_\Lambda(x\gamma)| \cdot \sum_{\gamma \in \Gamma/\Gamma \cap P} |\varphi'_{\Lambda'}(\rho(x\gamma))|$$

$$\le C \cdot C' \cdot \|\rho\|^{N'} \cdot \|x\|^{N+N'};$$

hence the proposition follows.   □

We set

$$E_{\rho,\varphi,\varphi'}(\Lambda, \Lambda', x) = \sum_{\gamma \in \Gamma/\Gamma \cap P} (\varphi_\Lambda \otimes (\varphi'_{\Lambda'} \circ \rho))(x\gamma) \tag{5.13}$$

for $x \in G$, which is an Eisenstein series for Γ associated to ρ, φ_Λ, and $\varphi'_{\Lambda'}$ in the sense of Definition 5.17 with $\Gamma_\infty = \Gamma \cap MU = \Gamma \cap P$ (see [37, p. 6]).

Theorem 5.20 *The Eisenstein series $E_{\rho,\varphi,\varphi'}(\Lambda, \Lambda', x)$ given by (5.13) is a mixed automorphic form for Γ of type (ρ, σ, σ').*

Proof. Recall that φ_Λ can be regarded as a function $\varphi_\Lambda : G \to V$ on G satisfying $\varphi_\Lambda(kmau) = \sigma(k)\varphi_\Lambda(m)$. Thus we have

$$\varphi_\Lambda(kg) = \sigma(k)\varphi_\Lambda(g)$$

for all $k \in K$ and $g \in G$. Similarly, we consider $\varphi'_{\Lambda'}$ as a V'-valued function on G' satisfying

$$\varphi'_{\Lambda'}(k'g') = \sigma'(k')\varphi'_{\Lambda'}(g')$$

for all $k' \in K'$ and $g' \in G'$. Thus we see that

$$(\varphi_\Lambda \otimes (\varphi'_{\Lambda'} \circ \rho))(kg) = (\sigma \otimes (\sigma' \circ \rho|_K))(k)(\varphi_\Lambda \otimes (\varphi'_{\Lambda'} \circ \rho))(g)$$

for all $k \in K$ and $g \in G$, which implies

$$E_{\rho,\varphi,\varphi'}(\Lambda, \Lambda', kg) = (\sigma \otimes (\sigma' \circ \rho|_K))(k)E_{\rho,\varphi,\varphi'}(\Lambda, \Lambda', g)$$

by (5.13). On the other hand, using (5.13) and Proposition 5.19, we obtain

$$\left| E_{\rho,\varphi,\varphi'}(\Lambda, \Lambda', x) \right| \leq \sum_{\gamma \in \Gamma/\Gamma \cap P} |(\varphi_\Lambda \otimes (\varphi'_{\Lambda'} \circ \rho))(x\gamma)| \leq C_0 \|x\|^{N_0}$$

for some $C_0, N_0 > 0$, and consequently the function $x \mapsto E_{\rho,\varphi,\varphi'}(\Lambda, \Lambda', x)$ is slowly increasing. Using (5.12), we have

$$(Y \cdot \varphi_\Lambda)(g) = \varphi_\Lambda(g)\chi(Y), \qquad (Y' \cdot \varphi'_{\Lambda'})(g') = \varphi'_{\Lambda'}(g')\chi'(Y')$$

for $Y \in Z(\mathfrak{g})$, $Y' \in Z(\mathfrak{g}')$, $g \in G$ and $g' \in G'$; hence it follows that φ_Λ is $Z(\mathfrak{g})$-finite and $\varphi'_{\Lambda'}$ is $Z(\mathfrak{g}')$-finite. Thus, as in the proof of Proposition 5.7, we see that the function $\varphi_\Lambda \otimes (\varphi'_{\Lambda'} \circ \rho)$ is $Z(\mathfrak{g})$-finite. Therefore $E_{\rho,\varphi,\varphi'}$ is also $Z(\mathfrak{g})$-finite, and the theorem follows. $\square$

5.4 Whittaker Vectors

We shall first extend the notion of mixed automorphic forms on semisimple Lie groups constructed in Section 5.2 to real reductive groups and describe Eisenstein series for such automorphic forms. We also discuss Whittaker vectors and Poincaré series for mixed automorphic forms on real reductive groups.

Let G be a real reductive group G of rank one, and let $G = NAK$ be its Iwasawa decomposition. Let $P = MAN$ be the associated Langlands decomposition of a minimal parabolic subgroup P of G. If $g \in G$, then we write

$$g = n(g)a(g)k(g)$$

with $n(g) \in N$, $a(g) \in A$ and $k(g) \in K$. Let $\mathfrak{g}, \mathfrak{k}, \mathfrak{m}, \mathfrak{a}, \mathfrak{n}$ denote the Lie algebras of G, K, M, A, N, respectively, and let

$$\rho(H) = \frac{1}{2} \operatorname{Tr}\big(\operatorname{ad}(H)|_{\mathfrak{n}}\big)$$

for $H \in \mathfrak{a}$.

Let $(\pi_{\xi,\nu}, H^{\xi,\nu})$ be the principal series representation of G corresponding to an element $\nu \in (\mathfrak{a}_{\mathbb{C}})^*$ and a unitary representation ξ of M in H_ξ. Thus $H^{\xi,\nu}$ is the space of all square-integrable functions $f : K \to H_\xi$ such that

$$f(mk) = \xi(m)f(k) \tag{5.14}$$

for $m \in M$ and $k \in K$ with

$$\pi_{\xi,\nu}(g)f(x) = a(xg)^{\nu+\rho}f(k(xg)) \tag{5.15}$$

for $g \in G$ and $x \in K$. Let $Z(\mathfrak{g})$ be the center of the universal enveloping algebra $U(\mathfrak{g})$ of $\mathfrak{g}$. Let V be a finite-dimensional complex inner product space, and let $\sigma : K \to GL(V)$ be a representation of K in V. Then an automorphic form on the real reductive group G for Γ and σ is a left Γ-invariant and $Z(\mathfrak{g})$-finite analytic function $f : G \to V$ satisfying the following conditions:

(i) $f(xk) = \sigma(k)f(x)$ for all $x \in G$ and $k \in K$,

(ii) f is slowly increasing, that is, there exist positive real numbers C and C' such that

$$\|f(x)\| \leq C' \cdot a|x|^C$$

for all $x \in G$, where $a|x|^C = e^{C|\rho(\log a(x))|}$.

Let G' be another real reductive group of rank one, and let K' and $P' = M'A'N'$ be a maximal compact subgroup and a Langlands decomposition of a minimal parabolic subgroup of G', respectively. Let $\varphi : G \to G'$ be a homomorphism such that $\varphi(K) \subset K'$, $\varphi(P) \subset P'$, $\varphi(A) \subset A'$ and $\varphi(N) \subset N'$, and let Γ' be a discrete subgroup of G' of finite covolume with $\varphi(\Gamma) \subset \Gamma'$ satisfying the assumptions in [66, §2]. We also consider a representation $\sigma' : K' \to GL(V')$ of K' in a finite-dimensional complex vector space V', so that $\sigma \otimes (\sigma' \circ \varphi)$ is a representation of K in $V \otimes V'$.

Definition 5.21 *A mixed automorphic form for Γ of type $(\varphi, \sigma, \sigma')$ on the real reductive group G is an automorphic form $f : G \to V \otimes V'$ for Γ and the representation $\sigma \otimes (\sigma' \circ \varphi)$ of K.*

Proposition 5.22 *Let $f : G \to V$ (resp. $f' : G' \to V'$) be an automorphic form for Γ (resp. Γ') and σ (resp σ'). Then the function $f \otimes (f' \circ \varphi) : G \to V \otimes V'$ is an automorphic form for Γ and $\sigma \otimes (\sigma' \circ \varphi)$.*

Proof. The proof is essentially the same as in Proposition 5.7. $\square$

Given an element $\nu' \in (\mathfrak{a}'_{\mathbb{C}})^*$ and a unitary representation ξ' of M' in $H_{\xi'}$, let $(H^{\xi',\nu'}, \pi_{\xi',\nu'})$ be the associated principal series representation of G'. Let $(\widehat{H}_\varphi^{\xi,\xi',\nu,\nu'}, \widehat{\pi}_{\xi,\xi',\nu,\nu'}^\varphi)$ be the induced representation of G associated to the representation $\xi \otimes (\xi' \circ \varphi)$ of M in $H_\xi \otimes H_{\xi'}$ and an element $\nu + (\nu' + \rho') \circ d\varphi$

of $(\mathfrak{a}_{\mathbb{C}})^*$, where $d\varphi : \mathfrak{a}_{\mathbb{C}} \to \mathfrak{a}'_{\mathbb{C}}$ is the linear map corresponding to the map $\varphi|_A : A \to A'$. Thus, using (5.14) and (5.15), we see that $\widehat{H}_\varphi^{\xi,\xi',\nu,\nu'}$ consists of square-integrable functions $\psi : K \to H_\xi \otimes H_{\xi'}$ such that

$$\psi(mk) = (\xi \otimes (\xi' \circ \varphi))(m)\psi(k)$$

for all $m \in M$ and $k \in K$, and we have

$$\widehat{\pi}_{\xi,\xi',\nu,\nu'}^\varphi(g)\psi(x) = a(xg)^{\nu+(\nu'+\rho')\circ d\varphi+\rho}\psi(k(xg))$$

for all $g \in G$ and $x \in K$.

Now we denote by $H_\varphi^{\xi,\xi',\nu,\nu'}$ the complex linear space generated by functions of the form

$$f \otimes (f' \circ \varphi) : K \to H_\xi \otimes H_{\xi'}$$

for some $f \in H^{\xi,\nu}$ and $f' \in H^{\xi',\nu'}$, and define the action of G on $H_\varphi^{\xi,\xi',\nu,\nu'}$ by

$$(\pi_{\xi,\xi',\nu,\nu'}^\varphi(g))(f \otimes (f' \circ \varphi)) = \big((\pi_{\xi,\nu} \otimes (\pi_{\xi',\nu'} \circ \varphi))(g)(f \otimes f')\big) \circ (1 \otimes \varphi) \quad (5.16)$$

for all $g \in G$. Thus, if $\psi = \sum_{i=1}^m f_i \otimes (f_i' \circ \varphi) \in H_\varphi^{\xi,\xi',\nu,\nu'}$ with $f_i \in H^{\xi,\nu}$ and $f_i' \in H^{\xi',\nu'}$, then by (5.14) and (5.15) we have

$$\psi(mk) = \sum_{i=1}^m f_i(mk) \otimes f_i'(\varphi(mk)) = \sum_{i=1}^m \xi(m)f_i(k) \otimes \xi'(\varphi(m))f_i'(\varphi(k))$$

$$= \sum_{i=1}^m (\xi \otimes (\xi' \circ \varphi))(m)\big(f_i \otimes (f_i' \circ \varphi)\big)(k) = (\xi \otimes (\xi' \circ \varphi))(m)\psi(k)$$

for all $m \in M$ and $k \in K$, and

$$\big(\pi_{\xi,\xi',\nu,\nu'}^\varphi(g)\psi\big)(x) = \sum_{i=1}^m (\pi_{\xi,\nu}(g)f_i)(x) \otimes (\pi_{\xi',\nu'}(\varphi(g))f_i')(\varphi(x))$$

$$= \sum_{i=1}^m a(xg)^{\nu+\rho}f_i(k(xg)) \otimes a'(\varphi(xg))^{\nu'+\rho'}f_i'(k'(\varphi(xg)))$$

$$= \sum_{i=1}^m a(xg)^{\nu+\rho}f_i(k(xg)) \otimes a'(\varphi(xg))^{\nu'+\rho'}f_i'(\varphi(k(xg)))$$

$$= a(xg)^{\nu+\rho}a'(\varphi(xg))^{\nu'+\rho'}\sum_{i=1}^m (f_i \otimes (f_i' \circ \varphi))(k(xg))$$

$$= a(xg)^{\nu+\rho}\varphi(a(xg))^{\nu'+\rho'}\sum_{i=1}^m (f_i \otimes (f_i' \circ \varphi))(k(xg))$$

$$= a(xg)^{\nu+\rho+(\nu'+\rho')\circ d\varphi}\sum_{i=1}^m (f_i \otimes (f_i' \circ \varphi))(k(xg))$$

$$= a(xg)^{\nu+\rho+(\nu'+\rho')\circ d\varphi}\psi(k(xg))$$

for all $g \in G$ and $x \in K$. Therefore $(H_\varphi^{\xi,\xi',\nu,\nu'}, \pi_{\xi,\xi',\nu,\nu'}^\varphi)$ is in fact a subrepresentation of $(\widehat{H}_\varphi^{\xi,\xi',\nu,\nu'}, \widehat{\pi}_{\xi,\xi',\nu,\nu'}^\varphi)$, and it has the structure of a $(\mathfrak{g}, K)$-module.

We define the linear map

$$\delta_{\xi,\xi',\nu,\nu'}^\varphi : H_\varphi^{\xi,\xi',\nu,\nu'} \to H_\xi \otimes H_{\xi'}$$

by

$$\delta_{\xi,\xi',\nu,\nu'}^\varphi \left(f \otimes (f' \circ \varphi) \right) = \left(f \otimes (f' \circ \varphi) \right)(1) \tag{5.17}$$

for $f \in H^{\xi,\nu}$ and $f' \in H^{\xi',\nu'}$. Let $H_K^{\xi,\nu} \subset H^{\xi,\nu}$ (resp. $H_{K'}^{\xi',\nu'} \subset H^{\xi',\nu'}$) denote the subspace of K-finite (resp. K'-finite) vectors, and let $H_{\varphi,K,K'}^{\xi,\xi',\nu,\nu'}$ be the subspace of $H_\varphi^{\xi,\xi',\nu,\nu'}$ generated by elements of the form $f \otimes (f' \circ \varphi)$ with $f \in H_K^{\xi,\nu}$ and $f' \in H_{K'}^{\xi',\nu'}$. For $\psi \in H_{\varphi,K,K'}^{\xi,\xi',\nu,\nu'}$, we set

$$E_\varphi(P,\xi,\xi',\nu,\nu')(\psi) = \sum_{\gamma \in \Gamma_N \backslash \Gamma} \delta_{\xi,\xi',\nu,\nu'}^\varphi \left(\pi_{\xi,\xi',\nu,\nu'}^\varphi(\gamma)\psi \right). \tag{5.18}$$

Now we consider an element $\psi \in H_{\varphi,K,K'}^{\xi,\xi',\nu,\nu'}$ given by

$$\psi = \sum_{i=1}^m f_i \otimes (f_i' \circ \varphi) \tag{5.19}$$

with $f_i \in H_K^{\xi,\nu}$ and $f_i' \in H_{K'}^{\xi',\nu'}$, then by (5.17) and (5.18) we see that $E_\varphi(P,\xi,\xi',\nu,\nu')(\psi)$ is an element of $H_\xi \otimes H_{\xi'}$ given by

$$E_\varphi(P,\xi,\xi',\nu,\nu')(\psi) \tag{5.20}$$

$$= \sum_{\gamma \in \Gamma_N \backslash \Gamma} \delta_{\xi,\xi',\nu,\nu'}^\varphi \left(\pi_{\xi,\xi',\nu,\nu'}^\varphi(\gamma g)\psi \right)$$

$$= \sum_{\gamma \in \Gamma_N \backslash \Gamma} \sum_{i=1}^m (\pi_{\xi,\nu}(\gamma)f_i)(1) \otimes (\pi_{\xi',\nu'}(\varphi(\gamma))f_i')(\varphi(1)).$$

For $g \in G$ and $\eta \in (H_\xi \otimes H_{\xi'})^*$, we set

$$E_\varphi(P,\xi,\xi',\nu,\nu',\psi)(g) = E_\varphi(P,\xi,\xi',\nu,\nu')(\pi_{\xi,\xi',\nu,\nu'}^\varphi(g)\psi), \tag{5.21}$$

$$E_\varphi^\eta(P,\xi,\xi',\nu,\nu',\psi)(g) = \eta\left(E_\varphi(P,\xi,\xi',\nu,\nu',\psi)(g) \right), \tag{5.22}$$

where $E_\varphi(P,\xi,\xi',\nu,\nu')$ is as in (5.18). If $\psi \in H_{\varphi,K,K'}^{\xi,\xi',\nu,\nu'}$ is as in (5.19), let $V_\xi^\psi \subset H_\xi$ (resp. $V_{\xi'}^\psi \subset H_{\xi'}$) be the subspace spanned by the set

$$\{f_i(k) \mid k \in K, 1 \le i \le m\} \qquad (\text{resp.} \quad \{f_i'(k') \mid k' \in K', 1 \le i \le m\}).$$

Since the function f_i (resp. f_i') is K-finite (resp. K'-finite), it follows that V_ξ^ψ (resp. $V_{\xi'}^\psi$) is a finite-dimensional complex vector space. Let σ_ξ^ψ (resp. $\sigma_{\xi'}^\psi$) be the representation of K (resp. K') on V_ξ^ψ (resp. $V_{\xi'}^\psi$) given by

$$\sigma_\xi^\psi(k)f_i(k_1) = f_i(k_1 k) \qquad (\text{resp.} \quad \sigma_{\xi'}^\psi(k')f_i'(k_1') = f_i'(k_1' k'))$$

for all $k, k_1 \in K$ (resp. $k', k_1' \in K'$). Then we have

$$E_\varphi(P, \xi, \xi', \nu, \nu', \psi)(g) \in V_\xi^\psi \otimes V_{\xi'}^\psi$$

for all $g \in G$.

Proposition 5.23 *Let* $\psi \in H_{\varphi,K,K'}^{\xi,\xi',\nu,\nu'}$ *be as in* (5.19)*. Then the function* $E_\varphi(P, \xi, \xi', \nu, \nu', \psi) : G \to V_\xi^\psi \otimes V_{\xi'}^\psi$ *given by* (5.21) *is a mixed automorphic form of type* $(\varphi, \sigma_\xi^\psi, \sigma_{\xi'}^\psi)$ *for* Γ.

Proof. From (5.18) and (5.21) we see easily that $E_\varphi(P, \xi, \xi', \nu, \nu', \psi)$ is left Γ-invariant. If $\psi \in H_{\varphi,K,K'}^{\xi,\xi',\nu,\nu'}$ is as in (5.19), then by (5.20) we have

$$E_\varphi(P, \xi, \xi', \nu, \nu', \psi)(g) = \sum_{i=1}^m \sum_{\gamma \in \Gamma_N \backslash \Gamma} a(xg)^{\nu+\rho} f_i(k(\gamma g)) \qquad (5.23)$$
$$\otimes a'(\varphi(\gamma g))^{\nu'+\rho'} f_i'(\varphi(k(\gamma g)))$$

for each $g \in G$. However, the sum

$$\sum_{\gamma \in \Gamma_N \backslash \Gamma} a(xg)^{\nu+\rho} f_i(k(\gamma g)) a'(\varphi(\gamma g))^{\nu'+\rho'} f_i'(\varphi(k(\gamma g)))$$

in (5.23) is an Eisenstein series in the sense of Definition 5.17, and therefore it is $Z(\mathfrak{g})$-finite and slowly increasing by Theorem 5.20. On the other hand, for $k_0 \in K$ we have

$$E_\varphi(P, \xi, \xi', \nu, \nu', \psi)(gk_0)$$
$$= \sum_{i=1}^m \sum_{\gamma \in \Gamma_N \backslash \Gamma} a(xgk_0)^{\nu+\rho} f_i(k(\gamma g k_0))$$
$$\otimes a'(\varphi(\gamma g k_0))^{\nu'+\rho'} f_i'(\varphi(k(\gamma g k_0)))$$
$$= \sum_{i=1}^m \sum_{\gamma \in \Gamma_N \backslash \Gamma} a(xg)^{\nu+\rho} f_i(k(\gamma g)k_0)$$
$$\otimes a'(\varphi(\gamma g)\varphi(k_0))^{\nu'+\rho'} f_i'(\varphi(k(\gamma g)\varphi(k_0)))$$
$$= \sum_{i=1}^m \sum_{\gamma \in \Gamma_N \backslash \Gamma} a(xg)^{\nu+\rho} f_i(k(\gamma g)k_0)$$
$$\otimes a'(\varphi(\gamma g)\varphi(k_0))^{\nu'+\rho'} f_i'(\varphi(k(\gamma g)\varphi(k_0)))$$
$$= \sum_{i=1}^m \sum_{\gamma \in \Gamma_N \backslash \Gamma} a(xg)^{\nu+\rho} \sigma_\xi^\psi(k_0) f_i(k(\gamma g))$$
$$\otimes a'(\varphi(\gamma g)\varphi(k_0))^{\nu'+\rho'} \sigma_{\xi'}^\psi(\varphi(k_0)) f_i'(\varphi(k(\gamma g))).$$

Hence we obtain

$$E_\varphi(P, \xi, \xi', \nu, \nu', \psi)(gk_0) = \left(\sigma_\xi^\psi \otimes (\sigma_{\xi'}^\psi \circ \varphi)\right)(k_0) E_\varphi(P, \xi, \xi', \nu, \nu', \psi)(g),$$

and therefore the proposition follows.   □

From Proposition 5.23 it follows that the function $E_\varphi^\eta(P, \xi, \xi', \nu, \nu', \psi)$: $G \to \mathbb{C}$ given by (5.22) is also a mixed automorphic form for Γ and that both of the functions $E_\varphi(P, \xi, \xi', \nu, \nu', \psi)$ and $E_\varphi^\eta(P, \xi, \xi', \nu, \nu', \psi)$ are Eisenstein series for mixed automorphic forms for Γ in the sense of Definition 5.17. We denote by $\mathcal{A}(\Gamma \backslash G, H_\xi \otimes H_{\xi'})$ the space of $(H_\xi \otimes H_{\xi'})$-valued automorphic forms for Γ. Then $\mathcal{A}(\Gamma \backslash G, H_\xi \otimes H_{\xi'})$ is a $(\mathfrak{g}, K)$-module, and the map

$$H_{\varphi, K, K'}^{\xi, \xi', \nu, \nu'} \to \mathcal{A}(\Gamma \backslash G, H_\xi \otimes H_{\xi'}), \quad \psi \mapsto E_\varphi(P, \xi, \xi', \nu, \nu', \psi)$$

is a homomorphism of $(\mathfrak{g}, K)$-modules.

We shall now discuss the construction of Whittaker vectors and describe Poincaré series for mixed automorphic forms. Let $G = NAK$, $P = MAN$ and other objects be as before. We fix $H_0 \in \mathfrak{a}$ such that $\alpha(H_0) = 1$ and set $a_t = \exp(tH_0)$ for $t \in \mathbb{R}$. If $\nu \in (\mathfrak{a}_\mathbb{C})^*$ and if $a = \exp H$ with $H \in \mathfrak{a}$, then we write $a^\nu = e^{\nu(H)}$ as usual. Let $H_\infty^{\xi, \nu}$ be the space of C^∞-vectors in $H^{\xi, \nu}$, and define the map $\delta_{\xi, \nu} : H_\infty^{\xi, \nu} \to H_\xi$ by $\delta_{\xi, \nu}(f) = f(1)$, which is a continuous (P, M)-homomorphism with M acting by ξ, $\mathfrak{a}$ by $\nu + \rho$, and $\mathfrak{n}$ by 0. For $\lambda \geq 1$ denote by $S_\lambda(H^{\xi, \nu}) \subset H^{\xi, \nu}$ the associated Gevrey space in the sense of Goodman and Wallach [30], which satisfies the inclusion relations

$$H_K^{\xi, \nu} \subset S_\lambda(H^{\xi, \nu}) \subset H_\infty^{\xi, \nu}.$$

We fix a nontrivial character χ on N.

Theorem 5.24 *There exists a weakly holomorphic family of continuous maps $\mathcal{W}(\xi, \nu) : S_\lambda(H^{\xi, \nu}) \to H_\xi$ for $1 \leq \lambda \leq 3/2$ satisfying the following conditions:*
(i) $\mathcal{W}(\xi, \nu)(\pi_{\xi, \nu}(n)v) = \chi(n)\mathcal{W}(\xi, \nu)(v)$ for all $n \in N$ and $v \in S_\lambda(H^\xi)$.
(ii) There exists a nonzero holomorphic function $I_\xi : (\mathfrak{a}_\mathbb{C})^* \to \mathbb{C}$ such that

$$\lim_{t \to -\infty} a_t^{-(\nu + \rho)} \mathcal{W}(\xi, \nu)(\pi_{\xi, \nu}(a_t)v) = I_\xi(\nu)\delta_{\xi, \nu}(v)$$

uniformly on compact subsets of $(\mathfrak{a}_\mathbb{C})^*$.

Proof. See [93, Theorem 1.1].   □

Let $(\widehat{H}_\varphi^{\xi, \xi', \nu, \nu'}, \widehat{\pi}_{\xi, \xi', \nu, \nu'}^\varphi)$ and $(H_\varphi^{\xi, \xi', \nu, \nu'}, \pi_{\xi, \xi', \nu, \nu'}^\varphi)$ be the representations of G described above. For $\lambda \geq 1$ let $S_\lambda(\widehat{H}_\varphi^{\xi, \xi', \nu, \nu'}) \subset \widehat{H}_\varphi^{\xi, \xi', \nu, \nu'}$ be the associated Gevrey space, and set

$$S_\lambda(H_\varphi^{\xi, \xi', \nu, \nu'}) = S_\lambda(\widehat{H}_\varphi^{\xi, \xi', \nu, \nu'}) \cap H_\varphi^{\xi, \xi', \nu, \nu'}. \tag{5.24}$$

Thus, if $H_{\varphi,\infty}^{\xi,\xi',\nu,\nu'}$ denotes the subspace of $H_\varphi^{\xi,\xi',\nu,\nu'}$ generated by elements of the form $f \otimes (f' \circ \varphi)$ with $f \in H_\infty^{\xi,\nu}$ and $f' \in H_\infty^{\xi',\nu'}$, then we have

$$H_{\varphi,K,K'}^{\xi,\xi',\nu,\nu'} \subset S_\lambda(H_\varphi^{\xi,\xi',\nu,\nu'}) \subset H_{\varphi,\infty}^{\xi,\xi',\nu,\nu'}.$$

Now the *Whittaker vector* for mixed automorphic forms is given by the next lemma.

Lemma 5.25 *Let χ be a nontrivial character of N, and let $S_\lambda(H_\varphi^{\xi,\xi',\nu,\nu'})$ be as in (5.24). If $1 \le \lambda \le 3/2$, then there exists a linear map*

$$\mathcal{W}^\varphi(\xi,\xi',\nu,\nu') : S_\lambda(H_\varphi^{\xi,\xi',\nu,\nu'}) \to H_\xi \otimes H_{\xi'}$$

satisfying the following conditions:

(i) $\mathcal{W}^\varphi(\xi,\xi',\nu,\nu')(\pi_{\xi,\xi',\nu,\nu'}^\varphi(n))\psi = \chi(n)\mathcal{W}^\varphi(\xi,\xi',\nu,\nu')\psi$ *for all $n \in N$ and $\psi \in S_\lambda(H_\varphi^{\xi,\xi',\nu,\nu'})$.*

(ii) There is a nonzero holomorphic function $I_{\xi,\xi'}^\varphi : \mathfrak{a}_\mathbb{C}^* \to \mathbb{C}$ such that

$$\lim_{t \to -\infty} a_t^{-(\nu+(\nu'+\rho')\circ d\varphi+\rho)} \mathcal{W}^\varphi(\xi,\xi',\nu,\nu')(\widehat{\pi}_{\xi,\xi',\nu,\nu'}^\varphi(a_t))\psi$$
$$= I_{\xi,\xi'}^\varphi(\nu + (\nu'+\rho') \circ d\varphi)\delta_{\xi,\xi',\nu,\nu'}^\varphi(\psi)$$

for all $\psi \in S_\lambda(H_\varphi^{\xi,\xi',\nu,\nu'})$.

Proof. Applying Theorem 5.24 to the representation $(\widehat{H}_\varphi^{\xi,\xi',\nu,\nu'}, \widehat{\pi}_{\xi,\xi',\nu,\nu'}^\varphi)$ of G associated to the representation $\xi \otimes (\xi \circ \varphi)$ of M and $\nu + (\nu'+\rho') \circ d\varphi \in (\mathfrak{a}_\mathbb{C})^*$, we obtain the linear map

$$\widehat{\mathcal{W}}^\varphi(\xi,\xi',\nu,\nu') : S_\lambda(\widehat{H}_\varphi^{\xi,\xi',\nu,\nu'}) \to H_\xi \otimes H_{\xi'}$$

such that

$$\widehat{\mathcal{W}}^\varphi(\xi,\xi',\nu,\nu')(\widehat{\pi}_{\xi,\xi',\nu,\nu'}^\varphi(n))\psi = \chi(n)\widehat{\mathcal{W}}^\varphi(\xi,\xi',\nu,\nu')\psi$$

for all $n \in N$ and $\psi \in S_\lambda(\widehat{H}_\varphi^{\xi,\xi',\nu,\nu'})$, and

$$\lim_{t \to -\infty} a_t^{-(\nu+(\nu'+\rho')\circ d\varphi+\rho)} \widehat{\mathcal{W}}^\varphi(\xi,\xi',\nu,\nu')(\widehat{\pi}_{\xi,\xi',\nu,\nu'}^\varphi(a_t))\psi$$
$$= I_{\xi,\xi'}^\varphi(\nu + (\nu'+\rho') \circ d\varphi)\delta_{\xi,\xi',\nu,\nu'}^\varphi(\psi)$$

for some holomorphic function $I_{\xi,\xi'}^\varphi : (\mathfrak{a}_\mathbb{C})^* \to \mathbb{C}$. Since $H_\varphi^{\xi,\xi',\nu,\nu'}$ is a subrepresentation of $\widehat{H}_\varphi^{\xi,\xi',\nu,\nu'}$ of G, the lemma follows by restricting $\widehat{\mathcal{W}}^\varphi(\xi,\xi',\nu,\nu')$ to $H_\varphi^{\xi,\xi',\nu,\nu'}$ of $\widehat{H}_\varphi^{\xi,\xi',\nu,\nu'}$. $\square$

Given $\psi \in S_\lambda(H_\varphi^{\xi,\xi',\nu,\nu'})$, let $V_{\varphi,\xi,\xi'}^\psi$ be the subspace of $H_\xi \otimes H_{\xi'}$ spanned by the set

$$\{\mathcal{W}^\varphi(\xi,\xi',\nu,\nu')\pi_{\xi,\xi',\nu,\nu'}^\varphi(g)\psi \mid g \in G\},$$

and let $\sigma_{\varphi,\xi,\xi'}^\psi$ be the representation of K on $V_{\varphi,\xi,\xi'}^\psi$ given by

$$\sigma_{\varphi,\xi,\xi'}^\psi(k)\mathcal{W}^\varphi(\xi,\xi',\nu,\nu')\pi_{\xi,\xi',\nu,\nu'}^\varphi(g)\psi = \mathcal{W}^\varphi(\xi,\xi',\nu,\nu')\pi_{\xi,\xi',\nu,\nu'}^\varphi(gk)\psi.$$

For $g \in G$ we set

$$\mathcal{P}_{\xi,\xi',\nu,\nu'}^{\varphi,\psi}(g) = \sum_{\gamma \in \Gamma_N \backslash \Gamma} \mathcal{W}^\varphi(\xi,\xi',\nu,\nu') \cdot (\pi_{\xi,\xi',\nu,\nu'}^\varphi(\gamma g))(\psi) \in V_{\varphi,\xi,\xi'}^\psi. \qquad (5.25)$$

In order to discuss the convergence of $\mathcal{P}_{\xi,\xi',\nu,\nu'}^{\varphi,\psi}$ let α be a simple root of (P, A), and fix an element $H_0 \in \mathfrak{a}$ such that $\alpha(H_0) = 1$. Let θ be a Cartan involution of $\mathfrak{g}$, and fix a nondegenerate G-invariant real-valued bilinear form B on $\mathfrak{g}$ such that $B(H_0, H_0) = 1$ and $-B(X, \theta X) < 0$ for all $X \in \mathfrak{g}$. Let $\langle , \rangle$ be the bilinear form on $(\mathfrak{a}_{\mathbb{C}})^*$ that is dual to $B|_{\mathfrak{a}_{\mathbb{C}} \times \mathfrak{a}_{\mathbb{C}}}$, and set

$$\mathfrak{a}_{\mathbb{C}}^*(\lambda)^+ = \{\mu \in (\mathfrak{a}_{\mathbb{C}})^* \mid \mathrm{Re}\langle \mu, \alpha \rangle > \langle \lambda, \alpha \rangle\}$$

for $\lambda \in \mathfrak{a}^*$.

Theorem 5.26 *Assume that $\nu + (\nu' + \rho') \circ d\varphi \in \mathfrak{a}_{\mathbb{C}}^*(\rho)^+$. Then the function $\mathcal{P}_{\xi,\xi',\nu,\nu'}^{\varphi,\psi}$ on G given by (5.25) satisfies the following conditions:*

(i) The series in (5.25) defining $\mathcal{P}_{\xi,\xi',\nu,\nu'}^{\varphi,\psi}$ converges uniformly on compact subsets of G.

(ii) $\mathcal{P}_{\xi,\xi',\nu,\nu'}^{\varphi,\psi}$ is left Γ-invariant.

(iii) $\mathcal{P}_{\xi,\xi',\nu,\nu'}^{\varphi,\psi}(gk) = \sigma_{\varphi,\xi,\xi'}^\psi(k)\mathcal{P}_{\xi,\xi',\nu,\nu'}^{\varphi,\psi}(g)$ for all $g \in G$ and $k \in K$.

Proof. Let $\widehat{\mathcal{W}}^\varphi(\xi,\xi',\nu,\nu')$ be as in the proof of Lemma 5.25. Then by Lemma 2.1 in [93] the statement in (i) is true for $\widehat{\mathcal{W}}^\varphi(\xi,\xi',\nu,\nu')$. Thus (i) follows from the fact that $\mathcal{W}^\varphi(\xi,\xi',\nu,\nu')$ is the restriction of $\widehat{\mathcal{W}}^\varphi(\xi,\xi',\nu,\nu')$ to $S_\lambda(H_\varphi^{\xi,\xi',\nu,\nu'})$. As for (iii), we have

$$\mathcal{P}_{\xi,\xi',\nu,\nu'}^{\varphi,\psi}(gk) = \sum_{\gamma \in \Gamma_N \backslash \Gamma} \mathcal{W}^\varphi(\xi,\xi',\nu,\nu')(\pi_{\xi,\xi',\nu,\nu'}^\varphi(\gamma gk))(\psi)$$

$$= \sum_{\gamma \in \Gamma_N \backslash \Gamma} \sigma_{\varphi,\xi,\xi'}^\psi(k)\mathcal{W}^\varphi(\xi,\xi',\nu,\nu')((\pi_{\xi,\xi',\nu,\nu'}^\varphi(\gamma g)(\psi))$$

$$= \sigma_{\varphi,\xi,\xi'}^\psi(k)\mathcal{P}_{\xi,\xi',\nu,\nu'}^{\varphi,\psi}(g)$$

for all $k \in K$. Since (ii) is clear from (5.25), the proof of the theorem is complete. $\qquad \square$

From Theorem 5.26 it follows that the function

$$\mathcal{P}^{\varphi,\psi}_{\xi,\xi',\nu,\nu'} : G \to V^{\psi}_{\varphi,\xi,\xi'}$$

given by (5.25) satisfies all the conditions for an automorphic form for Γ and $\sigma^{\psi}_{\varphi,\xi,\xi'}$ except the condition that it is slowly increasing. Since $\sigma^{\psi}_{\varphi,\xi,\xi'}$ can be considered as an analogue of $\sigma^{\psi}_{\xi} \otimes (\sigma^{\psi}_{\xi'} \circ \varphi)$ in the proof of Proposition 5.23, the series $\mathcal{P}^{\varphi,\psi}_{\xi,\xi',\nu,\nu'}(g)$ can be regarded as a generalized Poincaré series for mixed automorphic forms, and it provides an analogue of the Poincaré series of Miatello and Wallach [93].

5.5 Fourier Coefficients of Eisenstein Series

In this section we express the Fourier coefficients of the Eisenstein series for mixed automorphic forms described in Section 5.4 in terms of Jacquet integrals by using Whittaker vectors.

We shall use the same notations as in the previous sections. Let χ be a nontrivial character of N, and set

$$\mathrm{Wh}^{\chi,\varphi}_{\xi,\xi',\nu,\nu'} = \{\mu \in (H^{\xi,\xi',\nu,\nu'}_{\varphi})^* \mid \mu \circ (\pi^{\varphi}_{\xi,\xi',\nu,\nu'}(n)) \tag{5.26}$$
$$= \chi(n)\mu \text{ for all } n \in N\}.$$

Thus, if $\eta : H_{\xi} \times H_{\xi'} \to \mathbb{C}$ is a linear map and $\mathcal{W}^{\varphi}(\xi,\xi',\nu,\nu')$ is the Whittaker vector described in Lemma 5.25, then we have

$$\eta \circ (\mathcal{W}^{\varphi}(\xi,\xi',\nu,\nu')) \in \mathrm{Wh}^{\chi,\varphi}_{\xi,\xi',\nu,\nu'}.$$

Let s^* be an element of the normalizer M^* of A in K, and define the linear map

$$J^{\chi,\varphi}_{\xi,\xi',\nu,\nu'} : H^{\xi,\xi',\nu,\nu'}_{\varphi,\infty} \to H_{\xi} \otimes H_{\xi'}$$

by the integral

$$J^{\chi,\varphi}_{\xi,\xi',\nu,\nu'} = \int_N \chi(n)^{-1}\delta^{\varphi}_{\xi,\xi',\nu,\nu'} \circ (\pi^{\varphi}_{\xi,\xi',\nu,\nu'}(s^*n))dn, \tag{5.27}$$

where $\pi^{\varphi}_{\xi,\xi',\nu,\nu'}$ and $\delta^{\varphi}_{\xi,\xi',\nu,\nu'}$ are as in (5.16) and (5.17), respectively. The integral in (5.27) can be considered as an analogue of the *Jacquet integral* (cf. [45]) for mixed automorphic forms. Note that, if s^* is replaced by ms^* with $m \in M$, then $J^{\chi,\varphi}_{\xi,\xi',\nu,\nu'}$ should be replaced by $(\xi \otimes (\xi' \circ \varphi))(m)J^{\chi,\varphi}_{\xi,\xi',\nu,\nu'}$.

Let $\mu \in \mathrm{Wh}^{\chi,\varphi}_{\xi,\xi',\nu,\nu'}$, and extend χ to a map $\chi : G \to \mathbb{C}^{\times}$ by setting $\chi(nak) = \chi(n)$ for all $n \in N$, $a \in A$ and $k \in K$. Then μ is a quasi-invariant distribution on N with multiplier

$$G \times N \to \mathbb{C}^{\times}, \quad (g,n) \mapsto \chi(g)^{-1}$$

in the sense of [122, §5.2].

Lemma 5.27 *If* $\mu \in \mathrm{Wh}^{\chi,\varphi}_{\xi,\xi',\nu,\nu'}$, *then there exists a linear map* $S : H_\xi \otimes H_{\xi'} \to \mathbb{C}$ *such that*

$$\mu = S \circ (J^{\chi,\chi',\varphi}_{\xi,\xi',\nu,\nu'}),$$

where $J^{\chi,\chi',\varphi}_{\xi,\xi',\nu,\nu'}$ *is as in* (5.27).

Proof. Since μ is a quasi-invariant distribution on N with multiplier $(g,n) \mapsto \chi(g)^{-1}$, from [122, Theorem 5.2.2.1] it follows that there is a unique element $\widetilde{z} \in (H_\xi \otimes H_{\xi'})^*$ such that

$$\chi(g)^{-1}\widetilde{z} = \widetilde{z}$$

for all $g \in AK = N\backslash G$ and

$$\mu(f) = \left\langle \int_N \chi(n)^{-1} f(n) dn, \widetilde{z} \right\rangle$$

for $f \in H^{\xi,\xi',\nu,\nu'}_{\varphi,\infty}$. However, since $a(n) = 1$ and $a'(\varphi(n)) = 1$, we have

$$f(n) = (\pi^\varphi_{\xi,\xi',\nu,\nu'}(n))f(1) = \delta^\varphi_{\xi,\xi',\nu,\nu'}(\pi^\varphi_{\xi,\xi',\nu,\nu'}(n)f).$$

Thus the map $S : H_\xi \otimes H_{\xi'} \to \mathbb{C}$ defined by $S(v) = \langle v, \widetilde{z} \rangle$ for each $v \in H_\xi \otimes H_{\xi'}$ satisfies the desired condition. □

Let s be a nontrivial element of the Weyl group $W(G, A)$ of (G, A), and let $s^* \in K$ be an element of the normalizer M^* of A in K such that $\mathrm{ad}\,(s^*)|_{\mathfrak{a}} = s$. Then we set

$$(A_s(\xi, \xi', \nu, \nu')\psi)(k) = \int_N \delta^\varphi_{\xi,\xi',\nu,\nu'}(\pi^\varphi_{\xi,\xi',\nu,\nu'}(s^*nk)\psi)\,dn \tag{5.28}$$

for $\psi \in H^{\xi,\xi',\nu,\nu'}_{\varphi,\infty}$ and $k \in K$, and define the representation ξ^s of M and the element $s\nu$ of $(\mathfrak{a}_\mathbb{C})^*$ by

$$\xi^s(m) = \xi((s^*)^{-1}ms^*), \quad s\nu = (\mathrm{ad}\,(s^{*-1})(H)) \tag{5.29}$$

for all $m \in M$ and $H \in (\mathfrak{a}_\mathbb{C})^*$. Similarly, we can consider the representation $\xi'^{\varphi(s)}$ of M' and the element $\varphi(s)\nu'$ of $(\mathfrak{a}'_\mathbb{C})^*$.

Proposition 5.28 *For each* $\mu \in \mathrm{Wh}^{\chi,\varphi}_{\xi,\xi',\nu,\nu'}$ *there exist elements* $\eta, \eta' \in (H_\xi \otimes H'_\xi)^*$ *such that*

$$\mu = \eta \circ \Xi_{\xi,\xi'}(\nu + (\nu' + \rho') \circ d\varphi)^{-1}\mathcal{W}^\varphi(\xi, \xi', \nu, \nu')$$
$$+ \eta' \circ \Xi_{\xi^s,\xi'^{\varphi(s)}}(s\nu + (\varphi(s)\nu' + \rho') \circ d\varphi)^{-1}$$
$$\mathcal{W}^\varphi(\xi^s, \xi'^{\varphi(s)}, s\nu, \varphi(s)\nu') \circ A_s(\xi, \xi', \nu, \nu'),$$

where $A_s(\xi, \xi', \nu, \nu')$, ξ^s *and* $s\nu$ *are as in* (5.28) *and* (5.29).

Proof. Let $\eta_1, \dots, \eta_d$ be a basis for $\widehat{H}_\varphi^{\xi,\xi',\nu,\nu'}$, and let

$$\widehat{\mathrm{Wh}}_{\xi,\xi',\nu,\nu'}^{\chi,\varphi} = \{\widehat{\mu} \in (\widehat{H}_\varphi^{\xi,\xi',\nu,\nu'})^* \mid \widehat{\mu} \circ (\widehat{\pi}_{\xi,\xi',\nu,\nu'}^{\varphi}(n))$$
$$= \chi(n)\widehat{\mu} \text{ for all } n \in N\}.$$

Then, as in the proof of Theorem A.1.10 in [93], the $2d$ functionals

$$\eta_1 \circ \Xi_{\xi,\xi'}(\nu + (\nu' + \rho') \circ d\varphi)^{-1} \mathcal{W}^\varphi(\xi,\xi',\nu,\nu'),$$
$$\eta_1 \circ \Xi_{\xi^s,\xi'^{\varphi(s)}}(s\nu + (\varphi(s)\nu' + \rho') \circ d\varphi)^{-1}$$
$$\mathcal{W}^\varphi(\xi^s,\xi'^{\varphi(s)},s\nu,\varphi(s)\nu') \circ A_s(\xi,\xi',\nu,\nu'),$$
$$\dots\dots\dots\dots,$$
$$\eta_d \circ \Xi_{\xi,\xi'}(\nu + (\nu' + \rho') \circ d\varphi)^{-1} \mathcal{W}^\varphi(\xi,\xi',\nu,\nu'),$$
$$\eta_d \circ \Xi_{\xi^s,\xi'^{\varphi(s)}}(s\nu + (\varphi(s)\nu' + \rho') \circ d\varphi)^{-1}$$
$$\mathcal{W}^\varphi(\xi^s,\xi'^{\varphi(s)},s\nu,\varphi(s)\nu')A_s(\xi,\xi',\nu,\nu')$$

form a basis for $\widehat{\mathrm{Wh}}_{\xi,\xi',\nu,\nu'}^{\chi,\varphi}$. Thus, if $\widehat{\mu} \in \widehat{\mathrm{Wh}}_{\xi,\xi',\nu,\nu'}^{\chi,\varphi}$, there exist elements $\widehat{\eta}, \widehat{\eta}' \in (H_\xi \otimes H_{\xi'})^*$ such that

$$\widehat{\mu} = \widehat{\eta} \circ \Xi_{\xi,\xi'}(\nu + (\nu' + \rho') \circ d\varphi)^{-1} \mathcal{W}^\varphi(\xi,\xi',\nu,\nu')$$
$$+ \widehat{\eta}' \circ \Xi_{\xi^s,\xi'^{\varphi(s)}}(s\nu + (\varphi(s)\nu' + \rho') \circ d\varphi)^{-1}$$
$$\mathcal{W}^\varphi(\xi^s,\xi'^{\varphi(s)},s\nu,\varphi(s)\nu') \circ A_s(\xi,\xi',\nu,\nu').$$

Now the proposition follows from the fact that each $\mu \in \mathrm{Wh}_{\xi,\xi',\nu,\nu'}^{\chi,\varphi}$ is a restriction of an element $\widehat{\mu} \in \widehat{\mathrm{Wh}}_{\xi,\xi',\nu,\nu'}^{\chi,\varphi}$. $\square$

Let $\mathcal{A}(\Gamma\backslash G)$ be the space of $\mathbb{C}$-valued automorphic forms for Γ, and let $V \subset \mathcal{A}(\Gamma\backslash G)$ be a $(\mathfrak{g}, K)$-module. If $T : H_{\varphi,K,K'}^{\xi,\xi',\nu,\nu'} \to V$ is a homomorphism of $(\mathfrak{g}, K)$-modules, we set

$$I_g(\psi) = \int_{\Gamma\cap N\backslash N} \chi(n)^{-1}\big(T(\psi)\big)(ng)dn \tag{5.30}$$

for $\psi \in H_{\varphi,K,K'}^{\xi,\xi',\nu,\nu'}$ and $g \in G$.

Lemma 5.29 *If $J_{\xi,\xi',\nu,\nu'}^{\chi,\varphi}$ is as in (5.27), then for each $\psi \in H_{\varphi,K,K'}^{\xi,\xi',\nu,\nu'}$ and $g \in G$ the integral in (5.30) can be written in the form*

$$I_g(\psi) = \eta\big(J_{\xi,\xi',\nu,\nu'}^{\chi,\varphi}\big(\pi_{\xi,\xi',\nu,\nu'}^{\varphi}(g)\psi\big)\big)$$

for some element $\eta \in (H_\xi \otimes H_{\xi'})^*$.

Proof. Since T is a homomorphism of $(\mathfrak{g}, K)$-modules and the action of G on $V \subset \mathcal{A}(\Gamma \backslash G)$ is a right regular representation, if $n_0 \in N$, then by using (5.30) for the identity element 1 of G we obtain

$$I_1\left(\pi^\varphi_{\xi,\xi',\nu,\nu'}(n_0)\psi\right) = \int_{\Gamma \cap N \backslash N} \chi(n)^{-1}(T(\psi))(nn_0)dn$$

$$= \chi(n_0) \int_{\Gamma \cap N \backslash N} \chi(n)^{-1}(T(\psi))(n)dn = \chi(n_0)I_1(\psi).$$

Thus, using Lemma 5.27, we see that there is an element $\eta \in (H_\xi \otimes H_{\xi'})^*$ such that

$$I_1(\psi) = \eta(J^{\chi,\varphi}_{\xi,\xi',\nu,\nu'}(\psi)).$$

On the other hand, using the relation

$$T\left(\pi^\varphi_{\xi,\xi',\nu,\nu'}(g)\psi\right)(x) = (T\psi)(xg)$$

for all $g, x \in G$, we obtain

$$I_g(\psi) = \int_{\Gamma \cap N \backslash N} \chi(n)^{-1}T\left(\pi^\varphi_{\xi,\xi',\nu,\nu'}(g)\psi\right)(n)dn = I_1\left(\pi^\varphi_{\xi,\xi',\nu,\nu'}(g)\psi\right),$$

and hence the lemma follows. □

Let $Q = M_Q A_Q N_Q$ be the Langlands decomposition of another parabolic subgroup Q of G, so that there is an element $k \in K$ with $Q = kPk^{-1}$. Since we are now dealing with two parabolic subgroups P and Q, we shall use P and Q as subscripts on the left for various objects associated to respective parabolic subgroups, for example $_PH^{\xi,\xi',\nu,\nu'}_{\varphi,K,K'}$ for $H^{\xi,\xi',\nu,\nu'}_{\varphi,K,K'}$. Given a function ψ on G, we set

$$(L_k\psi)(x) = \psi(k^{-1}x) \tag{5.31}$$

for all $k, x \in G$. Let ξ^k be the representation of M_Q and $k\nu$ the element of $(\mathfrak{a}_Q)^*_\mathbb{C}$ corresponding to ξ and ν, respectively, where $\mathfrak{a}_Q$ is the Lie algebra of A_Q.

Lemma 5.30 *If $Q = kPk^{-1}$ with $k \in K$, then the operator L_k in (5.31) determines a linear map from $_PH^{\xi,\xi',\nu,\nu'}_\varphi$ to $_QH^{\xi^k,\xi'^{\varphi(k)},k\nu,\varphi(k)\nu'}_\varphi$.*

Proof. Let $\psi = \sum_{i=1}^m f_i \otimes (f'_i \circ \varphi) \in {}_PH^{\xi,\xi',\nu,\nu'}_\varphi$ with $f_i \in {}_PH^{\xi,\nu}$ and $f'_i \in {}_PH^{\xi',\nu'}$. Then by (5.31) we see that

$$L_k(\psi)(x) = \sum_{i=1}^m f_i(k^{-1}x) \otimes f'_i(\varphi(k)^{-1}\varphi(x))$$

$$= \sum_{i=1}^m \left((L_k f_i) \otimes ((L_{\varphi(k)} f'_i) \circ \varphi)\right)(x)$$

for all $x \in G$. However, we have

$$L_k f_i \in {}_Q H^{\xi^k, k\nu}, \quad L_{\varphi(k)} f_i' \in {}_Q H^{\xi'^{\varphi(k)}, \varphi(k)\nu'},$$

and therefore $L_k(\psi)(x) \in {}_Q H_\varphi^{\xi^k, \xi'^{\varphi(k)}, k\nu, \varphi(k)\nu'}$; hence the lemma follows. $\square$

Now we describe the Fourier coefficient of the Eisenstein series

$$E_\varphi^\eta(P, \xi, \xi', \nu, \nu', \psi)(g)$$

for mixed automorphic forms in (5.22) corresponding to a character of N_Q in the next theorem.

Theorem 5.31 *Let $E_\varphi^\eta(P, \xi, \xi', \nu, \nu', \psi)(g)$ be the Eisenstein series in (5.22) associated to a linear map $\eta : H_\xi \otimes H_{\xi'} \to \mathbb{C}$ and an element $\psi \in {}_P H_{\varphi, K, K'}^{\xi, \xi', \nu, \nu'}$, and let χ_Q be a character on N_Q. Then there exists a meromorphic function*

$$\Psi_{\varphi, \xi, \xi'}^{P, Q} : (\mathfrak{a}_Q)_{\mathbb{C}}^* \to \operatorname{Hom}_{\mathbb{C}}((H_\xi \otimes H_{\xi'})^*, (H_{\xi^k} \otimes H_{\xi'^{\varphi(k)}})^*)$$

such that

$$\int_{\Gamma \cap N_Q \backslash N_Q} \chi_Q(n)^{-1}(E_\varphi^\eta(P, \xi, \xi', \nu, \nu', \psi)(ng))dn$$
$$= \left(\Psi_{\varphi, \xi, \xi'}^{P, Q}(sk\nu + (\varphi(sk)\nu' + \rho') \circ d\varphi)\right)(\eta)$$
$$\times \left(J_{\xi^k, \xi'^{\varphi(k)}, k\nu, \varphi(k)\nu'}^{\chi_Q, \varphi}\left(\pi_{\xi^k, \xi'^{\varphi(k)}, k\nu, \varphi(k)\nu'}^{\varphi}(g)L_k\psi\right)\right)$$

for all $g \in G$.

Proof. If χ_P is the character of N_P corresponding to χ_Q, then the Jacquet integral $J_{\xi, \xi', \nu, \nu'}^{\chi_P, \varphi}$ on N_P corresponds to $J_{\xi^k, \xi'^{\varphi(k)}}^{\chi_Q, \varphi}$ on N_Q. Therefore by Lemma 5.29 there is a linear map $\Delta : H_\xi \otimes H_{\xi'} \to \mathbb{C}$ such that

$$\int_{\Gamma \cap N_Q \backslash N_Q} \chi_Q(n)^{-1}(E_\varphi^\eta(P, \xi, \xi', \nu, \nu', \psi)(ng))dn$$
$$= \Delta\left(J_{\xi^k, \xi'^{\varphi(k)}, k\nu, \varphi(k)\nu'}^{\chi_Q, \varphi}\left(\pi_{\xi^k, \xi'^{\varphi(k)}, k\nu, \varphi(k)\nu'}^{\varphi}(g)L_k\psi\right)\right).$$

On the other hand, if we consider the case of $Q = P$ and define $\Lambda(\eta) \in \operatorname{Wh}_{\xi, \xi', \nu, \nu'}^{\chi, \varphi}$ by

$$\int_{\Gamma \cap N_P \backslash N_P} \chi_P(n)^{-1}(E_\varphi^\eta(P, \xi, \xi', \nu, \nu', \psi)(ng))dn$$
$$= \int_{\Gamma \cap N_P \backslash N_P} \chi_P(n)^{-1}\eta\left(\sum_{\gamma \in \Gamma_N \backslash \Gamma} \delta_{\xi, \xi', \nu, \nu'}^\varphi(\pi_{\xi, \xi', \nu, \nu'}^\varphi(\gamma ng)\psi)\right)dn$$
$$= \Lambda(\eta)(\pi_{\xi, \xi', \nu, \nu'}^\varphi(g)\psi),$$

then $\Lambda(\eta)$ is an element of $\mathrm{Wh}^{\chi,\varphi}_{\xi,\xi',\nu,\nu'}$ in (5.26) and Λ is a linear function of η; hence by Proposition 5.28 there exist meromorphic functions

$$\Xi_1^{P,Q}, \Xi_2^{P,Q} : (\mathfrak{a}_Q)^*_{\mathbb{C}} \to \mathrm{Hom}_{\mathbb{C}}((H_\xi \otimes H_{\xi'})^*, (H_{\xi^k} \otimes H_{\xi'\varphi(k)})^*)$$

such that

$$\int_{\Gamma \cap N_Q \backslash N_Q} \chi_Q(n)^{-1}(E_\varphi^\eta(P, \xi, \xi', \nu, \nu', \psi)(ng))dn$$

$$= \Xi_1^{P,Q}(k\nu + (\varphi(k)\nu' + \rho') \circ d\varphi)(\eta)$$

$$\times (\mathcal{W}^\varphi(\xi^k, \xi'^{\varphi(k)}, k\nu, \varphi(k)\nu')(\pi^\varphi_{\xi^k, \xi'^{\varphi(k)}, k\nu, \varphi(k)}))\nu'(g)L_k\psi)$$

$$+ \Xi_2^{P,Q}(sk\nu + (\varphi(sk)\nu' + \rho') \circ d\varphi)(\eta)$$

$$\times (\mathcal{W}^\varphi((\xi^k)^s, (\xi'^{\varphi(k)})^{\varphi(s)}, sk\nu, \varphi(sk)\nu')$$

$$\circ A_s(\xi^k, \xi'^{\varphi(k)}, k\nu, \varphi(k)\nu')(\pi^\varphi_{\xi^k, \xi'^{\varphi(k)}, k\nu, \varphi(k)}\nu'(g)L_k\psi).$$

However, it can be shown that

$$A_s(\xi^k, \xi'^{\varphi(k)}, k\nu, \varphi(k)\nu') \circ \pi^\varphi_{\xi^k, \xi'^{\varphi(k)}, k\nu, \varphi(k)\nu'}$$

$$= \pi^\varphi_{(\xi^k)^s, (\xi'^{\varphi(k)})^{\varphi(s)}, sk\nu, \varphi(sk)\nu'} \circ A_s(\xi^k, \xi'^{\varphi(k)}, k\nu, \varphi(k)\nu').$$

Using this and applying Lemma 5.25 for $g = a_t$, we have

$$\Xi_1^{P,Q}(k\nu + (\varphi(k)\nu' + \rho') \circ d\varphi)(\eta)$$

$$(\mathcal{W}^\varphi(\xi^k, \xi'^{\varphi(k)}, k\nu, \varphi(k)\nu')(\pi^\varphi_{\xi^k, \xi'^{\varphi(k)}, k\nu, \varphi(k)}\nu'(a_t)L_k\psi)$$

$$+ \Xi_2^{P,Q}(sk\nu + (\varphi(sk)\nu' + \rho') \circ d\varphi)(\eta)$$

$$\times (\mathcal{W}^\varphi((\xi^k)^s, (\xi'^{\varphi(k)})^{\varphi(s)}, sk\nu, \varphi(sk)\nu')$$

$$\circ A_s(\xi^k, \xi'^{\varphi(k)}, k\nu, \varphi(k)\nu')(\pi^\varphi_{\xi^k, \xi'^{\varphi(k)}, k\nu, \varphi(k)}\nu'(a_t)L_k\psi)$$

$$\approx \Xi_1^{P,Q}(k\nu + (\varphi(k)\nu' + \rho') \circ d\varphi)(\eta)a_t^{k\nu + (\varphi(k)\nu' + \rho')\circ d\varphi + \rho}$$

$$\times I^\varphi_{\xi^k, \xi'^{\varphi(k)}}(\nu + (\nu' + \rho') \circ d\varphi)\delta^\varphi_{\xi^k, \xi'^{\varphi(k)}, k\nu, \varphi(k)}\nu'(L_k\psi)$$

$$+ \Xi_2^{P,Q}(sk\nu + (\varphi(sk)\nu' + \rho') \circ d\varphi)(\eta)$$

$$\times a_t^{k\nu + (\varphi(k)\nu' + \rho')\circ d\varphi + \rho}I^\varphi_{\xi^{sk}, \xi'^{\varphi(sk)}}(sk\nu + (\varphi(sk)\nu' + \rho') \circ d\varphi)$$

$$\times \delta^\varphi_{(\xi^k)^s, (\xi'^{\varphi(k)})^{\varphi(s)}, sk\nu, \varphi(sk)\nu'}(A_s(\xi^k, \xi'^{\varphi(k)}, k\nu, \varphi(k)\nu')(L_k\psi))$$

as $t \to \infty$. On the other hand, we have

$$\delta^\varphi_{(\xi^k)^s, (\xi'^{\varphi(k)})^{\varphi(s)}, sk\nu, \varphi(sk)\nu'}(A_s(\xi^k, \xi'^{\varphi(k)}, k\nu, \varphi(k)\nu')(L_k\psi))$$

$$= J^{\chi_Q, \varphi}_{(\xi^k)^s, (\xi'^{\varphi(k)})^{\varphi(s)}, sk\nu, \varphi(sk)\nu'}(L_k\psi).$$

Hence we obtain

$$
\begin{aligned}
\Delta\Big(J^{\chi_Q,\varphi}_{\xi^k,\xi'\varphi(k),k\nu,\varphi(k)\nu'} &\big(\pi^{\varphi}_{\xi^k,\xi'\varphi(k),k\nu,\varphi(k)\nu'}(a_t) L_k \psi \big) \Big) \\
= \Delta\Big(J^{\chi_Q,\varphi}_{\xi^k,\xi'\varphi(k),k\nu,\varphi(k)\nu'} & a_t^{k\nu+(\varphi(k)\nu'+\rho')\circ d\varphi+\rho} L_k \psi \Big) \\
\approx \Xi_1^{P,Q}(k\nu + (\varphi(k)\nu' &+ \rho') \circ d\varphi)(\eta) a_t^{k\nu+(\varphi(k)\nu'+\rho')\circ d\varphi+\rho} \\
\times\, I^{\varphi}_{\xi^k,\xi'\varphi(k)} &(k\nu + (\varphi(k)\nu' + \rho') \circ d\varphi) \delta^{\varphi}_{\xi^k,\xi'\varphi(k),k\nu,\varphi(k))\nu'}(L_k \psi) \\
+\, \Xi_2^{P,Q}(sk\nu + (\varphi(sk)\nu' &+ \rho') \circ d\varphi)(\eta) a_t^{sk\nu+(\varphi(sk)\nu'+\rho')\circ d\varphi+\rho} \\
\times\, I^{\varphi}_{\xi^{sk},\xi'\varphi(sk)} &(sk\nu + (\varphi(sk)\nu' + \rho') \circ d\varphi) \\
\times\, J^{\chi_Q,\varphi}_{(\xi^k)^s,(\xi'\varphi(k))\varphi(s),sk\nu,\varphi(sk)\nu'} &(L_k \psi)
\end{aligned}
$$

as $t \to \infty$. By comparing the coefficients, we have

$$
\begin{aligned}
\Delta = I^{\varphi}_{\xi^{sk},\xi'\varphi(sk)} &(sk\nu + (\varphi(sk)\nu' + \rho') \circ d\varphi) \\
&\times \Xi_2^{P,Q}(sk\nu + (\varphi(sk)\nu' + \rho') \circ d\varphi)(\eta).
\end{aligned}
$$

Thus the function $\Psi^{P,Q}_{\varphi,\xi,\xi'}$ on $(\mathfrak{a}_Q)^*_{\mathbb{C}}$ given by

$$
\Psi^{P,Q}_{\varphi,\xi,\xi'}(\nu) = I^{\varphi}_{\xi^{sk},\xi'\varphi(sk)}(\nu) \cdot \Xi_2^{P,Q}(\nu)
$$

is a meromorphic function, and the proof of the theorem is complete. $\qquad\square$

6

Families of Abelian Varieties

In earlier chapters we studied elliptic varieties and their connections with mixed automorphic forms. An elliptic variety can be described by a family of abelian varieties parametrized by an algebraic curve. The abelian varieties involved were products of elliptic curves. In this chapter we consider more general families of abelian varieties parametrized by an arithmetic quotient of a Hermitian symmetric domain.

Let $\mathcal{H}_n$ be the Siegel upper half space of degree n on which the symplectic group $Sp(n, \mathbb{R})$ acts as usual. If Γ' is an arithmetic subgroup of $Sp(n, \mathbb{R})$, then the associated quotient space $\Gamma' \backslash \mathcal{H}_n$ can be regarded as the moduli space for a certain family of polarized abelian varieties, known as a universal family (see e.g. [24, 42, 63]). Such a family of abelian varieties can be considered as a fiber variety over the Siegel modular variety $X' = \Gamma' \backslash \mathcal{H}_n$, and the geometry of a Siegel modular variety and the associated universal family of abelian varieties is closely connected with various topics in number theory including the theory of Siegel modular forms, theta functions and Jacobi forms.

Let $G = \mathbb{G}(\mathbb{R})$ be a semisimple Lie group of Hermitian type that can be realized as the set of real points of a linear algebraic group $\mathbb{G}$ defined over $\mathbb{Q}$. Thus the quotient $\mathcal{D} = G/K$ of G by a maximal compact subgroup K has the structure of a Hermitian symmetric domain. Let $\tau : \mathcal{D} \to \mathcal{H}_n$ be a holomorphic map, and let $\rho : G \to Sp(n, \mathbb{R})$ be a homomorphism of Lie groups such that $\tau(gz) = \rho(g)\tau(z)$ for all $z \in \mathcal{D}$ and $g \in G$. Let Γ be a torsion-free arithmetic subgroup of G such that $\rho(\Gamma) \subset \Gamma'$, and let $X = \Gamma \backslash \mathcal{D}$ be the corresponding arithmetic variety. Then the holomorphic map τ induces a morphism $\tau_X : X \to X'$ of arithmetic varieties, and by pulling the fiber variety over X' back via τ_X we obtain a fiber variety over X whose fibers are again polarized abelian varieties (see Section 6.1 for details). Such fiber varieties over an arithmetic variety are called Kuga fiber varieties (see [61, 108]), and various geometric and arithmetic aspects of Kuga fiber varieties have been investigated in numerous papers (see e.g. [1, 2, 31, 62, 69, 74, 84, 96, 108, 113]). A Kuga fiber variety is also an example of a mixed Shimura variety in more modern language (cf. [94]). Various objects connected with Siegel modular varieties and the associated universal families of abelian varieties can be generalized to the corresponding objects connected with more general locally symmetric varieties and the associated Kuga fiber varieties.

In Section 6.1 we review the construction of Kuga fiber varieties associated to equivariant holomorphic maps of symmetric domains. In Section 6.2 we describe canonical automorphy factors and kernel functions for semisimple Lie groups of Hermitian type as well as for generalized Jacobi groups. Section 6.3 is concerned with the interpretation of holomorphic forms of the highest degree on a Kuga fiber variety as mixed automorphic forms on symmetric domains involving canonical automorphy factors. The construction of an embedding of a Kuga fiber variety into a complex projective space is discussed in Section 6.4. This involves Jacobi forms of the type that will be considered in Chapter 7.

6.1 Kuga Fiber Varieties

In this section we review the construction of Kuga fiber varieties associated to equivariant holomorphic maps of symmetric domains. They are fiber bundles over locally symmetric spaces whose fibers are polarized abelian varieties. More details can be found in [61] and [108].

Let G be a Zariski-connected semisimple real algebraic group of Hermitian type defined over $\mathbb{Q}$. Thus G is the set of real points $\mathbb{G}(\mathbb{R})$ of a semisimple algebraic group $\mathbb{G}$ defined over $\mathbb{Q}$, and the associated Riemannian symmetric space $\mathcal{D} = G/K$, where K is a maximal compact subgroup of G, has a G-invariant complex structure. Such a space can be identified with a bounded symmetric domain in $\mathbb{C}^k$ for some k(see e.g. [36]), and is called a *Hermitian symmetric domain*. Let G' be another group of the same type, and let $\mathcal{D}'$ be the associated Hermitian symmetric domain. We assume that there exist a holomorphic map $\tau : \mathcal{D} \to \mathcal{D}'$ and a homomorphism $\rho : G \to G'$ of Lie groups such that

$$\tau(gz) = \rho(g)\tau(z)$$

for all $g \in G$ and $z \in \mathcal{D}$. In this case we say that τ is *equivariant* with respect to ρ or that (τ, ρ) is an *equivariant pair*. To construct Kuga fiber varieties we need to consider equivariant pairs when G' is a symplectic group. Such equivariant pairs were classified by Satake (see [108]).

Example 6.1 *Let W be a real vector space of dimension ν defined over $\mathbb{Q}$, and let S be a nondegenerate symmetric bilinear form of signature (p, q) for some positive integer p defined over $\mathbb{Q}$. Let $T(W) = \bigoplus_{r=0}^{\infty} W^{\otimes r}$ be the tensor algebra of W, and let $\mathfrak{A}_S$ be the two-sided ideal of $T(W)$ generated by the set*

$$\{x \otimes x - S(x, x) \mid x \in W\}.$$

Then the Clifford algebra *of (W, S) is given by*

$$\mathcal{C} = \mathcal{C}(V, S) = T(W)/\mathfrak{A}_S.$$

Let $\{e_1, \ldots, e_\nu\}$ with $\nu = p + q$ be an orthogonal basis of V such that

$$S(e_i, e_j) = \delta_{ij}\alpha_i$$

for some $\alpha_1, \ldots, \alpha_\nu \in \mathbb{R}$, where δ_{ij} is the Kronecker delta. If W is identified with its image in $\mathcal{C}$, then it is known that $\dim \mathcal{C} = 2^\nu$ and that the set

$$\{1\} \cup \{e_{i_1} \cdots e_{i_\nu} \mid 1 \le i_1 < \cdots < i_r \le \nu, \ 1 \le r \le \nu\}$$

is a basis of $\mathcal{C}$. Thus $\mathcal{C}$ is an associative algebra generated by $e_1, \ldots, e_\nu$ satisfying the conditions

$$e_i^2 = \alpha_i, \quad e_i e_j + e_j e_i = 0$$

for $1 \le i, j \le \nu$ with $i \ne j$. We set

$$\mathcal{C}^+ = \langle e_{i_1} \cdots e_{i_\nu} \mid 1 \le i_1 < \cdots < i_r \le \nu, \ r \ \text{even} \rangle_\mathbb{R},$$
$$\mathcal{C}^- = \langle e_{i_1} \cdots e_{i_\nu} \mid 1 \le i_1 < \cdots < i_r \le \nu, \ r \ \text{odd} \rangle_\mathbb{R}.$$

Then $\mathcal{C}^+$ is a subalgebra of $\mathcal{C}$ of dimension $2^{\nu-1}$, and we have

$$\mathcal{C} = \mathcal{C}^+ \oplus \mathcal{C}^-, \quad (\mathcal{C}^+)^2 = (\mathcal{C}^-)^2 = \mathcal{C}^+, \quad \mathcal{C}^+\mathcal{C}^- = \mathcal{C}^-\mathcal{C}^+ = \mathcal{C}^-.$$

Let ι be the canonical involution of $\mathcal{C}$ defined by $e_i^\iota = e_i$ for $1 \le i \le n$. Then the spin group is given by

$$\mathrm{Spin}(W, S) = \{g \in \mathcal{C}^+ \mid g^\iota g = 1, \ gWg^{-1} = W\}.$$

Given $\mathrm{Spin}(W, S)$, we set

$$\phi(g)x = gxg^{-1}$$

for all $x \in W$. Then we have $\phi(g) \in SO(W, S)$, and the map

$$\phi : \mathrm{Spin}(W, S) \to SO(W, S)$$

is a two-fold covering of $SO(W, S)$. Let a be an element $\mathcal{C}^+$ with $a^\iota = -a$, and let b_1 and b_2 be elements of $\mathcal{C}^+$ and $\mathcal{C}^\pm$, respectively, such that

$$b_1^2 + (-1)^{q(q+1)/2}b_2^2 = -1, \quad b_1 b_2 + b_2 b_1 = 0,$$

and the bilinear map

$$(x, y) \mapsto tr(b_1 a x^\iota y) + tr(b_2 a x^\iota e_- y)$$

for $x, y \in \mathcal{C}^+$ is symmetric and positive definite. We set

$$A(x, y) = tr(a x^\iota y), \quad I(x) = x b_1 + e_- x b_2.$$

for all $x, y \in \mathcal{C}^+$, where $e_- = e_{p+1} \cdots e_\nu$. Then A is a nondegenerate alternating bilinear form on $\mathcal{C}^+$, and I is a complex structure on $\mathcal{C}^+$ such that $(x, y) \mapsto A(x, Iy)$ is symmetric and positive definite and

$$A(gx, gy) = A(x, y), \quad I(gx) = gI(x)$$

for all $g \in \mathrm{Spin}(W, S)$ and $x, y \in \mathcal{C}^+$ (see [106, Section 2]). Thus we see that the left multiplication map

$$\widetilde{\rho}(g) : x \mapsto gx$$

determines a homomorphism

$$\widetilde{\rho} : \mathrm{Spin}(W, S) \to Sp(\mathcal{C}^+, A).$$

Now we assume that $p = 2$, that is the signature of S is $(2, q)$. Then it is known (see [106]) that the symmetric space $\mathcal{D} = G/K$ associated to a maximal compact subgroup of the spin group $G = \mathrm{Spin}(W, S)$ has a G-invariant complex structure. We choose a basis of $\mathcal{C}^+$ in such a way that $Sp(\mathcal{C}^+, A)$ can be identified with $Sp(2^\nu, \mathbb{R})$ with $\nu = 2 + q$ and denote by

$$\rho : \mathrm{Spin}(W, S) \to Sp(2^\nu, \mathbb{R})$$

the homomorphism induced by $\widetilde{\rho}$. Then we can consider a holomorphic map $\tau : \mathcal{D} \to \mathcal{H}_{2\nu}$ be that is equivariant with respect to ρ.

Let V be a real vector space of dimension $2n$ defined over $\mathbb{Q}$. A *complex structure* on V is an element $I \in GL(V)$ such that $I^2 = -1_V$ with 1_V denoting the identity map on V. Equipped with such a complex structure I, the real vector space V can be converted to a complex vector space if the complex multiplication operation is defined by

$$(a + bi) \cdot v = av + bIv \tag{6.1}$$

for all $a + bi \in \mathbb{C}$ and $v \in V$. Let β be a nondegenerate alternating bilinear form on V defined over $\mathbb{Q}$. Then the symplectic group

$$Sp(V, \beta) = \{g \in GL(V) \mid \beta(gv, gv') = \beta(v, v') \text{ for all } v, v' \in V\}$$

is of Hermitian type, and the associated Hermitian symmetric domain can be identified with the set $\mathcal{H} = \mathcal{H}(V, \beta)$ of all complex structures I on V such that the bilinear form $V \times V \to \mathbb{R}$, $(v, v') \mapsto \beta(v, Iv')$ is symmetric and positive definite. The group $Sp(V, \beta)$ acts on $\mathcal{H}$ by

$$g \cdot I = gIg^{-1}$$

for all $g \in Sp(V, \beta)$ and $I \in \mathcal{H}$. Let $\{e_1, \dots, e_{2n}\}$ be a symplectic basis of (V, β), that is, a basis of V satisfying the condition

$$\beta(e_i, e_j) = \begin{cases} 1 & \text{if } i = j + n, \\ -1 & \text{if } i = j - n, \\ 0 & \text{otherwise,} \end{cases}$$

for $1 \le i, j \le 2n$. Then with respect to such a basis $Sp(V, \beta)$ can be identified with the real symplectic group $Sp(n, \mathbb{R})$ of degree n. We also note that $\mathcal{H}$ can be identified with the *Siegel upper half space*

$$\mathcal{H}_n = \{Z \in M_n(\mathbb{C}) \mid {}^tZ = Z, \quad \operatorname{Im} Z \gg 0\}$$

of degree n consisting of $n \times n$ complex matrices with positive definite imaginary part and that the symplectic group $Sp(n, \mathbb{R})$ on $\mathcal{H}_n$ by

$$g(Z) = (AZ + B)(CZ + D)^{-1} \tag{6.2}$$

for all $Z \in \mathcal{H}_n$ and $g = \left(\begin{smallmatrix} A & B \\ C & D \end{smallmatrix}\right) \in Sp(n, \mathbb{R})$.

We now consider an equivariant pair (τ, ρ) for the special case of $G' = Sp(V, \beta)$. Thus we have the homomorphism $\rho : G \to Sp(V, \beta)$ and the holomorphic map $\tau : \mathcal{D} \to \mathcal{H}_n$ satisfying the condition $\tau(gz) = \rho(g)\tau(z)$ for all $g \in G$ and $z \in \mathcal{D}$. Let $G \ltimes_\rho V$ be the *semidirect product* of G and V with respect to the action of G on V via ρ. This means that $G \ltimes_\rho V$ consists of the elements $(g, v) \in G \times V$ and its multiplication operation is given by

$$(g, v) \cdot (g', v') = (gg', \rho(g)v' + v) \tag{6.3}$$

for $g, g' \in G$ and $v, v' \in V$. Then $G \ltimes_\rho V$ acts on $\mathcal{D} \times V$ by

$$(g, v) \cdot (z, w) = (gz, \rho(g)w + v) \tag{6.4}$$

for all $(g, v) \in G \ltimes_\rho V$ and $(z, w) \in \mathcal{D} \times V$. Let Γ be a torsion-free arithmetic subgroup of G. Then the corresponding quotient space $X = \Gamma \backslash \mathcal{D}$ has the structure of a complex manifold as well as the one of a locally symmetric space. It is also called an arithmetic variety taking into account the fact that it can be regarded as a quasi-projective complex algebraic variety (cf. [5]). Let L be a lattice in V with $V_{\mathbb{Q}} = L \otimes_{\mathbb{Z}} \mathbb{Q}$ such that

$$\beta(L, L) \subset \mathbb{Z}, \quad \rho(\Gamma)L \subset L. \tag{6.5}$$

If $(\gamma, \ell), (\gamma', \ell') \in G \ltimes_\rho V$ with $\gamma, \gamma' \in \Gamma$ and $\ell, \ell' \in L$, then by (6.3) we have

$$(\gamma, \ell) \cdot (\gamma', \ell') = (\gamma\gamma', \rho(\gamma)\ell' + \ell).$$

Using this and the condition $\rho(\Gamma)L \subset L$, we see that $(\gamma, \ell) \cdot (\gamma', \ell') \in \Gamma \times L$; hence we obtain the subgroup $\Gamma \ltimes_\rho L$ of $G \ltimes_\rho V$. Thus the action of $G \ltimes_\rho V$ in (6.4) induces the action of $\Gamma \ltimes_\rho L$ on $\mathcal{D} \times V$. We denote the associated quotient space by

$$Y = \Gamma \ltimes_\rho L \backslash \mathcal{D} \times V. \tag{6.6}$$

Then the natural projection map $\mathcal{D} \times V \to \mathcal{D}$ induces the map $\pi : Y \to X$, which has the structure of a fiber bundle over X whose fiber is isomorphic to the quotient space V/L.

We want to discuss next the complex structure on the fiber bundle Y over X given by (6.6). Let z_0 be a fixed element of $\mathcal{D}$, and let I_0 be the complex structure on V corresponding to the element $\tau(z_0)$ of $\mathcal{H}_n$. Let $V_{\mathbb{C}} = V \otimes_{\mathbb{R}} \mathbb{C}$ be the complexification of V, and denote by V_+ and V_- the subspaces of $V_{\mathbb{C}}$ defined by

$$V_\pm = \{v \in V_{\mathbb{C}} \mid I_0 v = \pm i v\}, \tag{6.7}$$

so that we have

$$V_{\mathbb{C}} = V_+ \oplus V_-, \quad V_+ = \overline{V}_-. \tag{6.8}$$

Using the fact that $\mathcal{H}_n$ can be identified with the set of complex structures on V, we see that each element $z \in \mathcal{D}$ determines a complex vector space $(V, I_{\tau(z)})$, where $I_{\tau(z)}$ is the complex structure on V corresponding to $\tau(z) \in \mathcal{H}_n$. Then each element v in $(V, I_{\tau(z)})$ determines an element

$$\xi(z, v) = v_z = v_+ - \tau(z) v_- = v_+ - I_{\tau(z)} v_- \tag{6.9}$$

of the subspace V_+ of $V_{\mathbb{C}}$, where the elements $v_\pm$ denote the $V_\pm$-components of $v \in V \subset V_{\mathbb{C}} = V_+ \oplus V_-$. We consider the map

$$\eta : \mathcal{D} \times V \to \mathcal{D} \times V_+$$

defined by

$$\eta(z, v) = (z, \xi(z, v)) \tag{6.10}$$

for all $(z, v) \in \mathcal{D} \times V$. Then it can be shown that η is a bijection, and therefore the action in (6.4) determines an action of $G \ltimes_\rho V$ on $\mathcal{D} \times V_+$ given by

$$(g, v) \cdot (z, u) = \eta((g, v) \cdot (\eta^{-1}(z, u))) \tag{6.11}$$

for all $(g, v) \in G \ltimes_\rho V$ and $(z, u) \in \mathcal{D} \times V_+$.

We consider the natural projection map $\widetilde{\pi} : \mathcal{D} \times V \to \mathcal{D}$ as the trivial vector bundle over $\mathcal{D}$ with fiber V. Then by using the isomorphism $\eta : \mathcal{D} \times V \cong \mathcal{D} \times V_+$ given by (6.10) the complex structure on $\mathcal{D} \times V_+$ can be carried over to a complex structure $\mathcal{I}$ on the vector bundle $\mathcal{D} \times V$ over $\mathcal{D}$. Note that the complex structure on $\mathcal{D} \times V_+$ is determined by the G-invariant complex structure on the Hermitian symmetric domain $\mathcal{D}$ and the complex structure $I_0 = I_{\tau(z_0)}$ on V. Let $\widetilde{\pi}' : \mathcal{D} \times V_+ \to \mathcal{D}$ be the natural projection map, which may be regarded as the trivial vector bundle over $\mathcal{D}$ with fiber V_+. If $z \in \mathcal{D}$, the fibers of $\widetilde{\pi}$ and $\widetilde{\pi}'$ over z can be written as

$$\widetilde{\pi}^{-1}(z) = \{z\} \times V \cong V, \quad \widetilde{\pi}'^{-1}(z) = \{z\} \times V_+ \cong V_+;$$

hence we have

$$\eta \mid_{\widetilde{\pi}^{-1}(z)} = \{z\} \times V_+.$$

Noting that the complex structure on V_+ is I_0, we see that

$$(\eta \mid_{\widetilde{\pi}^{-1}(z)}) \circ (\mathcal{I} \mid_{\widetilde{\pi}^{-1}(z)}) = I_0 \circ (\eta \mid_{\widetilde{\pi}^{-1}(z)})$$

for the fiber $\widetilde{\pi}^{-1}(z)$ of the bundle $\widetilde{\pi} : \mathcal{D} \times V \to \mathcal{D}$ over each $z \in \mathcal{D}$.

Proposition 6.2 *Let g be an element of G such that*

$$\rho(g) = \begin{pmatrix} A & B \\ C & D \end{pmatrix} \in Sp(V, \beta) \tag{6.12}$$

with respect to the decomposition $V_{\mathbb{C}} = V_+ \oplus V_-$ of $V_{\mathbb{C}} = V \otimes_{\mathbb{R}} \mathbb{C}$ in (6.8). Then the action in (6.11) can be written as

$$(g, v) \cdot (z, \xi(z, w)) = (gz, {}^t(C\tau(z) + D)^{-1}\xi(z, w) + \xi(gz, v)) \tag{6.13}$$

for all $z \in \mathcal{D}$ and $v, w \in V$.

Proof. Given $g \in G$, $z \in \mathcal{D}$ and $v, w \in V$, using (6.3), (6.9) and (6.11), we see that

$$(g, v) \cdot (z, \xi(z, w)) = (gz, \xi(gz, \rho(g)w + v)) \tag{6.14}$$
$$= (gz, \xi(gz, \rho(g)w)) + \xi(gz, v)).$$

If $\rho(g)$ is as in (6.12), we have

$$\rho(g)w = \begin{pmatrix} A & B \\ C & D \end{pmatrix} \begin{pmatrix} w_+ \\ w_- \end{pmatrix} = \begin{pmatrix} Aw_+ + Bw_- \\ Cw_+ + Dw_- \end{pmatrix};$$

hence we obtain

$$(\rho(g)w)_+ = Aw_+ + Bw_-, \quad (\rho(g)w)_- = Cw_+ + Dw_-.$$

Using this and (6.9), we have

$$\xi(gz, \rho(g)w) = (Aw_+ + Bw_-) - \tau(gz)(Cw_+ + Dw_-) \tag{6.15}$$
$$= (A - \tau(gz)C)w_+ + (B - \tau(gz)D)w_-$$
$$= (1, -\tau(gz)) \begin{pmatrix} A & B \\ C & D \end{pmatrix} \begin{pmatrix} w_+ \\ w_- \end{pmatrix}$$
$$= (1, -\tau(gz))\rho(g)w.$$

Since $\rho(g) \in Sp(V, \beta)$, its inverse is given by

$$\rho(g)^{-1} = \begin{pmatrix} A & B \\ C & D \end{pmatrix}^{-1} = \begin{pmatrix} {}^tD & -{}^tB \\ -{}^tC & {}^tA \end{pmatrix}.$$

Using this, (6.2), and the fact that the matrices $\tau(z), \tau(gz) \in \mathcal{H}_n$ are symmetric, we see that

$$(1, -\tau(gz)) = (1, -{}^t(\tau(gz))) \tag{6.16}$$
$$= (1, -{}^t(C\tau(z) + D)^{-1\,t}(A\tau(z) + B))$$
$$= {}^t(C\tau(z) + D)^{-1}(\tau(z)\,{}^tC + {}^tD, -\tau(z)\,{}^tA - {}^tB)$$
$$= {}^t(C\tau(z) + D)^{-1}(1, -\tau(z)) \begin{pmatrix} {}^tD & -{}^tB \\ -{}^tC & {}^tA \end{pmatrix}$$
$$= {}^t(C\tau(z) + D)^{-1}(1, -\tau(z))\rho(g)^{-1}.$$

By substituting this into (6.15) we obtain

$$\xi(gz, \rho(g)w) = {}^t(C\tau(z) + D)^{-1}(1, -\tau(z))w \qquad (6.17)$$
$$= {}^t(C\tau(z) + D)^{-1}\xi(z, w);$$

hence the proposition follows by combining this with (6.14).   □

Corollary 6.3 *The complex structure $\mathcal{I}$ on $\mathcal{D} \times V$ described above is invariant under the action of $G \ltimes_\rho V$.*

Proof. Let $z = (z_1, \ldots, z_k)$ and $\xi = (\xi_1, \ldots, \xi_n)$ be global complex coordinate systems for $\mathcal{D}$ and V, respectively, associated to the complex structure $\mathcal{I}$ on $\mathcal{D} \times V$. Given $(g, v) \in G \ltimes_\rho V$, we denote by

$$z \circ (g, v) = (z_1 \circ (g, v), \ldots, z_k \circ (g, v)),$$

$$\xi \circ (g, v) = (\xi_1 \circ (g, v), \ldots, \xi_n \circ (g, v))$$

the corresponding transformed coordinate systems. Then by (6.13) we have

$$(z \circ (g, v))(z, w) = gz, \qquad (6.18)$$

$$(\xi \circ (g, v))(z, w) = {}^t(C\tau(z) + D)^{-1}\xi(z, w) + \xi(gz, v). \qquad (6.19)$$

Since the complex structure on $\mathcal{D}$ determined by $\mathcal{I}$ is G-invariant, the formula (6.18) shows that the transformed coordinates $z_j \circ (g, v)$ are holomorphic functions of $z_1, \ldots, z_k$. Similarly, the coordinates z_j are holomorphic functions of $z_1 \circ (g, v), \ldots, z_k \circ (g, v)$. On the other hand, from (6.19) we see that the transformed coordinates $\xi_j \circ (g, v)$ are holomorphic functions of z and ξ and that the coordinates ξ_j are holomorphic functions of z and $\xi \circ (g, v)$; hence the lemma follows.   □

By Corollary 6.3 the complex structure $\mathcal{I}$ on $\mathcal{D} \times V$ induces the complex structure $\mathcal{I}_Y$ on the fiber bundle Y over X in (6.6); hence its fiber V/L becomes a complex torus. In addition, the alternating bilinear form β on V determines the structure of a polarized abelian variety on the complex torus V/L. Thus Y may be regarded as a family of abelian varieties parametrized by the locally symmetric space X and is known as a *Kuga fiber variety*.

Lemma 6.4 *The complex structure $\mathcal{I}$ on $\mathcal{D} \times V$ described above satisfies*

$$\mathcal{I}\,|_{\tilde{\pi}^{-1}(z)} = I_{\tau(z)}$$

for each $z \in \mathcal{D}$, where $I_{\tau(z)}$ is the complex structure on V corresponding to the element $\tau(z) \in \mathcal{H}_n$.

Proof. Since the complex structure on V_+ is I_0, it suffices to show that

$$\xi(z, I_{\tau(z)}w) = I_0\xi(z, w)$$

for all $w \in V$. Let z_0 be the element of $\mathcal{D}$ such that the complex structure I_0 corresponds to $\tau(z_0) \in \mathcal{H}_n$ as before, and let $g \in G$ be the element with $z = gz_0$. Then we have

$$I_{\tau(z)} = I_{\rho(g)\tau(z_0)} = \rho(g)I_{\tau(z_0)}\rho(g)^{-1} = \rho(g)I_0\rho(g)^{-1}.$$

Using this, (6.9) and (6.16), we have

$$
\begin{aligned}
\xi(z, I_{\tau(z)}w) &= (1, -\tau(z))I_{\tau(z)}w \\
&= (1, -\tau(gz_0))\rho(g)I_0\rho(g)^{-1}w \\
&= {}^t(C\tau(z) + D)^{-1}(1, -\tau(z_0))\rho(g)^{-1}\rho(g)I_0\rho(g)^{-1}w \\
&= I_0{}^t(C\tau(z) + D)^{-1}(1, -\tau(z_0))\rho(g)^{-1}w \\
&= I_0\xi(z, w),
\end{aligned}
$$

and therefore the lemma follows. $\square$

Using the notation in (6.9), the action of $G \ltimes_\rho V$ on $\mathcal{D} \times V_+$ given by (6.11) can be written in the form

$$(g, v) \cdot (z, u) = (gz, {}^t(C\tau(z) + D)^{-1}u + v_{gz}) \tag{6.20}$$

for all $v \in V$, $(z, u) \in \mathcal{D} \times V_+$ and $g \in G$ with $\rho(g) = (\begin{smallmatrix} A & B \\ C & D \end{smallmatrix}) \in Sp(V, \beta)$. Thus the Kuga fiber variety in (6.6) can be written as

$$Y = \Gamma \ltimes_\rho L \backslash \mathcal{D} \times V_+, \tag{6.21}$$

where the quotient is taken with respect to the action given by (6.20).

Given $g \in G$, we define the map $\phi_g : V \to V$ by

$$\phi_g(v) = \rho(g)v \tag{6.22}$$

for all $v \in V$. Then we see easily that ϕ_g is an $\mathbb{R}$-linear isomorphism. Using the fact that

$$I_{\tau(gz)} = I_{\rho(g)\tau(z)} = \rho(g)I_{\tau(z)}\rho(g)^{-1}$$

for each $z \in \mathcal{D}$, we have

$$
\begin{aligned}
(I_{\tau(gz)} \circ \phi_g)(v) &= \rho(g)I_{\tau(z)}\rho(g)^{-1}\rho(g)v \\
&= \rho(g)I_{\tau(z)}v = (\phi_g \circ I_{\tau(z)})(v)
\end{aligned}
$$

for all $z \in \mathcal{D}$ and $v \in V$. Thus it follows that

$$I_{\tau(gz)} \circ \phi_g = \phi_g \circ I_{\tau(z)} \tag{6.23}$$

for all $z \in \mathcal{D}$.

Lemma 6.5 *Given $z \in \mathcal{D}$ and $g \in G$, let $(V, I_{\tau(z)})$ and $(V, I_{\tau(gz)})$ be the complex vector spaces with respect to the complex structures $I_{\tau(z)}$ and $I_{\tau(gz)}$, respectively. Then the $\mathbb{R}$-linear isomorphism $\phi_g : V \to V$ given by (6.22) induces the $\mathbb{C}$-linear isomorphism*

$$\phi_g : (V, I_{\tau(z)}) \to (V, I_{\tau(gz)})$$

defined also by (6.22).

Proof. Given $a + bi \in \mathbb{C}$, using (6.1) and (6.23), we have

$$\phi_g((a + bi) \cdot v) = \phi_g(av + bI_{\tau(z)}v) = a\phi_g(v) + b(\phi_g \circ I_{\tau(z)})v$$
$$= a\phi_g(v) + bI_{\tau(gz)}\phi_g(v) = (a + bi) \cdot \phi_g(v)$$

for all $v \in V$; hence the lemma follows. □

By Lemma 6.5 the element of $(V, I_{\tau(gz)})$ corresponding to the element v in $(V, I_{\tau(z)})$ is $\phi_g(v) = \rho(g)v$. This means that as an element of $(V, I_{\tau(gz)})$ the element v_{gz} in (6.20) should be considered as

$$(\rho(g)v)_{gz} = \xi(gz, \rho(g)v)) = {}^t(C\tau(z) + D)^{-1}v_z,$$

where we used the calculation in (6.17). Therefore we can now rewrite (6.20) in the form

$$(g, v) \cdot (z, u) = (gz, {}^t(C\tau(z) + D)^{-1}(u + v_z)). \tag{6.24}$$

for all $(g, v) \in G \ltimes_\rho V$ and $(z, u) \in \mathcal{D} \times V_+$.

Example 6.6 *We consider the case where the real vector space is $V = \mathbb{R}^{2n} = \mathbb{R}^n \times \mathbb{R}^n$, so that $V_+ = \mathbb{C}^n$. Given $(r, s) \in \mathbb{R}^n \times \mathbb{R}^n = V$, we choose the complex structure on V in such a way that $(s, -r) \in V_+ \oplus V_- = V_\mathbb{C}$. Then by (6.9) we have*

$$(r, s)_z = s + \tau(z)r$$

for $z \in \mathcal{D}$; hence (6.24) can be written as

$$(g, (r, s)) \cdot (z, u) = (gz, {}^t(C\tau(z) + D)^{-1}(u + \tau(z)r + s))$$

for $u \in \mathbb{C}^n$ and $g \in G$ with $\rho(g) = \left(\begin{smallmatrix} A & B \\ C & D \end{smallmatrix}\right) \in Sp(V, \beta)$.

In order to consider another interpretation of the Kuga fiber variety Y in (6.21), we consider the case of $G_0 = Sp(V, \beta)$, $\mathcal{D}_0 = \mathcal{H}_n$ and $\Gamma_0 = Sp(L, \beta)$, and assume that $\rho_0 : G_0 \to Sp(V, \beta)$ and $\tau_0 : \mathcal{D}_0 \to \mathcal{H}_n$ are the respective identity maps. If we denote by Y_0 the Kuga fiber variety determined by the equivariant pair (τ_0, ρ_0) and the discrete subgroup $\Gamma_0 \ltimes_{\rho_0} L$ of $G_0 \ltimes_{\rho_0} V$, then we obtain the complex torus bundle $\pi_0 : Y_0 \to X_0$ over the Siegel modular variety $X_0 = \Gamma_0 \backslash \mathcal{H}_n$. Let Y be the Kuga fiber variety in (6.6) or (6.21) associated to the equivariant pair (τ, ρ) and the discrete group $\Gamma \ltimes_\rho L$ considered before. Using (6.5), we see that $\rho(\Gamma) \subset \Gamma_0$; hence the map $\tau : \mathcal{D} \to \mathcal{H}_n$ induces the map $\tau_X : X \to X_0$. Then the Kuga fiber variety Y can also be obtained by pulling the bundle Y_0 back via τ_X.

6.2 Automorphy Factors and Kernel Functions

In this section we review the notion of canonical automorphy factors and canonical kernel functions for semisimple Lie groups of Hermitian type and those for generalized Jacobi groups by following closely the descriptions of Satake given in [108] (see also [99]). A generalized Jacobi group can be constructed by using a Heisenberg group and an equivariant pair which determines a Kuga fiber variety considered in Section 6.1. Canonical automorphy factors for generalized Jacobi groups will be used later to define Jacobi forms.

Let $G = \mathbb{G}(\mathbb{R})$ be a semisimple real algebraic group of Hermitian type as in Section 6.1, and let K be a maximal compact subgroup of G. Then the Riemannian symmetric space $\mathcal{D} = G/K$ has a G-invariant complex structure $\mathcal{I}$ which determines a complex structure on the tangent space $T_z(\mathcal{D})$ for each $z \in \mathcal{D}$. Let $\mathfrak{k}$ be the Lie algebra of K, and let $\mathfrak{g} = \mathfrak{k} + \mathfrak{p}$ be the corresponding Cartan decomposition of $\mathfrak{g}$. Let $z_0 \in \mathcal{D}$ be the fixed point of K, and let $\mathcal{I}_0$ be the complex structure on $T_{z_0}(\mathcal{D}) = \mathfrak{p}$. We set

$$\mathfrak{p}_{\pm} = \{X \in \mathfrak{p}_{\mathbb{C}} \mid \mathcal{I}_0(X) = \pm i X\}, \tag{6.25}$$

and denote by P_+ and P_- the $\mathbb{C}$-subgroups of $G_{\mathbb{C}}$ corresponding to the subspaces $\mathfrak{p}_+$ and $\mathfrak{p}_-$, respectively, of $\mathfrak{g}_{\mathbb{C}}$. Then we have

$$P_+ \cap K_{\mathbb{C}} P_- = \{1\}, \quad G \subset P_+ K_{\mathbb{C}} P_-, \quad G \cap K_{\mathbb{C}} P_- = K \tag{6.26}$$

(see for example [108, Lemma II.4.2], [99]). If $g \in P_+ K_{\mathbb{C}} P_- \subset G_{\mathbb{C}}$, we denote by $(g)_+ \in P_+$, $(g)_0 \in K_{\mathbb{C}}$ and $(g)_- \in P_-$ the components of g such that

$$g = (g)_+ (g)_0 (g)_-.$$

We denote by $(G_{\mathbb{C}} \times \mathfrak{p}_+)_*$ the subset of $G_{\mathbb{C}} \times \mathfrak{p}_+$ consisting of elements (g, z) such that $g \exp z \in P_+ K_{\mathbb{C}} P_-$. Then the *canonical automorphy factor* is the map $J : (G_{\mathbb{C}} \times \mathfrak{p}_+)_* \to K_{\mathbb{C}}$ defined by

$$J(g, z) = (g \exp z)_0 \tag{6.27}$$

for $(g, z) \in (G_{\mathbb{C}} \times \mathfrak{p}_+)_*$. If $(g, z) \in (G_{\mathbb{C}} \times \mathfrak{p}_+)_*$, we also define the element $g(z) \in \mathfrak{p}_+$ by

$$\exp g(z) = (g \exp z)_+. \tag{6.28}$$

Furthermore, for $z, z' \in \mathfrak{p}_+$ with $(\exp \overline{z}')^{-1} \exp z \in P_+ K_{\mathbb{C}} P_-$, we set

$$\kappa(z, z') = J((\exp \overline{z}')^{-1}, z)^{-1} = (((\exp \overline{z}')^{-1} \exp z)_0)^{-1} \in K_{\mathbb{C}}. \tag{6.29}$$

Thus we obtain a $K_{\mathbb{C}}$-valued function $\kappa(\cdot, \cdot)$ defined on an open subset of $\mathfrak{p}_+ \times \mathfrak{p}_+$ called the *canonical kernel function* for G, and it satisfies the relations

$$\kappa(z', z) = \overline{\kappa(z, z')}^{-1}, \quad \kappa(z_0, z) = \kappa(z, z_0) = 1,$$

$$\kappa(g(z), \overline{g}(z')) = J(g,z)\kappa(z,z')\overline{J(\overline{g}, w)}^{-1}$$

for $z, z' \in \mathfrak{p}_+$ and $g \in G_{\mathbb{C}}$ for which $\kappa(z,z')$ and $\kappa(g(z), \overline{g}(z'))$ are defined (see e.g. [108, Section II.5]).

Let (V, β) be the real symplectic space of dimension $2n$ defined over $\mathbb{Q}$ as is described in Section 6.1. We extend β to a bilinear form on $V_{\mathbb{C}}$ and denote by $\beta_{I_0} : V_{\mathbb{C}} \times V_{\mathbb{C}} \to \mathbb{C}$ the bilinear map defined by

$$\beta_{I_0}(v, v') = \beta(v, I_0 v')$$

for all $v, v' \in V_{\mathbb{C}}$. Then we have $\beta_{I_0} = i\beta$ on $V_+ \times V_+$ and $\beta_{I_0} = -i\beta$ on $V_- \times V_-$. Thus each of $\beta_{I_0}|_{V_+ \times V_+}$ and $\beta_{I_0}|_{V_- \times V_-}$ is both symmetric and alternating; hence we have

$$\beta_{I_0}|_{V_+ \times V_+} = 0, \quad \beta_{I_0}|_{V_- \times V_-} = 0.$$

Let $\mathcal{H} = \mathcal{H}(V, \beta)$ be as in Section 6.1, which can be identified with the Siegel upper half space $\mathcal{H}_n$ of degree n. Since $I_0 \in \mathcal{H}$, from the definition of $\mathcal{H}$ it follows that both $\beta_{I_0}|_{V_+ \times V_-}$ and $\beta_{I_0}|_{V_- \times V_+}$ are positive definite. Therefore we can identify V_- with the dual V_+^* of V_+. Now we define a Hermitian form $\widetilde{\beta} : V_{\mathbb{C}} \times V_{\mathbb{C}} \to \mathbb{C}$ on $V_{\mathbb{C}}$ by

$$\widetilde{\beta}(v, v') = i\beta(\overline{v}, v')$$

for all $v, v' \in V_{\mathbb{C}}$. Then $\widetilde{\beta}$ is positive definite on $V_+ \times V_+$, negative definite on $V_- \times V_-$, and is zero on $V_+ \times V_-$ and $V_- \times V_+$.

Let $\{u_1, \ldots, u_n\}$ be an orthonormal basis of V_+ with respect to the restriction of the positive definite form $\widetilde{\beta}$ to $V_+ \times V_+$, and define the elements $u_{n+1}, \ldots, u_{2n}$ by $u_{j+n} = \overline{u}_j$ for $1 \leq j \leq n$. Then we have

$$i\beta(u_{n+j}, u_j) = i\beta(\overline{u}_j, u_j) = \widetilde{\beta}(u_j, u_j) = 1,$$

$$i\beta(u_k, u_{n+k}) = i\beta(u_k, \overline{u}_k) = -i\beta(\overline{u}_k, u_k) = -\widetilde{\beta}(u_k, u_k) = 1$$

for $1 \leq j, k \leq n$. On the other hand, using Lemma 1.1(iii), we obtain

$$i\beta(u_j, u_k) = \widetilde{\beta}(\overline{u}_j, u_k) = 0,$$

$$i\beta(u_{n+j}, u_{n+k}) = i\beta(\overline{u}_j, \overline{u}_k) = \widetilde{\beta}(u_j, \overline{u}_k) = 0$$

for $1 \leq j, k \leq n$. Thus $\{u_1, \ldots, u_{2n}\}$ is a symplectic basis for $(V_{\mathbb{C}}, i\beta)$, and we have $Sp(V_{\mathbb{C}}, i\beta) = Sp(V_{\mathbb{C}}, \beta)$.

Now we discuss the canonical automorphy factor for $G = Sp(V, \beta)$. We shall regard the elements of $Sp(V, \beta)$ and the elements of its Lie algebra as matrices using the basis $\{u_1, \ldots, u_{2n}\}$ of $V_{\mathbb{C}}$ described above. Indeed, we have $G = Sp(n, \mathbb{R})$ with respect to this basis. Thus, for example, $\mathfrak{p}_+$ and P_+ can be written in the form

$$\mathfrak{p}_+ = \left\{ \begin{pmatrix} 0 & Z \\ 0 & 0 \end{pmatrix} \, \middle| \, Z \in S_n(\mathbb{C}) \right\},$$

$$P_+ = \exp \mathfrak{p}_+ = \left\{ \begin{pmatrix} 1 & Z \\ 0 & 1 \end{pmatrix} \, \middle| \, Z \in S_n(\mathbb{C}) \right\},$$

where $S_n(\mathbb{C})$ denotes the set of complex symmetric $n \times n$ matrices. Similarly, we have

$$\mathfrak{p}_- = \left\{ \begin{pmatrix} 0 & 0 \\ \overline{Z} & 0 \end{pmatrix} \, \middle| \, Z \in S_n(\mathbb{C}) \right\},$$

$$P_- = \exp \mathfrak{p}_- = \left\{ \begin{pmatrix} 1 & 0 \\ \overline{Z} & 1 \end{pmatrix} \, \middle| \, Z \in S_n(\mathbb{C}) \right\}.$$

We shall identify $\mathfrak{p}_+$ with $S_m(\mathbb{C})$ using the correspondence $\begin{pmatrix} 0 & Z \\ 0 & 0 \end{pmatrix} \mapsto Z$. Thus we may write

$$\exp Z = \begin{pmatrix} 1 & Z \\ 0 & 1 \end{pmatrix} \in P_+$$

for $Z \in \mathfrak{p}_+ = S_m(\mathbb{C})$, and $G_{\mathbb{C}}$ acts on $\mathfrak{p}_+$ by

$$gZ = \exp^{-1}((g \exp Z)_+) \in \mathfrak{p}_+$$

for all $g \in G_{\mathbb{C}}$, where $\exp^{-1}(W)$ for $W \in P_+$ denotes the $(1,2)$-block of the 2×2 block matrix W. Let g' be an element of $G_{\mathbb{C}} = Sp(V_{\mathbb{C}}, \beta)$ whose matrix representation is of the form $g' = \begin{pmatrix} A' & B' \\ C' & D' \end{pmatrix}$. Then $g' \in P_+ K_{\mathbb{C}} P_-$ if and only if D' is nonsingular, and in this case its decomposition is given by

$$g' = \begin{pmatrix} 1 & B'D'^{-1} \\ 0 & 1 \end{pmatrix} \begin{pmatrix} {}^t D'^{-1} & 0 \\ 0 & D' \end{pmatrix} \begin{pmatrix} 1 & 0 \\ D'^{-1}C' & 1 \end{pmatrix}.$$

Lemma 6.7 *Let $g = \begin{pmatrix} A & B \\ C & D \end{pmatrix} \in G_{\mathbb{C}}$, $Z \in \mathfrak{p}_+ = S_n(\mathbb{C})$, and let*

$$J^S : (G_{\mathbb{C}} \times \mathfrak{p}_+)_* \to K_{\mathbb{C}}$$

be the canonical automorphy factor for the symplectic group $G = Sp(V, \beta)$. If $CZ + D$ is nonsingular, then we have

$$gZ = (AZ + B)(CZ + D)^{-1},$$

$$J^S(g, Z) = \begin{pmatrix} {}^t(CZ + D)^{-1} & 0 \\ 0 & (CZ + D) \end{pmatrix}. \tag{6.30}$$

Proof. Given $g \in G_{\mathbb{C}}$ and $Z \in \mathfrak{p}_+$ as above, we have

$$g \exp Z = \begin{pmatrix} A & B \\ C & D \end{pmatrix} \begin{pmatrix} 1 & Z \\ 0 & 1 \end{pmatrix} = \begin{pmatrix} A & AZ + B \\ C & CZ + D \end{pmatrix}.$$

Since $CZ + D$ is nonsingular, we see that the element $g \exp Z$ belongs to $P_+ K_{\mathbb{C}} P_-$ and that its decomposition is given by

$$g \exp Z = (g \exp Z)_+ (g \exp Z)_0 (g \exp Z)_-,$$

$$(g \exp Z)_+ = \begin{pmatrix} 1 & (AZ + B)(CZ + D)^{-1} \\ 0 & 1 \end{pmatrix},$$

$$(g \exp Z)_0 = \begin{pmatrix} {}^t(CZ + D)^{-1} & 0 \\ 0 & (CZ + D) \end{pmatrix},$$

$$(g \exp Z)_- = \begin{pmatrix} 1 & 0 \\ (CZ + D)^{-1}C & 1 \end{pmatrix}.$$

Thus we have

$$J^S(g, Z) = (g \exp p)_0 = \begin{pmatrix} {}^t(CZ + D)^{-1} & 0 \\ 0 & (CZ + D) \end{pmatrix},$$

and gZ is the $(1,2)$-block $(AZ + B)(CZ + D)^{-1}$ of the matrix $(g \exp p)_+$. □

Since the Siegel upper half space $\mathcal{H}_n$ is as the set of complex symmetric $n \times n$ matrices with positive definite imaginary part, there is a natural embedding of $\mathcal{H}_n$ into $\mathfrak{p}_+ = S_m(\mathbb{C})$. If $g = (\begin{smallmatrix} A & B \\ C & D \end{smallmatrix}) \in Sp(V, \beta)$ and $Z \in \mathcal{H}_n$, then $CZ + D$ is nonsingular; hence, using Lemma 6.7 we obtain the usual action $Z \mapsto (AZ + B)(CZ + D)^{-1}$ of G on $\mathcal{H}_n$. On the other hand, given $g \in G$ and $Z \in \mathcal{H}_n$, the associated complex $n \times n$ matrix $J^S(g, Z)$ can be regarded as a linear map of $V_{\mathbb{C}}$ into itself.

Corollary 6.8 *(i) Let $Z \in \mathcal{H}_n$ and*

$$g = \begin{pmatrix} A & B \\ C & D \end{pmatrix} \in Sp(V, \beta).$$

Then the restriction $J_+^S(g, Z)$ of the linear map $J^S(g, Z) : V_{\mathbb{C}} \to V_{\mathbb{C}}$ to the subspace V_+ of $V_{\mathbb{C}}$ is given by

$$J_+^S(g, Z) = {}^t(CZ + D)^{-1}. \tag{6.31}$$

(ii) Let κ^S be the canonical kernel function for the symplectic group $Sp(V, \beta)$, and let κ_+^S be its restriction to V_+ of the type described in (i). Then we have

$$\kappa_+^S(Z, Z') = 1 - Z\overline{Z}' \tag{6.32}$$

for all $Z, Z' \in \mathcal{H}_n$, where 1 denotes the $n \times n$ identity matrix.

Proof. The formula (6.31) follows immediately from (6.30). On the other hand, applying Lemma 6.7 to the group

$$(\exp \overline{Z}')^{-1} = \begin{pmatrix} 1 & 0 \\ -\overline{Z}' & 1 \end{pmatrix}$$

with $Z' \in \mathcal{H}_n$, we see that

$$\kappa_+^S(Z, Z') = \begin{pmatrix} 1 - Z\overline{Z}' & 0 \\ 0 & 1 - \overline{Z}'Z \end{pmatrix},$$

where we used (6.29) and the fact that

$${}^t(1 - \overline{Z}'Z) = 1 - {}^tZ^t\overline{Z}' = 1 - Z\overline{Z}'.$$

This implies (6.32), and hence the proof of the lemma is complete. □

Let (τ, ρ) be an equivariant pair which determines Kuga fiber varieties considered in Section 6.1. Thus $\tau : \mathcal{D} \to \mathcal{H}_n$ is a holomorphic map that is equivariant with respect to the homomorphism $\rho : G \to Sp(V, \beta)$ of Lie groups. We denote by $\widetilde{G}$ the group of all elements of $G \times V \times \mathbb{R}$ whose multiplication operation is defined by

$$(g, v, t)(g', v', t') = (gg', \rho(g)v' + v, t + t' + \beta(v, \rho(g)v')/2) \qquad (6.33)$$

for all $(g, v, t), (g', v', t') \in G \times V \times \mathbb{R}$. Thus the subgroup $\{0\} \times V \times \mathbb{R}$ of $\widetilde{G}$ is the *Heisenberg group* associated to the symplectic space (V, β). The group $\widetilde{G}$ is the group of Harish-Chandra type in the sense of Satake [108] and can be considered as a *generalized Jacobi group* since it reduces to a usual Jacobi group when ρ is the identity map on $Sp(V, \beta)$ (see for example [8, 124]). Let K be a maximal compact subgroup of G with $\mathcal{D} = G/K$, and let $\mathfrak{g}$ and $\mathfrak{k}$ be the Lie algebras of G and K, respectively. If $\mathfrak{g} = \mathfrak{k} + \mathfrak{p}$ is the corresponding Cartan decomposition of $\mathfrak{g}$, we recall that $\mathfrak{p}_+$ and $\mathfrak{p}_-$ are subspaces of $\mathfrak{p}_{\mathbb{C}}$ defined by (6.25). If the subspaces V_+ and V_- of $V_{\mathbb{C}}$ are as in (6.7), we set

$$\widetilde{\mathfrak{p}}_+ = \mathfrak{p}_+ \oplus V_+, \quad \widetilde{\mathfrak{p}}_- = \mathfrak{p}_+ \oplus V_-, \qquad (6.34)$$

and let $\widetilde{P}_+, \widetilde{P}_-$ be the corresponding subgroup of $\widetilde{G}_{\mathbb{C}} = G_{\mathbb{C}} \times V_{\mathbb{C}} \times \mathbb{C}$, respectively. If $\widetilde{K}_{\mathbb{C}} = K_{\mathbb{C}} \times \{0\} \times \mathbb{C}$, we have $\widetilde{G} \subset \widetilde{P}_+\widetilde{K}_{\mathbb{C}}\widetilde{P}_-$. Thus each element $\widetilde{g} \in \widetilde{G}$ has a decomposition of the form

$$\widetilde{g} = (\widetilde{g})_+ \cdot (\widetilde{g})_0 \cdot (\widetilde{g})_-$$

with $(\widetilde{g})_+ \in \widetilde{P}_+$, $(\widetilde{g})_0 \in \widetilde{K}_{\mathbb{C}}$ and $(\widetilde{g})_- \in \widetilde{P}_-$. The *canonical automorphy factor* $\widetilde{J}$ for the group $\widetilde{G}$ and the action of $\widetilde{G}$ on $\widetilde{\mathfrak{p}}_+ = \mathfrak{p}_+ \oplus V_+$ is defined by

$$\widetilde{J}((g, v, t), (z, w)) = ((g, v, t)\exp(z, w))_0, \qquad (6.35)$$

$$\exp((g, v, t) \cdot (z, w)) = ((g, v, t)\exp(z, w))_+, \qquad (6.36)$$

assuming that $(g, v, t)\exp(z, w) \in \widetilde{P}_+\widetilde{K}_{\mathbb{C}}\widetilde{P}_-$. Here $\exp(z, w)$ is an element of $\widetilde{G}$ and is given by

$$\exp(z, w) = (\exp z, w, 0) \tag{6.37}$$

for all $z \in \mathfrak{p}_+$ and $w \in V_+$. Since $\mathcal{D}$ is embedded into $\mathfrak{p}_+$, the quotient space $\widetilde{\mathcal{D}} = \widetilde{G}/\widetilde{K} = \mathcal{D} \times V_+$ can be embedded into $\widetilde{\mathfrak{p}}_+ = \mathfrak{p}_+ \oplus V_+$. Thus (6.36) defines the action of $\widetilde{G}$ on $\widetilde{\mathcal{D}}$. We also define the *canonical kernel function* $\widetilde{\kappa}(\cdot, \cdot) : \widetilde{\mathfrak{p}}_+ \times \widetilde{\mathfrak{p}}_+ \to \widetilde{K}_{\mathbb{C}}$ by

$$\widetilde{\kappa}((z, w), (z', w')) = \widetilde{J}((\exp \overline{(z', w')})^{-1}, (z, w))^{-1} \tag{6.38}$$
$$= ((\exp \overline{(z', w')})^{-1} \exp(z, w))_0^{-1}$$

for $(z, w), (z', w') \in \widetilde{\mathfrak{p}}_+$ such that

$$(\exp \overline{(z', w')})^{-1} \exp(z, w) \in \widetilde{P}_+ \widetilde{K}_{\mathbb{C}} \widetilde{P}_-. \tag{6.39}$$

The condition (6.39) is satisfied for $(z, w), (z', w') \in \widetilde{\mathcal{D}}$; hence we obtain a canonical kernel function on $\widetilde{\mathcal{D}} \times \widetilde{\mathcal{D}}$.

Lemma 6.9 *The canonical kernel function* $\widetilde{\kappa}((z, w), (z', w'))$ *is holomorphic in* (z, w) *and satisfies the relations*

$$\widetilde{\kappa}((z', w'), (z, w)) = \overline{\widetilde{\kappa}((z, w), (z', w'))}^{-1},$$

$$\widetilde{\kappa}(\widetilde{g} \cdot (z, w), \widetilde{g} \cdot (z', w')) = \widetilde{J}(\widetilde{g}, (z, w)) \widetilde{\kappa}((z, w), (z', w')) \overline{\widetilde{J}(\widetilde{g}, (z', w'))}^{-1}$$

for $(z, w), (z', w') \in \widetilde{\mathcal{D}}$ *and* $\widetilde{g} \in \widetilde{G}$.

Proof. The first relation follows immediately from (6.38). In order to verify the second relation, let $(z, w), (z', w') \in \widetilde{\mathcal{D}}$ and $\widetilde{g} \in \widetilde{G}$. Then, using (6.35) and (6.36), we see that

$$\widetilde{g} \exp(z, w) = \exp(\widetilde{g} \cdot (z, w)) \widetilde{J}(\widetilde{g}, (z, w)) p_1,$$

$$\widetilde{g} \exp(z', w') = \exp(\widetilde{g} \cdot (z', w')) \widetilde{J}(\widetilde{g}, (z', w')) p_2$$

for some $p_1, p_2 \in \widetilde{P}_-$. Since $\overline{\overline{g}} = \widetilde{g}$, we have

$$\exp\overline{(z', w')}^{-1} \exp(z, w) \tag{6.40}$$
$$= \exp \overline{(z', w')}^{-1} \overline{\widetilde{g}}^{-1} \widetilde{g}^{-1} \exp(z, w)$$
$$= \overline{p_2}^{-1} \overline{\widetilde{J}(\widetilde{g}, (z', w'))}^{-1} \exp \overline{(\widetilde{g} \cdot (z', w'))}^{-1} \exp(\widetilde{g} \cdot (z, w)) \widetilde{J}(\widetilde{g}, (z, w)) p_1,$$

which shows that

$$\exp \overline{(z', w')}^{-1} \exp(z, w) \in \widetilde{P}_+ \widetilde{K}_{\mathbb{C}} \widetilde{P}_-.$$

Thus, by (6.38) and (6.40) the $\widetilde{K}_{\mathbb{C}}$-component of $\exp \overline{(z', w')}^{-1} \exp(z, w)$ is given by

$$\widetilde{\kappa}((z,w),(z',w'))^{-1}$$
$$= \overline{\widetilde{J}(\widetilde{g},(z',w'))}^{-1} \exp\overline{(\widetilde{g} \cdot (z',w'))}^{-1} \exp(\widetilde{g} \cdot (z,w))\widetilde{J}(\widetilde{g},(z,w));$$

hence we obtain

$$\exp\overline{(\widetilde{g} \cdot (z',w'))}^{-1} \exp(\widetilde{g} \cdot (z,w))^{-1}$$
$$= \widetilde{J}(\widetilde{g},(z,w))\widetilde{\kappa}((z,w),(z',w'))\overline{\widetilde{J}(\widetilde{g},(z',w'))}^{-1}.$$

Now the second relation is obtained from this and (6.38).   □

We shall now describe below the action of $\widetilde{G}$ on $\widetilde{\mathcal{D}}$ and the canonical automorphy factor for $\widetilde{G}$ defined by (6.36) more explicitly. Given $(g,v,t) \in \widetilde{G}$ and $(z,w) \in \widetilde{\mathcal{D}} \subset \mathfrak{p}_+ \oplus V_+$, we set

$$(g,v,t) \cdot (z,w) = (z',w') \in \widetilde{\mathcal{D}}, \tag{6.41}$$

$$\widetilde{J}((g,v,t),(z,w)) = (J_1,0,J_2) \in \widetilde{K}_{\mathbb{C}} = K_{\mathbb{C}} \times \{0\} \times \mathbb{C}. \tag{6.42}$$

Using (6.35), (6.36), (6.37), (6.41) and (6.42), we obtain a decomposition of the form

$$(g,v,t)\exp(z,w) = (\exp z', w', 0)(J_1, 0, J_2)(p_-, w_-, 0) \in \widetilde{P}_+ \widetilde{K}_{\mathbb{C}} \widetilde{P}_- \tag{6.43}$$

for some $(p_-, w_-, 0) \in \widetilde{P}_-$. Using the multiplication rule on $\widetilde{G}$ in (6.33), we see that the right hand side of (6.43) reduces to

$$((\exp z')J_1, w', J_2)(p_-, w_-, 0) = (g'', v'', t''),$$

where

$$g'' = (\exp z')J_1 p_-,$$
$$v'' = w' + \rho((\exp z')J_1)w_-,$$
$$t'' = J_2 + \beta(w', \rho((\exp z')J_1)w_-)/2.$$

On the other hand, the left hand side of (6.43) can be written as

$$(g,v,t)\exp(z,w) = (g,v,t)(\exp z, w, 0)$$
$$= (g\exp z, v + \rho(g)w, t + \beta(v,\rho(g)w)/2).$$

Hence we obtain

$$g\exp z = (\exp z')J_1 p_-, \tag{6.44}$$

$$v + \rho(g)w = w' + (\exp z')J_1 w_-, \tag{6.45}$$

$$t + \beta(v,\rho(g)w)/2 = J_2 + \beta(w',\rho((\exp z')J_1)w_-)/2.$$

for $(g,v,t) \in \widetilde{G} = G \times V \times \mathbb{R}$ and $(z,w) \in \widetilde{\mathcal{D}} = \mathcal{D} \times V_+$.

Proposition 6.10 *Given* $(z, w) \in \tilde{\mathcal{D}}$, *we set*

$$\rho(J_1(z, w)) = \begin{pmatrix} J_+(z, w) & 0 \\ 0 & J_-(z, w) \end{pmatrix} \in Sp(V, \beta),$$

where we identified $Sp(V, \beta)$ *with* $Sp(n, \mathbb{R})$ *by fixing a symplectic basis of* V. *Then the action of* $\tilde{G}$ *on* $\tilde{\mathcal{D}}$ *is given by*

$$(g, v, t) \cdot (z, w) = (g(z), v_{gz} + J_+(z, w)w) \tag{6.46}$$

for all $(g, v, t) \in \tilde{G}$.

Proof. Consider the elements $(g, v, t) \in \tilde{G}$ and $(z, w) \in \tilde{\mathcal{D}}$, and assume that

$$(g, v, t) \cdot (z, w) = (z', w') \in \tilde{\mathcal{D}}$$

as in (6.41). Since $\exp g(z) = (g \exp z)_+$ by (6.28), from (6.44) we obtain

$$\exp z' = \exp g(z), \quad J_1 = (g \exp z)_-.$$

Hence it follows that $z' = g(z)$, and

$$J_1((g, v, t), (z, w)) = J(g, z),$$

where J is the canonical automorphy factor for the group G given in (6.27). Now we consider the matrix representations

$$\rho(\exp z) = \begin{pmatrix} 1 & \tau(z) \\ 0 & 1 \end{pmatrix}, \quad \rho(g) = \begin{pmatrix} A & B \\ C & D \end{pmatrix}, \quad \rho(J_1) = \begin{pmatrix} J_+ & 0 \\ 0 & J_- \end{pmatrix}$$

for $z \in \mathfrak{p}_-$ relative to the decomposition $V = V_+ \oplus V_-$. Applying ρ to both sides of the relation (6.44), we obtain

$$\begin{pmatrix} A & B \\ C & D \end{pmatrix} \begin{pmatrix} 1 & \tau(z) \\ 0 & 1 \end{pmatrix} = \begin{pmatrix} 1 & \tau(z') \\ 0 & 1 \end{pmatrix} \begin{pmatrix} J_+ & 0 \\ 0 & J_- \end{pmatrix} \begin{pmatrix} 1 & 0 \\ M & 1 \end{pmatrix}$$

for some matrix M. Thus we have

$$\begin{pmatrix} A & A\tau(z) + B \\ C & C\tau(z) + D \end{pmatrix} = \begin{pmatrix} J_+ + \tau(z')J_- M & \tau(z')J_- \\ J_- M & J_- \end{pmatrix}.$$

Hence we see that

$$J_+ = A - \tau(z')C, \quad J_- = C\tau(z) + D,$$

$$\tau(z') = (A\tau(z) + B)(C\tau(z) + D)^{-1}.$$

On the other hand the matrix form of the relation (6.45) is given by

$$\begin{pmatrix} v_+ \\ v_- \end{pmatrix} + \begin{pmatrix} A & B \\ C & D \end{pmatrix} \begin{pmatrix} w \\ 0 \end{pmatrix} = \begin{pmatrix} w' \\ 0 \end{pmatrix} + \begin{pmatrix} 1 & \tau(z') \\ 0 & 1 \end{pmatrix} \begin{pmatrix} J_+ & 0 \\ 0 & J_- \end{pmatrix} \begin{pmatrix} 0 \\ w_- \end{pmatrix},$$

which implies that

$$v_+ + Aw = w' + \tau(z')J_-w_-, \quad v_- + Cw = J_-w_-.$$

Therefore we obtain

$$w' = (v_+ + Aw) - \tau(z')(v_- + Cw)$$
$$= (v_+ - \tau(z')v_-) + (A - \tau(z')C)w = v_{z'} + J_+w;$$

hence we have $(z', w') = (g(z), v_{gz} + J_+w)$.   □

Corollary 6.11 *Let J_+^S be the restriction of the canonical automorphy factor for $Sp(V, \beta)$ to V_+ given in Corollary 6.8. Then, for $(g, v, t) \in \widetilde{G}$ and $(z, w) \in \widetilde{\mathcal{D}}$, we have*

$$(g, v, t) \cdot (z, w) = (gz, v_{gz} + J_+^S(\rho(g), \tau(z))w). \tag{6.47}$$

In particular, if

$$\rho(g) = \begin{pmatrix} A_\rho & B_\rho \\ C_\rho & D_\rho \end{pmatrix} \in Sp(V, \beta),$$

then we have

$$(g, v, t) \cdot (z, w) = (gz, v_{gz} + {}^t(C_\rho\tau(z) + D_\rho)^{-1}w). \tag{6.48}$$

Proof. From (6.27), for $g \in G$ and $z \in \mathcal{D}$, we have

$$\rho(J_1) = \rho(J_1(g, z)) = \rho((g \exp z)_0) = (\rho(g) \exp \tau(z))_0.$$

Thus we see that

$$\rho(J_1)(g, z) = J^S(\rho(g), \tau(z)),$$

where J^S is the canonical automorphy factor for the symplectic group $Sp(V, \beta)$ given in (6.30). Therefore, if J_+ is as in Proposition 6.10, we have

$$J_+(g, z) = J_+^S(\rho(g), \tau(z)). \tag{6.49}$$

Using this and Proposition 6.10, we obtain (6.47). Then (6.48) is obtained by using (6.31).   □

From the multiplication operation on $\widetilde{G}$ given in (6.33) we see that the induced operation on $G \times V$ by the natural projection $\widetilde{G} \to G \times V$ is exactly the one on $G \ltimes_\rho V$ considered in Section 6.1. On the other hand, by (6.49) the restriction of the action of $\widetilde{G}$ on $\mathcal{D} \times V_+$ given by (6.46) to $G \times V$ can be written in the form

$$(g, v) \cdot (z, w) = (gz, v_{gz} + J_+^S(\rho(g), \tau(z))w)$$
$$= (gz, v_{gz} + {}^t(C_\rho\tau(z) + D_\rho)^{-1}w).$$

for all $(z, w) \in \mathcal{D} \times V_+$ and $(g, v) \in G \times V \subset \widetilde{G}$ with $\rho(g) = \begin{pmatrix} A_\rho & B_\rho \\ C_\rho & D_\rho \end{pmatrix} \in Sp(V, \beta)$, which coincides with the action in (6.20) that was used for the construction of the Kuga fiber variety Y in (6.21).

Lemma 6.12 *If $(g, v, t) \cdot (z, w) = (z', w')$, then we have*

$$\beta(v, \rho(g)w) - \beta(v_{z'}, Cw) = \beta(v, J_1 w),$$

where the matrix C is the $(2, 1)$ block of $\rho(g)$ as in the proof of Lemma 6.10.

Proof. Using $v_{z'} = v_+ - \tau(z')v_-$ and the matrix representation of $\rho(g)$, we have

$$\beta(v, \rho(g)w) - \beta(v_{z'}, Cw) = \beta(v, Aw + Cw) - \beta\left(\left(\begin{smallmatrix} 1 & -\tau(z') \\ 0 & 1 \end{smallmatrix}\right) v + v_-, Cw\right).$$

Since $Cw \in V_-$ and $\beta|_{V_- \times V_-} = 0$, using the fact that β is invariant under $\rho(G)$, we obtain

$$\begin{aligned}
\beta(v, \rho(g)w) - \beta(v_{z'}, Cw) &= \beta(v, Aw) + \beta(v, Cw) - \beta\left(v, \left(\begin{smallmatrix} 1 & \tau(z') \\ 0 & 1 \end{smallmatrix}\right) Cw\right) \\
&= \beta(v, Aw) + \beta(v, Cw) - \beta(v, Cw + \tau(z')Cw) \\
&= \beta(v, Aw - \tau(z')Cw).
\end{aligned}$$

Now the lemma follows from the fact that $J_1 w = J_+ w = Aw - \tau(z')Cw$. $\square$

Proposition 6.13 *The canonical automorphy factor $\widetilde{J}$ for $\widetilde{G}$ is given by*

$$\widetilde{J}((g, v, t), (z, w)) = (J_1((g, v), (z, w)), 0, J_2((g, v, t), (z, w))) \tag{6.50}$$

for all $(g, v, t) \in \widetilde{G}$ and $(z, w) \in \widetilde{D}$, where J_1 is the canonical automorphy factor J for the group G given by (6.27) and

$$\begin{aligned}
J_2((g, v, t), (z, w)) = t &+ \beta(v, v_{gz})/2 + \beta(v, J_1 w) \tag{6.51} \\
&+ \beta(\rho(g)w, J_1 w)/2.
\end{aligned}$$

Proof. If C is as in Lemma 6.12, then we have $(v + \rho(g)w)_- = v_- + Cw$. Using this and the fact that $\beta|_{V_+ \times V_+} = 0$, we see that

$$\begin{aligned}
J_2 &= t + \beta(v, \rho(g)w)/2 - \beta(w', \rho((\exp z')J_1)w_-)/2 \\
&= t + \beta(v, \rho(g)w)/2 - \beta(w', v + \rho(g)w - w')/2 \\
&= t + \beta(v, \rho(g)w)/2 - \beta(v_{z'} + J_+w, v + \rho(g)w)/2 \\
&= t + \beta(v, \rho(g)w)/2 - \beta(v_{z'}, v)/2 - \beta(v_{z'}, \rho(g)w)/2 \\
&\qquad - \beta(J_+w, v_- + Cw)/2 \\
&= t + \beta(v, \rho(g)w)/2 - \beta(v_{z'}, v)/2 - \beta(v_{z'}, \rho(g)w)/2 \\
&\qquad - \beta(J_+w, v_-)/2 - \beta(J_+w, Cw)/2
\end{aligned}$$

Using Lemma 6.12, we thus obtain

$$\begin{aligned}
J_2 &= t + \beta(v, J_1 w)/2 - \beta(v_{z'}, v)/2 - \beta(J_+w, v_-)/2 - \beta(J_+w, Cw)/2 \\
&= t + \beta(v, J_1 w) - \beta(v_{z'}, v)/2 - \beta(J_+w, Cw)/2.
\end{aligned}$$

Since $\rho(g)w = Aw + Cw$ with $Aw \in V_+$, we have

$$\beta(J_+w, Cw) = \beta(J_+w, \rho(g)w) = \beta(J_1w, \rho(g)w) = -\beta(\rho(g)w, J_1w).$$

Therefore the proposition follows by using this and the relation $\beta(v_{z'}, v) = \beta(v_{gz}, v) = -\beta(v, v_{gz})$. $\qquad\square$

Now we define the complex-valued function $\mathcal{J} : \widetilde{G} \times \widetilde{\mathcal{D}} \to \mathbb{C}$ by

$$\begin{aligned}
\mathcal{J}((g, v, t), (z, w)) &= \mathbf{e}\left(J_2((g, v, t), (z, w))\right) \\
&= \mathbf{e}\left(t + \beta(v, v_{gz})/2 + \beta(v, J_1w) \right. \\
&\qquad\left. + \beta(\rho(g)w, J_1w)/2\right)
\end{aligned} \tag{6.52}$$

for all $(g, v, t) \in \widetilde{G}$, where $\mathbf{e}\left(\cdot\right) = e^{2\pi i(\cdot)}$.

Proposition 6.14 *The function $\mathcal{J}$ is an automorphy factor, that is, it satisfies the relation*

$$\mathcal{J}(\widetilde{g}\widetilde{g}', \widetilde{z}) = \mathcal{J}(\widetilde{g}, \widetilde{g}'\widetilde{z})\mathcal{J}(\widetilde{g}', \widetilde{z}) \tag{6.53}$$

for $\widetilde{g} = (g, v, t), \widetilde{g}' = (g'.v', t') \in \widetilde{G}$ and $\widetilde{z} = (z, w) \in \widetilde{\mathcal{D}}$.

Proof. Let $\widetilde{g}, \widetilde{g}' \in \widetilde{G}$ and $\widetilde{z} \in \widetilde{\mathcal{D}}$. Since the map

$$\widetilde{J} = (J_1, 0, J_2) : \widetilde{G} \times \widetilde{\mathcal{D}} \to \widetilde{K}_{\mathbb{C}}$$

is an automorphy factor, using the multiplication rule (6.33) in $\widetilde{G}$, we obtain

$$\begin{aligned}
(J_1(\widetilde{g}\widetilde{g}', \widetilde{z}), 0, J_2(\widetilde{g}\widetilde{g}', \widetilde{z})) &= (J_1(\widetilde{g}, \widetilde{g}'\widetilde{z}), 0, J_2(\widetilde{g}, \widetilde{g}'\widetilde{z}))(J_1(\widetilde{g}', \widetilde{z}), 0, J_2(\widetilde{g}', \widetilde{z})) \\
&= (J_1(\widetilde{g}, \widetilde{g}'\widetilde{z})J_1(\widetilde{g}', \widetilde{z}), 0, J_2(\widetilde{g}, \widetilde{g}'\widetilde{z}) + J_2(\widetilde{g}', \widetilde{z})).
\end{aligned}$$

Thus we have

$$J_2(\widetilde{g}\widetilde{g}', \widetilde{z}) = J_2(\widetilde{g}, \widetilde{g}'\widetilde{z}) + J_2(\widetilde{g}', \widetilde{z}),$$

and hence $\mathcal{J} = \mathbf{e}\left(J_2\right)$ satisfies the desired relation. $\qquad\square$

6.3 Mixed Automorphic Forms and Kuga Fiber Varieties

In this section we describe a connection between Kuga fiber varieties and mixed automorphic forms (cf. [71, 74, 73, 76]). To be more specific, we show that the holomorphic forms of the highest degree on a Kuga fiber variety can be interpreted as mixed automorphic forms on a Hermitian symmetric domain.

Let (τ, ρ) be an equivariant pair consisting of a homomorphism $\rho : G \to G'$ and a holomorphic map $\tau : \mathcal{D} \to \mathcal{D}'$ described in Section 6.1. Let Γ be a torsion-free arithmetic subgroups of G, and let $j : \Gamma \times \mathcal{D} \to \mathbb{C}^\times$ be an automorphy factor, which means that j satisfies the cocycle condition

$$j(\gamma_1, \gamma_2, z) = j(\gamma_1, \gamma_2 z) j(\gamma_2, z)$$

for all $z \in \mathcal{D}$ and $\gamma_1, \gamma_2 \in \Gamma$. Let Γ' be a torsion-free arithmetic subgroup of G' such that $\rho(\Gamma) \subset \Gamma'$, and let $j' : \Gamma' \times \mathcal{D} \to \mathbb{C}^\times$ be an automorphy factor.

Definition 6.15 *A mixed automorphic form on $\mathcal{D}$ of type (j, j', ρ, τ) is a holomorphic function $f : \mathcal{D} \to \mathbb{C}$ such that*

$$f(\gamma z) = j(\gamma, z) \cdot j'(\rho(\gamma), \tau(z)) \cdot f(z)$$

for all $z \in D$ and $\gamma, \gamma' \in \Gamma$.

Remark 6.16 *Note that the map $(g, z) \mapsto j(\gamma, z) \cdot j'(\rho(\gamma), \tau(z))$ is also an automorphy factor of Γ. Thus, if we denote this automorphy factor by $J_{j,j'}$, then a mixed automorphic form on $\mathcal{D}$ of type (j, j', ρ, τ) is simply an automorphic form on $\mathcal{D}$ of type $J_{j,j'}$ in the sense of Definition 6.19.*

Let Y be the Kuga fiber variety over $X = \Gamma \backslash \mathcal{D}$ associated to the equivariant pair (τ, ρ) with $G' = Sp(V, \beta)$ given by (6.21), and let Y^m be the m-fold fiber power of the fiber bundle $\pi : Y \to X$ over X, that is,

$$Y^m = \{(y_1, \ldots, y_m) \in Y \times \cdots \times Y \mid \pi(y_1) = \cdots = \pi(y_m)\}.$$

Thus we have

$$Y^m = \Gamma \times L^m \backslash \mathcal{D} \times V_+^m, \tag{6.54}$$

and each fiber of Y^m is isomorphic to the m-fold power $(V_+/L)^m$ of the polarized abelian variety V_+/L.

Let $J : G \times \mathcal{D} \to K_{\mathbb{C}}$ be the automorphy factor obtained by restricting the canonical automorphy factor $(G_{\mathbb{C}} \times \mathfrak{p}_+)_* \to K_{\mathbb{C}}$ given in (6.27) by identifying $\mathcal{D}$ as a subdomain of $\mathfrak{p}_+$, and set

$$j_H(g, z) = \det[\mathrm{Ad}_{\mathfrak{p}_+}(J(g, z))] \tag{6.55}$$

for all $g \in G$ and $z \in \mathcal{D}$, where $\mathrm{Ad}_{\mathfrak{p}_+}$ is the restriction of the adjoint representation of $G_{\mathbb{C}}$ to $\mathfrak{p}_+$. Then $j_H : G \times \mathcal{D} \to \mathbb{C}$ is an automorphy factor, and for each $g \in G$ the map $z \mapsto j_H(g, z)$ is simply the Jacobian map of the transformation $z \mapsto gz$ of $\mathcal{D}$. We also consider the automorphy factor $j_V : Sp(V, \beta) \times \mathcal{H}_n \to \mathbb{C}$ given by

$$j_V(g', Z) = \det(CZ + D) \tag{6.56}$$

for all $Z \in \mathcal{H}_n$ and

$$g' = \begin{pmatrix} A & B \\ C & D \end{pmatrix} \in Sp(V, \beta),$$

where we identified $Sp(V, \beta)$ with $Sp(n, \mathbb{R})$ by using a symplectic basis for V.

Theorem 6.17 *Let Y^m be the m-fold fiber power of the Kuga fiber variety Y over X given by (6.6), and let Ω^{k+mn} be the sheaf of holomorphic $(k + mn)$-forms on Y^m. Then the space $H^0(Y^m, \Omega^{k+mn})$ of sections of Ω^{k+mn} is canonically isomorphic to the space of mixed automorphic forms on $\mathcal{D}$ of type $(j_H^{-1}, j_V^m, \rho, \tau)$.*

Proof. We assume that the Hermitian symmetric domain $\mathcal{D}$ is realized as a bounded symmetric domain in $\mathbb{C}^k$ for some positive integer k, and let $z = (z_1, \ldots, z_k)$ be the global coordinate system for $\mathcal{D}$. Recalling that each fiber of Y^m is of the form $(V_+/L)^m$, let $\zeta^{(j)} = (\zeta_1^{(j)}, \ldots, \zeta_n^{(j)})$ be coordinate system for V_+ for $1 \leq j \leq m$. Let Φ be a holomorphic $(k + mn)$-form on Y^m. Then Φ can be regarded as a holomorphic $(k + mn)$-form on $\mathcal{D} \times V_+^m$ that is invariant under the action of $\Gamma \ltimes_\rho L^m$. Thus there is a holomorphic function $f_\Phi(z, \zeta)$ on $\mathcal{D} \times V_+^m$ such that

$$\Phi(z, \zeta) = f_\Phi(z, \zeta) dz \wedge d\zeta^{(1)} \wedge \cdots \wedge d\zeta^{(m)}, \tag{6.57}$$

where $z = (z_1, \ldots, z_k) \in \mathcal{D}$, $\zeta = (\zeta^{(1)}, \ldots, \zeta^{(m)})$ and

$$\zeta^{(j)} = (\zeta_1^{(j)}, \ldots, \zeta_n^{(j)}) \in V_+$$

for $1 \leq j \leq m$. Give an element $x_0 \in \mathcal{D}$, the restriction of the form Φ to the fiber $Y_{x_0}^n$ over x_0 is the holomorphic mn-form

$$\Phi(x_0, \zeta) = f_\Phi(x_0, \zeta) dz \wedge d\zeta^{(1)} \wedge \cdots \wedge d\zeta^{(m)},$$

where $\zeta \mapsto f_\Phi(x_0, \zeta)$ is a holomorphic function on $Y_{x_0}^m$. However, $Y_{x_0}^m$ is a complex torus of dimension mn, and therefore is compact. Since any holomorphic function on a compact complex manifold is constant, we see that f_Φ is a function of z only. Thus (6.57) can be written in the form

$$\Phi(z, \zeta) = \widetilde{f}_\Phi(z) dz \wedge d\zeta^{(1)} \wedge \cdots \wedge d\zeta^{(m)}, \tag{6.58}$$

where $\widetilde{f}$ is a holomorphic function on $\mathcal{D}$. In order to use the condition that Φ is invariant under the action of $\Gamma \ltimes_\rho L^m$, consider an element

$$(\gamma, l) = (\gamma, l_1, \ldots, l_m) \in \Gamma \ltimes_\rho L^m.$$

Then we have

$$dz \circ (\gamma, l) = j_H(\gamma, z) dz,$$

since $j_H(\gamma, *)$ is the Jacobian map for the transformation $z \mapsto \gamma z$ as stated above. On the other hand, by (6.20) the action of the element $(\gamma, l) \in \Gamma \ltimes_\rho L^m$ on

$$d\zeta^{(j)} = (d\zeta_1^{(j)}, \ldots, d\zeta_n^{(j)}) \in V_+ \tag{6.59}$$

is given by

$$d\zeta^{(j)} \circ (\gamma, l) = \bigwedge_{i=1}^{m} d\Big({}^{t}(C_\rho \tau(z) + D_\rho)^{-1}\zeta_i^{(j)} + (l_i)_{\gamma z} \Big) \qquad (6.60)$$

$$= \det(C_\rho \tau(z) + D_\rho)^{-1} \bigwedge_{i=1}^{m} d\zeta_i^{(j)}$$

$$= j_V(\rho(\gamma), \tau(z))^{-1} d\zeta^{(j)}$$

for $1 \le j \le m$, where

$$\rho(\gamma) = \begin{pmatrix} A_\rho & B_\rho \\ C_\rho & D_\rho \end{pmatrix} \in Sp(V, \beta).$$

Thus from (6.58), (6.59) and (6.60) we obtain

$$\Phi \circ (\gamma, l) = \widetilde{f}_\Phi(\gamma z) j_H(\gamma, z) j_V(\rho(\gamma), \tau(z))^{-m} dz \wedge d\zeta^{(1)} \wedge \cdots \wedge d\zeta^{(m)}.$$

Since Φ is invariant under $\Gamma \ltimes_\rho L^m$, by comparing the previous relation with (6.58) we see that

$$\widetilde{f}_\Phi(\gamma z) = j_H(\gamma, z)^{-1} j_V(\rho(\gamma), \tau(z))^m \widetilde{f}_\Phi(z)$$

for all $\gamma \in \Gamma$ and $z \in D$. On the other hand, given a mixed automorphic form f on $\mathcal{D}$ of type $(j_H^{-1}, j_V^m, \rho, \tau)$, we define the $(k+mn)$-form Φ_f on Y^m by

$$\Phi_f(z, \zeta) = f(z) dz \wedge d\zeta^{(1)} \wedge \cdots \wedge d\zeta^{(m)}.$$

Then for $(\gamma, l) = (\gamma, l_1, \ldots, l_m) \in \Gamma \ltimes_\rho L^m$ we have

$$(\Phi_f \circ (\gamma, l))(z, \zeta) = f(\gamma z) j_H(\gamma, z) j_V(\rho(\gamma), \tau(z))^{-m} dz \wedge d\zeta^{(1)} \wedge \cdots \wedge d\zeta^{(m)}$$

$$= f(z) dz \wedge d\zeta^{(1)} \wedge \cdots \wedge d\zeta^{(m)} = \Phi_f(z, \zeta).$$

Therefore the map $f \mapsto \Phi_f$ gives an isomorphism between the space of mixed automorphic forms on $\mathcal{D}$ of type $(j_H^{-1}, j_V^m, \rho, \tau)$ and the space $H^0(Y^m, \Omega^{k+mn})$ of holomorphic $(k+mn)$-forms on Y^m. $\qquad\qquad\square$

Remark 6.18 *Note that the cohomology group $H^0(Y^m, \Omega^{k+mn})$ is isomorphic to the Dolbeault cohomology group $H^{k+mn,0}(Y^m, \mathbb{C})$ of Y^m, which is the $(k+mn, 0)$-component in the Hodge decomposition of the de Rham cohomology group $H^{k+mn}(Y^m, \mathbb{C})$ of Y^m.*

6.4 Embeddings of Kuga Fiber Varieties

If Y is the Kuga fiber variety over X given by (6.6) or (6.21), it is well-known (cf. [5]) that the locally symmetric space $X = \Gamma \backslash \mathcal{D}$ has the structure of a quasi-projective algebraic variety over $\mathbb{C}$. In this section we show that Y is also a projective variety when Γ is cocompact by following Kuga [61] and Satake [108].

We shall first start with a review of the theory of embeddings of Kähler manifolds into complex projective spaces (see [108]). Let M be a complex manifold, and let $\{U_\alpha \mid \alpha \in \Xi\}$ be an open cover of M. A collection $\{f_{\alpha\beta}\}$ of nonzero holomorphic functions $f_{\alpha\beta} : U_\alpha \cap U_\beta \to \mathbb{C}$ is a *1-cocycle* of M with respect to the covering $\{U_\alpha\}$ if $f_{\alpha\alpha} = 1$ on U_α and $f_{\alpha\beta} \cdot f_{\beta\gamma} \cdot f_{\gamma\alpha} = 1$ on $U_\alpha \cap U_\beta \cap U_\gamma$. Two 1-cocycles $\{f_{\alpha\beta}\}$ and $\{f'_{\alpha\beta}\}$ are *cohomologous* if there are nonzero holomorphic functions $h_\alpha : U_\alpha \to \mathbb{C}$ such that

$$f'_{\alpha\beta} = h_a \cdot f_{\alpha\beta} \cdot h_\beta^{-1}$$

on $U_\alpha \cap U_\beta$. The set of 1-cocycles on M modulo the cohomologous relation forms the first cohomology group $H^1(M, \mathcal{O}^\times)$, where $\mathcal{O}^\times$ denotes the sheaf of nonzero holomorphic functions on M. On the other hand, a 1-cocycle $\{f_{\alpha\beta}\}$ determines a *line bundle* $\mathcal{L}$ over M obtained from the set

$$\coprod_{\alpha \in \Xi} U_\alpha \times \mathbb{C}$$

by identifying the elements $(z, \eta_\alpha) \in U_\alpha \times \mathbb{C}$ and $(z', \eta_\beta) \in U_\beta \times \mathbb{C}$ with $U_\alpha \cap U_\beta \neq \emptyset$, $z = z'$ and $\eta_\alpha = f_{\alpha\beta} \cdot \eta_\beta$. The isomorphism class of such a line bundle depends only on the cohomology class of the corresponding 1-cocycle, and there is a natural isomorphism between the group of complex line bundles over M and the cohomology group $H^1(M, \mathcal{O}^\times)$.

Let $\mathcal{L}$ be the line bundle on M determined by the cocycle $\{f_{\alpha\beta}\}$. If we set

$$k_{\alpha\beta\gamma} = -\frac{1}{2\pi i}(\log f_{\alpha\beta} + \log f_{\beta\gamma} + \log f_{\gamma\alpha}),$$

then the collection $\{k_{\alpha\beta\gamma}\}$ is a 2-cocycle and determines an element $c(\mathcal{L})$ of the second cohomology group $H^2(M, \mathbb{Z})$ with coefficients in $\mathbb{Z}$. The element $c(\mathcal{L}) \in H^2(M, \mathbb{Z})$ is called the *Chern class* of $\mathcal{L}$. We denote by $c_\mathbb{R}(\mathcal{L}) \in H^2(M, \mathbb{R})$ the *real Chern class* of $\mathcal{L}$, that is, the image of $c(\mathcal{L})$ in $H^2(M, \mathbb{R})$. Let $\{h_\alpha\}$ be a collection of C^∞ functions $h_\alpha : U_\alpha \to \mathbb{C}$ satisfying

$$h_\alpha(z) = h_\beta(z) \cdot |f_{\alpha\beta}(z)|^{-2}$$

for all $z \in U_\alpha \cap U_\beta$. Then the corresponding differential forms

$$h_\alpha(z) d\eta_\alpha \wedge d\overline{\eta}_\alpha$$

determine a Hermitian metric on each fiber of $\mathcal{L}$. We shall call $\{h_\alpha\}$ the *Hermitian structure* on $\mathcal{L}$. Given such a Hermitian structure, it is known that the real Chern class $c_\mathbb{R}(\mathcal{L})$ of $\mathcal{L}$ is given by

$$c_\mathbb{R}(\mathcal{L}) = \text{Cl}\left[\frac{1}{2\pi i} d' d'' \log h_\lambda\right], \tag{6.61}$$

where $\mathrm{Cl}[\cdot]$ denotes the cohomology class in $H^2(M, \mathbb{R})$ (see [108, p. 203]).

Let $\mathfrak{D}$ be a domain in $\mathbb{C}^m$, and let Δ be a group that acts on $\mathfrak{D}$ holomorphically, properly discontinuously, and without fixed points. Let $j : \Delta \times \mathfrak{D} \to \mathbb{C}^\times$ be an automorphy factor satisfying

$$j(\delta_1\delta_2, z) = j(\delta_1, \delta_2 z) \cdot j(\delta_2, z) \tag{6.62}$$

for all $\delta_1, \delta_2 \in \Delta$ and $z \in \mathfrak{D}$. We consider the action of Δ on $\mathfrak{D} \times \mathbb{C}$ given by

$$\delta \cdot (z, \zeta) = (\delta z, j(\delta, z)\zeta)$$

for $\delta \in \Delta$ and $(z, \zeta) \in \mathfrak{D} \times \mathbb{C}$. Then the quotient

$$\mathcal{L}(j) = \Delta \backslash \mathfrak{D} \times \mathbb{C} \tag{6.63}$$

with respect to this action has the structure of a line bundle over $M = \Delta \backslash \mathfrak{D}$, which is induced by the natural projection map $\mathfrak{D} \times \mathbb{C} \to \mathfrak{D}$.

Definition 6.19 *Let $j : \Delta \times \mathfrak{D} \to \mathbb{C}^\times$ be the automorphy factor described above. An* automorphic form of type j for Δ *is a holomorphic map $f : \mathfrak{D} \to \mathbb{C}$ that satisfies*

$$f(\delta z) = j(\delta, z)f(z)$$

for all $z \in \mathfrak{D}$ and $\delta \in \Delta$.

Lemma 6.20 *Let $\Gamma_0(M, \mathcal{L}(j))$ be the space of all sections of the line bundle $\mathcal{L}(j)$. Then each element of $\Gamma_0(M, \mathcal{L}(j))$ can be identified with an automorphic form of type j for Δ.*

Proof. Let $s : M \to \mathcal{L}(j)$ be an element of $\Gamma_0(M, \mathcal{L}(j))$. Then for each $z \in \mathfrak{D}$ we have

$$s(\Delta z) = [(z, w_z)] \in \mathcal{L}(j) = \Delta \backslash \mathfrak{D} \times \mathbb{C} \tag{6.64}$$

for some $w_z \in \mathbb{C}$, where $\Delta z \in M$ and $[(z, w_z)] \in \mathcal{L}(j)$ denote the elements corresponding to $z \in \mathfrak{D}$ and $(z, w_z) \in \mathfrak{D} \times \mathbb{C}$, respectively. We define the function $f_s : \mathfrak{D} \to \mathbb{C}$ by $f_s(z) = w_z$ for all $z \in \mathfrak{D}$. Using (6.64), for each $\delta \in \Delta$ we have

$$s(\Delta z) = s(\Delta \delta z) = [(\delta z, w_{\delta z})] = [\delta^{-1} \cdot (\delta z, w_{\delta z})] = (z, j(\delta^{-1}, \delta z)w_{\delta z})),$$

which implies that

$$f_s(z) = j(\delta^{-1}, \delta z)w_{\delta z} = j(\delta^{-1}, \delta z)f_s(\delta z) \tag{6.65}$$

for all $z \in D$ and $\delta \in \Delta$. However, from (6.62) we obtain

$$1 = j(\delta^{-1}\delta, z) = j(\delta^{-1}, \delta z) \cdot j(\delta, z).$$

Thus we have $j(\delta^{-1}, \delta z) = j(\delta, z)^{-1}$, and therefore (6.65) implies that f_s is an automorphic form of type j for Δ. $\qquad\square$

The 1-cocycle associated to the line bundle $\mathcal{L}(j)$ in (6.63) can be described as follows. Let $\mathfrak{D} \to M = \Delta \backslash \mathfrak{D}$, and choose an open cover $\{U_\alpha \mid \alpha \in \Xi\}$ of M such that $U_\alpha \cap U_\beta$ is connected for all $\alpha, \beta \in \Xi$ and each U_α satisfies the following conditions:

(i) Every connected component of $p^{-1}(U_\alpha)$ is homeomorphic to U_α.

(ii) If U'_α is a connected component of $p^{-1}(U_\alpha)$, then we have

$$p^{-1}(U_\alpha) = \coprod_{\delta \in \Delta} \delta \cdot U'_\alpha.$$

Let $\alpha, \beta \in \Xi$ with $U_\alpha \cap U_\beta \neq \emptyset$, and let U'_α, U'_β be the connected components of U_α, U_β, respectively. Then there is a unique element $\delta_{\alpha\beta} \in \Delta$ such that

$$U'_\alpha \cap \delta_{\alpha\beta} \cdot U'_\beta \neq \emptyset.$$

Thus we obtain a collection $\{f_{\alpha\beta}\}$ of functions $f_{\alpha\beta} : U_\alpha \cap U_\beta \to \mathbb{C}$ such that

$$f_{\alpha\beta}(z) = j(\delta_{\alpha\beta}, z_\beta) \tag{6.66}$$

for all $z \in U_\alpha \cap U_\beta$, where $z_\alpha = (p|_{U'_\beta})^{-1}(z)$. Using (6.62), we see that $\{f_{\alpha\beta}\}$ is a 1-cocycle on M, and it can be shown that $\mathcal{L}(j)$ is exactly the line bundle determined by this 1-cocycle.

Given an automorphy factor $j : \Delta \times \mathfrak{D} \to \mathbb{C}^\times$, a *kernel function* associated to j is a function $k : \mathfrak{D} \times \mathfrak{D} \to \mathbb{C}^\times$ that is holomorphic in the first argument and satisfies

$$k(z', z) = \overline{k(z, z')}, \quad k(\delta z, \delta z') = j(\delta, z) \cdot k(z, z') \cdot \overline{j(\delta, z')} \tag{6.67}$$

for all $z, z' \in \mathfrak{D}$ and $\delta \in \Delta$.

Lemma 6.21 *If* $k : \mathfrak{D} \times \mathfrak{D} \to \mathbb{C}^\times$ *is a kernel function such that* $k(z, z) > 0$ *for all* $z \in \mathfrak{D}$, *then we have*

$$c_\mathbb{R}(\mathcal{L}(j)) = \mathrm{Cl}\left[-\frac{1}{2\pi i} d'd'' \log k(z, z) \right]. \tag{6.68}$$

Proof. From (6.66) and (6.67) we obtain

$$k(\delta_{\alpha\beta} z_\beta, \delta_{\alpha\beta} z_\beta) = j(\delta_{\alpha\beta}, z_\beta) \cdot k(z_\beta, z_\beta) \cdot \overline{j(\delta_{\alpha\beta}, z_\beta)} \tag{6.69}$$
$$= k(z_\beta, z_\beta) \cdot |f_{\alpha\beta}(z_\beta)|^2$$

for all $z \in U_\alpha \cap U_\beta$. Since $\delta_{\alpha\beta} \in U'_\alpha$, we see that $\delta_{\alpha\beta} z_\beta = \delta_\alpha$; hence if we set

$$h_\alpha(z) = k(z_\alpha, z_\alpha)^{-1},$$

for all $z \in \mathfrak{D}$, then (6.69) becomes

$$h_\alpha(z) = h_\beta(z) \cdot |f_{\alpha\beta}(z)|^{-2}.$$

Therefore $\{h_\alpha\}$ is a Hermitian structure on the line bundle $\mathcal{L}(j)$; hence we obtain (6.68) by using (6.61). $\qquad \square$

Let M be a Hermitian manifold, that is, a complex manifold with a Hermitian metric ds^2 on M. Then we have

$$ds^2 = \sum s_{\alpha\beta} dz^\alpha d\overline{z}^\beta, \tag{6.70}$$

where $s_{\alpha\beta} = ds^2(\partial/\partial z^\alpha, \partial/\partial \overline{z}^\beta)$. Given such a Hermitian metric, the 2-form Φ given by

$$\Phi = \frac{i}{2} \sum s_{\alpha\beta} dz^\alpha \wedge d\overline{z}^\beta$$

is called the associated *fundamental 2-form*. If the real $(1,1)$-form Φ is closed, then ds^2 in (6.70) is called a *Kähler metric* and (M, ds^2) is called a *Kähler manifold*.

Theorem 6.22 (Kodaira) *Let (M, ds^2) be a compact Kähler manifold, and let Φ be the associated fundamental 2-form. If the cohomology class of Φ is in the image of the natural homomorphism $H^2(M, \mathbb{Z}) \to H^2(M, \mathbb{R})$, then M is algebraic. Furthermore, if $\mathcal{L}$ is a line bundle over M with $c_{\mathbb{R}}(\mathcal{L}) = \mathrm{Cl}[\Phi]$ and if $(\eta_0, \ldots, \eta_N)$ is a basis of the space of sections of $\mathcal{L}^{\otimes\nu}$ with ν a positive integer, then, for sufficiently large ν, the map*

$$z \mapsto (\eta_0(z), \ldots, \eta_N(z)), \quad M \to P^N(\mathbb{C})$$

provides an embedding of M into an algebraic variety in $P^N(\mathbb{C})$.

Proof. See e.g. [32, §1.4]. □

Proposition 6.23 *Let $k : \mathfrak{D} \times \mathfrak{D} \to \mathbb{C}^\times$ be a kernel function on $\mathfrak{D}$ associated to an automorphy factor $j : \Delta \times \mathfrak{D} \to \mathbb{C}^\times$ satisfying $k(z, z) > 0$ for all $z \in \mathfrak{D}$. Assume that the Hessian matrix*

$$\left(2 \frac{\partial^2}{\partial \overline{z}^\alpha \partial z^\beta} \log k(z, z) \right)$$

of $\log k(z, z)$ is definite. For a positive integer ν, let $(\phi_0, \ldots, \phi_N)$ is a basis of the space of automorphic forms on $\mathfrak{D}$ of type j^ν. Then for sufficiently large ν the map

$$z \mapsto (\phi_0(z), \ldots, \phi_N(z)), \quad M \to P^N(\mathbb{C}) \tag{6.71}$$

provides an embedding of $M = \Delta \backslash \mathfrak{D}$ into an algebraic variety in $P^N(\mathbb{C})$.

Proof. Since the Hessian matrix of $\log k(z, z)$ is definite, we obtain a Hermitian metric on $\mathfrak{D}$ given by

$$ds^2 = \frac{1}{\pi} \sum_{\alpha,\beta} \frac{\partial^2}{\partial \overline{z}^\alpha \partial z^\beta} \log k(z, z) d\overline{z}^\alpha dz^\beta,$$

which is Δ-invariant and Kähler. Thus it induces a Kähler metric on $M = \Delta \backslash \mathfrak{D}$ whose associated fundamental 2-form is given by

$$\Phi = \frac{i}{2\pi} d'd'' \log k(z, z).$$

Using Lemma 6.21, we see that the cohomology class of Φ is contained in the image of $H^2(M, \mathbb{Z})$. Therefore, using Theorem 6.22, we see that for sufficiently large ν a basis $(\phi_0, \ldots, \phi_N)$ of the space of sections of the line bundle $\mathcal{L}(j)^{\otimes \nu}$ determines an embedding of M onto an algebraic variety in $P^N(\mathbb{C})$ by the map (6.71). Hence the proposition follows by applying Lemma 6.20. $\square$

In order to construct an embedding of a Kuga fiber variety into a complex projective space, let G, $\mathcal{D}$, V, Γ, L, and the equivariant pair (τ, ρ) be as in Section 6.1. In particular, V is a real vector space of dimension $2n$. We fix $z_0 \in \mathcal{D}$ and consider the complex structure $I_0 = I_{\tau(z_0)}$ on V corresponding to $\tau(z_0) \in \mathcal{H}_n$ as before. We denote by V_+ and V_- the associated subspaces of $V_{\mathbb{C}}$ given by (6.7).

Lemma 6.24 *If $v = v_+ + v_- \in V \subset V_{\mathbb{C}}$ with $v_+ \in V_+$ and $v_- \in V_-$, then we have*

$$v_+ = \overline{v}_-, \quad v_- = \overline{v}_+, \tag{6.72}$$

where the decomposition $V_{\mathbb{C}} = V_+ \oplus V_-$ is with respect to the complex structure I_0 on V.

Proof. Using (6.7), we have

$$I_0 \overline{v}_+ = \overline{(I_0 v_+)} = \overline{(iv_+)} = -i\overline{v}_+,$$
$$I_0 \overline{v}_- = \overline{(I_0 v_-)} = \overline{(-iv_-)} = i\overline{v}_-;$$

hence it follows that $\overline{v}_+ \in V_-$ and $\overline{v}_- \in V_+$. However, since v belongs to the real vector space V, we have

$$v_+ + v_- = v = \overline{v} = \overline{v}_+ + \overline{v}_-.$$

Therefore we obtain $v_+ = \overline{v}_-$ and $v_- = \overline{v}_+$, which proves the lemma. $\square$

Example 6.25 *Let $V = \mathbb{R}^2$, so that $V_{\mathbb{C}} = \mathbb{C}^2$, and let $I_0 : \mathbb{R}^2 \to \mathbb{R}^2$ be the linear map defined by*

$$I_0 \begin{pmatrix} x_1 \\ x_2 \end{pmatrix} = \begin{pmatrix} x_2 \\ -x_1 \end{pmatrix}$$

for all $x_1, x_2 \in \mathbb{R}$. Then $I_0^2 = -1_V$; hence I_0 is a complex structure on $\mathbb{R}^2$. Consider the element

$$v = \begin{pmatrix} 1 \\ 3 \end{pmatrix} = v_+ + v_- \in V_+ \oplus V_- = \mathbb{C}^2.$$

Then it can be easily shown that

$$v_+ = \frac{1}{2} \begin{pmatrix} 1 - 3i \\ 3 + i \end{pmatrix}, \quad v_- = \frac{1}{2} \begin{pmatrix} 1 + 3i \\ 3 - i \end{pmatrix} = \overline{v}_+.$$

Let $\xi : \mathcal{D} \times V \to V_+$ be as in (6.9), and let $(z, w) \in \mathcal{D} \times V_{\mathbb{C}}$ with $w = \xi(z, v)$ for some $v \in V$. If $v = v_+ + v_- \in V$ with $v_+ \in V_+$ and $v_- \in V_-$, we set

$$d^z w = dv_+ - \tau(z) \cdot dv_- = dv_+ - \tau(z) \cdot d\bar{v}_+, \qquad (6.73)$$

where we used (6.72). We also set

$$d^z \bar{w} = d\bar{v}_+ - \overline{\tau(z)} \cdot dv_+. \qquad (6.74)$$

Using (6.7) and (6.72), we have

$$w = \xi(z, v) = v_z = v_+ - \tau(z)\bar{v}_+;$$

hence we see that

$$\begin{aligned}
dw &= dv_+ - d(\tau(z) \cdot \bar{v}_+) && (6.75)\\
&= dv_+ - (d\tau(z)) \cdot \bar{v}_+ - \tau(z) \cdot d\bar{v}_+ \\
&= d^z w - (d\tau(z)) \cdot \bar{v}_+.
\end{aligned}$$

Lemma 6.26 *Let d' and d'' be the holomorphic and antiholomorphic part of the differential operator d, and let $\kappa : \mathcal{D} \times \mathcal{D} \to K_{\mathbb{C}}$ be the canonical kernel function for G given by (6.29). Then we have*

$$d' v_+ = \rho(\kappa_1(z, z))^{-1} \cdot d^z w, \quad d'' v_+ = \rho(\kappa_1(z, z))^{-1} \cdot \tau(z) \cdot d^z \bar{w}.$$

Proof. From (6.29) and (6.32) it follows that

$$\rho(\kappa_1(z, z')) = 1 - \tau(z) \cdot \overline{\tau(z')} \qquad (6.76)$$

for $z, z' \in \mathcal{D}$. Using this, (6.73) and (6.74), we have

$$\begin{aligned}
d^z w + \tau(z) \cdot d^z \bar{w} &= dv_+ - \tau(z) \cdot d\bar{v}_+ + \tau(z) \cdot (d\bar{v}_+ - \overline{\tau(z)} \cdot dv_+) \\
&= (1 - \tau(z) \cdot \overline{\tau(z)}) \cdot dv_+ \\
&= \rho(\kappa(z, z)) \cdot dv_+
\end{aligned}$$

for $(z, w) \in \mathcal{D} \times V_{\mathbb{C}}$, which implies that

$$dv_+ = \rho(\kappa_1(z, z))^{-1} \cdot (d^z w + \tau(z) \cdot d^z \bar{w}).$$

However, from (6.75) we see that $d^z w$ is of type $(1, 0)$, and therefore $d^z \bar{w}$ is of type $(0, 1)$; hence the lemma follows. $\square$

Let $\widetilde{\kappa}(\cdot, \cdot)$ be the canonical kernel function for the group $\widetilde{G} = G \times V \times \mathbb{R}$ given by (6.38). Thus we have

$$\widetilde{\kappa}((z, w), (z', w')) = \widetilde{J}((\exp \overline{(z', w')})^{-1}, (z, w))^{-1}$$

for $(z, w), (z', w') \in \tilde{\mathcal{D}} = \mathcal{D} \times V_+$. Since $\tilde{J} = (J_1, 0, J_2) \in \tilde{K}_{\mathbb{C}}$, by restricting $\tilde{\kappa}$ to $\mathcal{D} \times \mathcal{D}$ we have

$$\tilde{\kappa}((z, w), (z', w'))$$
$$= (J_1((\exp \overline{(z', w')})^{-1}, (z, w))), 0, J_2((\exp \overline{(z', w')})^{-1}, (z, w)))^{-1}$$
$$= (J_1((\exp \overline{(z', w')})^{-1}, (z, w)))^{-1}, 0, -J_2((\exp \overline{(z', w')})^{-1}, (z, w))).$$

Using (6.37), we obtain

$$((\exp \overline{(z', w')})^{-1} = (\exp \overline{z}', \overline{w}', 0)^{-1} = ((\exp \overline{z}')^{-1}, -\overline{w}', 0),$$

and hence we see that

$$J_1((\exp \overline{(z', w')})^{-1}, (z, w)))^{-1} = J((\exp \overline{z}')^{-1}, z)^{-1} = \kappa(z, z'),$$

where J and κ are the canonical automorphy factor and the canonical kernel function for the group G given in (6.27) and (6.29), respectively. Thus we obtain

$$\tilde{\kappa}((z, w), (z', w')) = (\kappa(z, z'), 0, \kappa_2((z, w), (z', w'))),$$

where

$$\kappa_2((z, w), (z', w')) = -J_2((\exp \overline{(z', w')})^{-1}, (z, w)) \quad (6.77)$$
$$= \beta(\overline{w}', \rho(\kappa(z, z'))^{-1} \tau(z) \overline{w}')/2$$
$$\quad + \beta(\overline{w}', \rho(\kappa(z, z'))^{-1} w)$$
$$\quad + \beta(\overline{\tau(z')} w, \rho(\kappa(z, z'))^{-1} w)/2.$$

We set

$$\mathfrak{K}((z, w), (z', w')) = \mathbf{e}\left[\kappa_2((z, w), (z', w'))\right] \quad (6.78)$$

for $(z, w), (z', w') \in \tilde{\mathcal{D}}$. If $\mathcal{J} = \mathbf{e}\left[J_2\right]$ is as in Section 6.2, then it can be shown that

$$\mathfrak{K}((z', w'), (z, w)) = \overline{\mathfrak{K}((z, w), (z', w'))},$$
$$\mathfrak{K}(\tilde{g} \cdot (z, w), \tilde{g} \cdot (z', w')) = \mathcal{J}(\tilde{g}, (z, w)) \cdot \mathfrak{K}((z, w), (z', w')) \cdot \overline{\mathcal{J}(\tilde{g}, (z', w'))}$$

for all $(z, w), (z', w') \in \tilde{\mathcal{D}}$ and $\tilde{g} \in \tilde{G}$. Given elements $z \in \mathcal{D}$ and $w, w' \in V_+$, we set

$$\mathfrak{K}_z(w, w') = \mathfrak{K}((z, w), (z, w')). \quad (6.79)$$

Throughout the rest of this section we shall fix a basis of $V_{\mathbb{C}}$ so that the alternating bilinear form β on $V_{\mathbb{C}}$ satisfies

$$i\beta(\overline{w}, w) = {}^t\overline{w} \cdot w'$$

for all $w, w' \in V_+$.

Lemma 6.27 *For fixed $z \in \mathcal{D}$, we have*

$$d'd'' \log \mathfrak{K}_z(w, w) = 2\pi \cdot {}^t(d^z w) \wedge \rho(\kappa(z, z))^{-1} \cdot d^z \overline{w}$$

for all $w \in V_+$, where $d^z w$ and $d^z \overline{w}$ are as in (6.73) and (6.74), respectively.

Proof. Using (6.77), (6.78) and (6.79), we have

$$\log \mathfrak{K}_z(w, w) = \pi[{}^t\overline{w} \cdot \rho(\kappa)^{-1} \cdot \tau(z) \cdot \overline{w} + 2 \cdot {}^t\overline{w} \cdot \rho(\kappa)^{-1} \cdot w$$
$$+ {}^tw \cdot \overline{\tau(z)} \cdot \rho(\kappa)^{-1} \cdot w]$$
$$= \pi[{}^t\overline{w} \cdot \rho(\kappa)^{-1} \cdot (\tau(z) \cdot \overline{w} + w)$$
$$+ ({}^t\overline{w} + {}^tw \cdot \overline{\tau(z)}) \cdot \rho(\kappa)^{-1} \cdot w]$$

for all $w \in V_+$. Note that $w = \xi(z, v)$ for a unique element $v \in V$, where ξ is as in (6.9). Then we have

$$w = v_+ - \tau(z) \cdot \overline{v}_+, \quad {}^t\overline{w} = {}^t\overline{v}_+ - {}^tv_+ \cdot \overline{\tau(z)};$$

hence we see that

$$\tau(z) \cdot \overline{w} + w = (1 - \tau(z) \cdot \overline{\tau(z)})v_+ = \rho(\kappa) \cdot v_+,$$
$${}^t\overline{w} + {}^tw \cdot \overline{\tau(z)} = {}^t\overline{v}_+ \cdot \rho(\kappa),$$

where we used (6.76). Thus we obtain

$$\log \mathfrak{K}_z(w, w) = \pi[{}^t\overline{w} \cdot v_+ + {}^t\overline{v}_+ \cdot w] \tag{6.80}$$
$$= \pi[({}^t\overline{v}_+ - {}^tv_+ \cdot \overline{\tau(z)}) \cdot v_+ + {}^t\overline{v}_+ \cdot (v_+ - \tau(z) \cdot \overline{v}_+)]$$
$$= \pi[{}^t\overline{v}_+ \cdot v_+ - {}^tv_+ \cdot \overline{\tau(z)} \cdot \overline{v}_+ + {}^t\overline{v}_+ \cdot v_+ - {}^t\overline{v}_+ \cdot \tau(z) \cdot \overline{v}_+].$$

However, we have

$${}^t\overline{w} \cdot w = ({}^t\overline{v}_+ - {}^tv_+ \cdot \overline{\tau(z)}) \cdot (v_+ - \tau(z) \cdot \overline{v}_+)$$
$$= {}^t\overline{v}_+ \cdot v_+ - {}^tv_+ \cdot \overline{\tau(z)} \cdot \overline{v}_+ - {}^t\overline{v}_+ \cdot \overline{\tau(z)} \cdot v_+ + {}^t\overline{v}_+ \cdot \overline{\tau(z)} \cdot \tau(z) \cdot \overline{v}_+;$$

hence (6.80) can be written in the form

$$\log \mathfrak{K}_z(w, w) = \pi[{}^t\overline{v}_+ \cdot v_+ + {}^t\overline{w} \cdot w - {}^tv_+ \cdot \overline{\tau(z)} \cdot \tau(z) \cdot \overline{v}_+]$$
$$= \pi[{}^t\overline{v}_+ \cdot \rho(\kappa) \cdot v_+ + {}^t\overline{w} \cdot w].$$

Using this, Lemma 6.26, (6.75), and the relation $d''({}^t\overline{v}_+) = {}^t\overline{d'v_+}$, we have

$$d'' \log \mathfrak{K}_z(w,w) = \pi[({}^t\overline{d'v_+}) \cdot \rho(\kappa)v_+ + {}^t\overline{v}_+ \cdot (d''\rho(\kappa)) \cdot v_+$$
$$+ {}^t\overline{v}_+ \cdot \rho(\kappa) \cdot d''v_+ + ({}^t d\overline{w}) \cdot w]$$
$$= \pi[{}^t\overline{(\rho(\kappa)^{-1} \cdot d^z w)} \cdot \rho(\kappa) \cdot v_+ + {}^t\overline{v}_+ \cdot (-\tau(z) \cdot d\overline{\tau}) \cdot v_+$$
$$+ {}^t\overline{v}_+ \cdot \rho(\kappa) \cdot \rho(\kappa)^{-1} \cdot \tau(z) \cdot d^z\overline{w} + ({}^t d\overline{w}) \cdot w]$$
$$= \pi[{}^t d^z\overline{w} \cdot v_+ - {}^t\overline{v}_+ \cdot \tau(z) \cdot d\overline{\tau} \cdot v_+$$
$$+ {}^t\overline{v}_+ \cdot \tau(z) \cdot d^z\overline{w} + ({}^t d\overline{w}) \cdot w].$$
$$= \pi[({}^t d\overline{w} + {}^t v_+ \cdot {}^t d\overline{\tau}) \cdot v_+ - {}^t\overline{v}_+ \cdot \tau(z) \cdot d\overline{\tau} \cdot v_+$$
$$+ {}^t\overline{v}_+ \cdot \tau(z) \cdot (d\overline{w} + d\overline{\tau} \cdot v_+)$$
$$+ ({}^t d\overline{w}) \cdot (v_+ - \tau(z) \cdot \overline{v}_+)]$$
$$= \pi[2 \cdot {}^t d\overline{w} \cdot v_+ + {}^t v_+ \cdot {}^t d\overline{\tau} \cdot v_+].$$

From this and Lemma 6.26 we obtain

$$d'd'' \log \mathfrak{K}_z(w,w) = \pi[-2 \cdot {}^t d\overline{w} \wedge \rho(\kappa)^{-1} \cdot d^z w$$
$$- 2 \cdot {}^t v_+ \cdot {}^t d\overline{\tau} \wedge \rho(\kappa)^{-1} \cdot d^z w]$$
$$= \pi[-2 \cdot {}^t d^z\overline{w} \wedge \rho(\kappa)^{-1} \cdot d^z w],$$

which proves the lemma. □

As before, we regard the Hermitian symmetric domain $\mathcal{D}$ as a bounded domain in $\mathbb{C}^k$, and let $d\mu(z)$ be the Euclidean volume element of $\mathbb{C}^k$ given by

$$d\mu(z) = \left(\frac{i}{2}\right)^k \prod_{\alpha=1}^{k} dz_\alpha \wedge d\overline{z}_\alpha.$$

The the space $\mathcal{H}^2(\mathcal{D})$ of all square-integrable holomorphic functions on $\mathcal{D}$ is a Hilbert space with respect to the inner product

$$\langle f, g \rangle = \int_{\mathcal{D}} f(z)\overline{g(z)}d\mu(z)$$

for $f, g \in \mathcal{H}^2(\mathcal{D})$. If $\{\mu_i \mid i = 1, 2, \ldots\}$ is an orthonormal basis of $\mathcal{H}^2(\mathcal{D})$, then the *Bergman kernel function* $k_\mathcal{D} : \mathcal{D} \times \mathcal{D} \to \mathbb{C}$ is given by

$$k_\mathcal{D}(z, z') = \sum_{i=1}^{\infty} \mu_i(z)\overline{\mu_i(z')}$$

for $z, z' \in \mathcal{D}$. Let $\chi_0 : K_\mathbb{C} \to \mathbb{C}^\times$ be the character of $K_\mathbb{C}$ given by

$$\chi_0(k) = \det(\mathrm{Ad}_{\mathfrak{p}_+}(k))$$

for $k \in K_\mathbb{C}$. Then it can be shown that

$$k_{\mathcal{D}}(z, z')^{-1} = \mathrm{vol}(\mathcal{D}) \cdot \chi_0(\kappa(z, z'))$$

for all $z, z' \in \mathcal{D}$, where κ is the canonical kernel function of G and $\mathrm{vol}(\mathcal{D})$ is the Euclidean volume of $\mathcal{D}$. Furthermore, for a bounded domain $\mathcal{D}$ in $\mathbb{C}$, we have $k_{\mathcal{D}}(z, z) > 0$ for all $z \in D$ and the Hessian matrix $(h_{\alpha\beta})$ with

$$h_{\alpha\beta} = \frac{\partial^2}{\partial \overline{z}_\alpha \partial z_\beta} \log k_{\mathcal{D}}(z, z)$$

is a positive definite Hermitian matrix, where

$$\frac{\partial}{\partial z_\alpha} = \frac{1}{2}\left(\frac{\partial}{\partial \overline{x}_\alpha} - i\frac{\partial}{\partial \overline{y}_\alpha}\right), \quad \frac{\partial}{\partial \overline{z}_\alpha} = \frac{1}{2}\left(\frac{\partial}{\partial \overline{x}_\alpha} + i\frac{\partial}{\partial \overline{y}_\alpha}\right).$$

The associated Hermitian metric on $\mathcal{D}$ given by

$$ds_{\mathcal{D}}^2 = \sum_{\alpha, \beta} h_{\alpha\beta} d\overline{z}_\alpha dz_\beta \tag{6.81}$$

is called the *Bergman metric* of $\mathcal{D}$.

Theorem 6.28 *Assume that that the arithmetic variety $X = \Gamma \backslash \mathcal{D}$ is compact, and let $(\eta_0, \dots, \eta_N)$ be a basis of the space of holomorphic automorphic forms on $\mathcal{D} \times V_+$ for $\Gamma \times L$ relative to the automorphy factor*

$$(\Gamma \times L) \times (\mathcal{D} \times V_+) \to \mathbb{C}, \tag{6.82}$$

$$((\gamma, l), (z, w)) \mapsto (j_H(\gamma, z)^{-1} J((\gamma, l, 0), (z, w)))^\nu,$$

where j_H is as in (6.55). Then, for sufficiently large ν, the map

$$[(z, w)] \mapsto (\eta_0(z, w), \dots, \eta_N(z, w))$$

gives an embedding $Y \hookrightarrow P(\mathbb{C})^N$ of the Kuga fiber variety $Y = \Gamma \times L \backslash \mathcal{D} \times V_+$ to the complex projective space $P(\mathbb{C})^N$.

Proof. By Lemma 6.27 the differential form

$$\frac{1}{2\pi i} d' d'' \log \mathfrak{K}_z(w, w)$$

is the fundamental 2-form associated to the Hermitian metric

$$2(^t d^z \overline{w}) \rho(\kappa(z, z))^{-1} d^z w = 2i \cdot \beta(d^z \overline{w}, \rho(\kappa(z, z))^{-1} d^z w)$$

on the fiber V_+/L over $z \in \mathcal{D}$. Therefore, if we consider the Hermitian metric

$$ds^2 = \pi^{-1} \sum_{\alpha, \beta} \frac{\partial^2}{\partial \overline{z}_\alpha \partial z_\beta} \log k_{\mathcal{D}}(z, z) d\overline{z}_\alpha dz_\beta \tag{6.83}$$

$$+ 2(^t d^z \overline{w}) \rho(\kappa(z, z))^{-1} d^z w$$

on Y, then the fundamental 2-form associated to ds^2 is given by

$$\omega = \frac{i}{2\pi} d'd'' \log k_{\mathcal{D}}(z, z) + i(^t d^z w)\rho(\kappa(z, z))^{-1} d^z \overline{w}$$
$$= \frac{i}{2\pi} d'd'' \log k_{\mathcal{D}}(z, z) + \frac{i}{2\pi} d'd'' \log \mathfrak{K}_z(w, w),$$

which represents the real Chern class of the line bundle $\mathcal{L}(j_H^{-1} \cdot \mathcal{J})^{\otimes \nu}$ associated to the automorphy factor in (6.82). Thus the theorem follows by applying Proposition 6.23. □

Remark 6.29 *The automorphic forms $\eta_0, \ldots, \eta_N$ used in Theorem 6.28 to construct an embedding a Kuga fiber variety into a complex projective space are essentially Jacobi forms on symmetric domains which will be discussed in Chapter 7.*

7

Jacobi Forms

Jacobi forms on the Poincaré upper half plane share properties in common with both elliptic functions and modular forms in one variable, and they were systematically developed by Eichler and Zagier in [23]. They are functions defined on the product $\mathcal{H} \times \mathbb{C}$ of the Poincaré upper half plane $\mathcal{H}$ and the complex plane $\mathbb{C}$ which satisfy certain transformation formulas with respect to the action of a discrete subgroup Γ of $SL(2, \mathbb{R})$, and important examples of Jacobi forms include theta functions and Fourier coefficients of Siegel modular forms. Numerous papers have been devoted to the study of such Jacobi forms in connection with various topics in number theory (see e.g. [7], [9], [54], [116]). Jacobi forms of several variables have been studied most often on Siegel upper half spaces (cf. [123], [124]). Such Jacobi forms and their relations with Siegel modular forms and theta functions have also been studied extensively over the years (cf. [25], [49], [50], [123], [124]). A number of papers have appeared recently which deal with Jacobi forms on domains associated to orthogonal groups, and one notable such paper was written by Borcherds [12] (see also [11], [55], [59]). Borcherds gave a highly interesting construction of Jacobi forms and modular forms on such domains and investigated their connections with generalized Kac-Moody algebras. Jacobi forms for more general semisimple Lie groups were in fact considered before by Piatetskii-Shapiro in [102, Chapter 4]. Such Jacobi forms occur as coefficients of Fourier-Jacobi series of automorphic forms on symmetric domains. In this chapter we study Jacobi forms on Hermitian symmetric domains associated to equivariant holomorphic maps into Siegel upper half spaces. Such Jacobi forms can be used to construct a projective embedding of a Kuga fiber variety and can be identified with certain line bundles on a Kuga fiber variety. When the Hermitian symmetric domain $\mathcal{D}$ is the Poincaré upper half plane or a Siegel upper half space, the interpretation of Jacobi forms as sections of a line bundle was investigated by Kramer and Runge (see [57, 58, 105].

One of the important nilpotent Lie groups is the Heisenberg group whose irreducible representations were classified by Stone and von Neumann (see for example [44, 98, 121]). One way of realizing representations of the Heisenberg group is by using Fock representations, whose representation spaces are Hilbert spaces of functions on complex vector spaces with inner products associated to points on a Siegel upper half space (see [107]). Another topic that is treated in this chapter is a generalization of such Fock representations by using inner products associated to points on a Hermitian symmetric domain that is mapped into a Siegel upper half space by an equivariant holomorphic map. These representations of the Heisenberg group are given by an automorphy factor for Jacobi forms on symmetric domains considered in Chapter 6. We also introduce theta functions associated to an equivariant pair and study connections between such theta functions and Fock representations described above.

In Section 7.1 we construct circle bundles as well as line bundles over a Kuga fiber variety. We also introduce Jacobi forms on symmetric domains and discuss their connections with those bundles. Section 7.2 is about Fock representations of Heisenberg groups obtained by using automorphy factors for Jacobi forms on symmetric domains. We study some of the properties of such representations including the fact that they are unitary and irreducible. In Section 7.3 we introduce theta functions on Hermitian symmetric domains and show that certain types of such theta functions generate the eigenspace of the Fock representation associated to a quasi-character. Such theta functions provide examples of Jacobi forms. Vector-valued Jacobi forms on symmetric domains are discussed in Section 7.4 in connection with modular forms on symmetric domains.

7.1 Jacobi Forms on Symmetric Domains

In this section, we construct twisted torus bundles over locally symmetric spaces, or circle bundles over Kuga fiber varieties, associated to generalized Jacobi groups. We then define Jacobi forms associated to an equivariant pair, which generalize the usual Jacobi forms (see [124]), and discuss connections between such generalized Jacobi forms and twisted torus bundles. Similar Jacobi forms were also considered in [72, 77, 81].

Let $\widetilde{G} = G \times V \times \mathbb{R}$ be the generalized Jacobi group in Section 6.2 associated to an equivariant pair (τ, ρ) which acts on the space $\widetilde{\mathcal{D}} = \widetilde{G}/\widetilde{K} = \mathcal{D} \times V_+$. Other notations will also be the same as in Section 6.2. In particular, $\widetilde{P}_+$ and $\widetilde{P}_-$ are the subgroup of $\widetilde{G}_{\mathbb{C}} = G_{\mathbb{C}} \times V_{\mathbb{C}} \times \mathbb{C}$ corresponding to the subspaces $\mathfrak{p}_+$ and $\mathfrak{p}_-$, respectively, of $\widetilde{\mathfrak{p}}_{\mathbb{C}} = \mathfrak{p}_{\mathbb{C}} \oplus V_{\mathbb{C}}$ given by (6.34). Recall that $\widetilde{G} \subset \widetilde{P}_+ \widetilde{K}_{\mathbb{C}} \widetilde{P}_-$ with $\widetilde{K}_{\mathbb{C}} = K_{\mathbb{C}} \times \{0\} \times \mathbb{C}$; hence each element $\widetilde{g} \in \widetilde{G}$ has a decomposition of the form

$$\widetilde{g} = (\widetilde{g})_+ \cdot (\widetilde{g})_0 \cdot (\widetilde{g})_-$$

with $(\widetilde{g})_+ \in \widetilde{P}_+$, $(\widetilde{g})_0 \in \widetilde{K}_{\mathbb{C}}$ and $(\widetilde{g})_- \in \widetilde{P}_-$. We also recall that the $\mathbb{C}$-subgroups P_+ and P_- of $G_{\mathbb{C}}$ corresponding to the subspaces $\mathfrak{p}_+$ and $\mathfrak{p}_-$, respectively, of $\mathfrak{g}_{\mathbb{C}}$ in (6.25) satisfy the relations in (6.26). From the relation $G \cap K_{\mathbb{C}} P_- = K$ in (6.26) we have

$$\mathcal{D} = G/K = G/(G \cap K_{\mathbb{C}}P_-) = GK_{\mathbb{C}}P_-/K_{\mathbb{C}}P_-.$$

Using this and the condition $G \subset P_+ K_{\mathbb{C}} P_-$ in (6.26), we obtain the natural embedding

$$\mathcal{D} \hookrightarrow P_+ K_{\mathbb{C}} P_-/K_{\mathbb{C}} P_- \hookrightarrow G_{\mathbb{C}}/K_{\mathbb{C}} P_-. \tag{7.1}$$

However, from the relation $P_+ \cap K_{\mathbb{C}} P_- = \{1\}$ in (6.26) we see that

$$P_+ K_{\mathbb{C}} P_-/K_{\mathbb{C}} P_- = P_+/(P_+ \cap K_{\mathbb{C}} P_-) = P_+ \cong \mathfrak{p}_+. \tag{7.2}$$

From (7.1) and (7.2) we obtain

$$\mathcal{D} \hookrightarrow \mathfrak{p}_+ \hookrightarrow G_{\mathbb{C}}/K_{\mathbb{C}} P_-, \tag{7.3}$$

where the second embedding is given by the exponential map.

We note that the exponential map on $\mathfrak{p}_+ \oplus V_+ \oplus \mathbb{C}$ is given as follows. Given an element $(z, w, u) \in \mathfrak{p}_+ \oplus V_+ \oplus \mathbb{C}$, we denote by $\exp(z, w, u)$ the element of $\widetilde{G}$ defined by

$$\exp(z, w, u) = (\exp z, w, u).$$

In particular, we have

$$\exp u = \exp(0, 0, u) = (1, 0, u), \quad \exp(z, w) = \exp(z, w, 0) = (\exp z, w, 0),$$

which agrees with (6.37). Thus, using (6.33), we see that

$$\exp(z, w, u) = \exp(z, w) \exp u. \tag{7.4}$$

The embeddings in (7.3) induces

$$\widetilde{\mathcal{D}} = \mathcal{D} \oplus V_+ \hookrightarrow \widetilde{\mathfrak{p}}_+ = \mathfrak{p}_+ \oplus V_+ \hookrightarrow \widetilde{G}_{\mathbb{C}}/\widetilde{K}_{\mathbb{C}} \widetilde{P}_-.$$

On the other hand, since the elements of $\exp \mathbb{C}$ and the elements of $\widetilde{P}_+$ commute, the exponential map determines the natural embedding

$$\widetilde{\mathfrak{p}}_+ \oplus \mathbb{C} = \mathfrak{p}_+ \oplus V_+ \oplus \mathbb{C} \hookrightarrow \widetilde{G}_{\mathbb{C}}/K_{\mathbb{C}} \widetilde{P}_-.$$

Thus by using the embedding $\mathcal{D} \hookrightarrow \mathfrak{p}_+$ we obtain a commutative diagram

$$
\begin{array}{ccc}
\mathcal{D} \times V_+ \times \mathbb{C} & \longrightarrow & \widetilde{G}_{\mathbb{C}}/K_{\mathbb{C}} \widetilde{P}_- \\
\downarrow & & \downarrow \\
\mathcal{D} \times V_+ & \longrightarrow & \widetilde{G}_{\mathbb{C}}/\widetilde{K}_{\mathbb{C}} \widetilde{P}_- \\
\downarrow & & \downarrow \\
\mathcal{D} & \longrightarrow & G_{\mathbb{C}}/K_{\mathbb{C}} \widetilde{P}_-,
\end{array}
$$

where the horizontal arrows are the natural embeddings given by the exponential map and the vertical arrows are the natural projection maps.

Now we define an action of $\widetilde{G}$ on $\mathcal{D} \times V_+ \times \mathbb{C}$ by requiring that

$$(g, v, t) \cdot (z, w, u) = (z', w', u')$$

if and only if

$$(g, v, t) \exp(z, w, u) \in \exp(z', w', u') \widetilde{K}_\mathbb{C} \widetilde{P}_- \qquad (7.5)$$

for $(g, v, t) \in G \times V \times \mathbb{R}$ and $(z, w, u), (z', w', u') \in \mathcal{D} \times V_+ \times \mathbb{C}$. More specifically, the action is defined by the condition

$$(g, v, t) \exp(z, w, u) = \exp(z', w', u') \widetilde{k} \widetilde{p}_- \qquad (7.6)$$

for all $(g, v, t) \in \widetilde{G}$ and $(z, w, u) \in \mathcal{D} \times V_+ \times \mathbb{C}$, where $\widetilde{k} \in \widetilde{K}_\mathbb{C}$ and $\widetilde{p}_- \in \widetilde{P}_-$ with $\widetilde{k} = (k, 0, 0)$ for some $k \in K_\mathbb{C}$. Using (7.4), the condition (7.6) can be written in the form

$$(g, v, t) \exp(z, w) \exp u = \exp(z', w') \exp u' \cdot \widetilde{k} \widetilde{p}_-.$$

Thus, by considering the natural projection map

$$\mathcal{D} \times V_+ \times \mathbb{C} \to \mathcal{D} \times V_+,$$

we see that the action of $\widetilde{G}$ on $\mathcal{D} \times V_+$ is given by

$$(g, v, t) \exp(z, w) = \exp(z', w') \exp(u' - u) \widetilde{k} \widetilde{p}_-,$$

which implies that

$$((g, v, t) \exp(z, w))_0 = \exp(u' - u) \widetilde{k} = (k, 0, u' - u).$$

However, using (6.35) and (6.50), we see that

$$((g, v, t) \exp(z, w))_0 = (J_1, 0, J_2),$$

where $J_1 = J(g, z)$ with J being the canonical automorphy factor for G given by (6.27) and $J_2 = J_2((g, v, t), (z, w))$ is as in (6.51). Therefore we obtain

$$k = J_1, \quad u' = u + J_2.$$

On the other hand, by comparing the conditions (6.36) and (7.5) we see that the corresponding actions of (g, v, t) on (z, w) coincide. Hence it follows that

$$(g, v, t) \cdot (z, w, u) = (gz, v_{gz} + J_+ w, u + J_2), \qquad (7.7)$$

where $J_+ = J_+(g, z)$ is as in (6.49). For convenience we recall the formulas

$$J_2 = J_2((g, v, t), (z, w)) = t + \beta(v, v_{gz})/2 + \beta(v, J_1 w) + \beta(\rho(g)w, J_1 w)/2,$$

$$J_+ = J_+(g, z) = J_+^S(\rho(g), \tau(z)) = {}^t(C_\rho \tau(z) + D_\rho)^{-1} \qquad (7.8)$$

for $g \in G$ $\rho(g) = \begin{pmatrix} A_\rho & B_\rho \\ C_\rho & D_\rho \end{pmatrix} \in Sp(V, \beta)$, where J_+^S is the restriction of the canonical automorphy factor J^S for $Sp(V, \beta)$ to V_+.

Now we restrict the holomorphic action of $\widetilde{G}$ on $\mathcal{D} \times V_+ \times \mathbb{C}$ to the real analytic action of $\widetilde{G}$ on $\mathcal{D} \times V_+ \times \mathbb{R}$. Let the arithmetic subgroup $\Gamma \subset G$ and the lattice $L \subset V_+$ be as in Section 6.1, and consider the quotient

$$\mathfrak{C} = \Gamma \times L \times \mathbb{Z} \backslash \mathcal{D} \times V_+ \times \mathbb{R} \qquad (7.9)$$

of $\mathcal{D} \times V_+ \times \mathbb{R}$ by the action of $\Gamma \times L \times \mathbb{Z} \subset \widetilde{G}$ given in (7.7). We shall identify $\mathfrak{C}$ with the quotient

$$\Gamma \times L \backslash \mathcal{D} \times V_+ \times (\mathbb{R}/\mathbb{Z})$$

by using the map

$$(z, w, u) \mapsto (z, w, \mathbf{e}\,[u]), \quad \mathcal{D} \times V_+ \times \mathbb{R} \to \mathcal{D} \times V_+ \times (\mathbb{R}/\mathbb{Z}),$$

where we identify $\mathbb{R}/\mathbb{Z}$ with the unit circle $\{z \in \mathbb{C} \mid |z| = 1\}$ in $\mathbb{C}$. Then the action of $\Gamma \times L$ on $\mathcal{D} \times V_+ \times (\mathbb{R}/\mathbb{Z})$ is given by

$$(\gamma, \ell) \cdot (z, w, \lambda) = (\gamma z, \ell_{\gamma z} + J_+ w, \mathcal{J}((\gamma, \ell, 0), (z, w))\lambda) \qquad (7.10)$$

for all $(\gamma, \ell) \in \Gamma \times L$ and $(z, w, \lambda) \in \mathcal{D} \times V_+ \times (\mathbb{R}/\mathbb{Z})$, and the natural projection map

$$\mathcal{D} \times V_+ \times (\mathbb{R}/\mathbb{Z}) \to \mathcal{D} \times V_+$$

equips $\mathfrak{C}$ with the structure of a fiber bundle $\pi_{\mathcal{J}} : \mathfrak{C} \to Y$ over the Kuga fiber variety $Y = \Gamma \times L \backslash \mathcal{D} \times V_+$ whose fiber is isomorphic to the circle $\mathbb{R}/\mathbb{Z}$. Thus $\mathfrak{C}$ is a *circle bundle* over Y. On the other hand, by composing $\pi_{\mathcal{J}}$ with $\pi : Y \to X$ we can also consider $\mathfrak{C}$ as a *twisted torus bundle* over the arithmetic variety $X = \Gamma \backslash \mathcal{D}$ in the sense of [72].

Let $j : G \times \mathcal{D} \to \mathbb{C}$ be an automorphy factor satisfying as usual the cocycle condition

$$j(gg', z) = j(g, g'z)j(g', z) \qquad (7.11)$$

for all $z \in \mathcal{D}$ and $g, g' \in G$. Given a function $f : \mathcal{D} \times V_+ \to \mathbb{C}$, an element $(g, v) \in G \times V$ and a nonnegative integer ν, we set

$$(f \mid_{j,\nu}^{\rho,\tau} (g, v))(z, w) = j(g, z)^{-1} \mathcal{J}((g, v, 0), (z, w))^{-\nu} \qquad (7.12)$$
$$\times f(gz, v_{gz} + J_+ w)$$
$$= j(g, z)^{-1} \mathbf{e}^\nu \, (\beta(v, v_{gz})/2$$
$$+ \beta(v, J_1 w) + \beta(\rho(g)w, J_1 w)/2) \qquad (7.13)$$
$$\times f(gz, v_{\gamma z} + J_+ w)$$

for all $(z, w) \in \mathcal{D} \times V_+$, where $\mathbf{e}^\nu(\cdot) = \mathbf{e}(\nu(\cdot)) = e^{2\pi i \nu(\cdot)}$ and $\mathcal{J} = \mathbf{e}(J_2)$ is as in (6.52).

Lemma 7.1 *If $f : \mathcal{D} \times V_+ \to \mathbb{C}$ is a function and ν is a nonnegative integer, we have*

$$(f \mid_{j,\nu}^{\rho,\tau} (g,v)) \mid_{j,\nu}^{\rho,\tau} (g',v') = f \mid_{j,\nu}^{\rho,\tau} ((g,v)(g',v'))$$

for all $(g,v), (g',v') \in G \times V$, where the product $(g,v)(g',v')$ is as in (6.3).

Proof. Using (6.46) and (7.12), we have

$$((f \mid_{j,\nu}^{\rho,\tau}(g,v)) \mid_{j,\nu}^{\rho,\tau} (g',v'))(z,w)$$
$$= j(g',z)^{-1} \mathcal{J}((g',v',0),(z,w))^{-\nu}(f \mid_{j,\nu}^{\rho,\tau} (g,v))((g',v',0) \cdot (z,w))$$
$$= j(g',z)^{-1} j(g,g'z)^{-1} \mathcal{J}((g',v',0),(z,w))^{-\nu}$$
$$\times \mathcal{J}((g,v,0),(g',v',0) \cdot (z,w))^{-\nu} f((g,v,0)(g',v',0) \cdot (z,w))$$

for all $(z,w) \in \mathcal{D} \times V_+$, which can be shown to be equal to

$$(f \mid_{j,\nu}^{\rho,\tau} ((g,v)(g',v')))(z,w)$$

by using the cocycle conditions for $\mathcal{J}$ and j in (6.53) and (7.11), respectively. $\square$

Definition 7.2 *A holomorphic function $f : \mathcal{D} \times V_+ \to \mathbb{C}$ is a Jacobi form of weight j and index ν for (Γ, L, ρ, τ) if it satisfies*

$$f \mid_{j,\nu}^{\rho,\tau} (\gamma, \ell) = f \qquad (7.14)$$

for all $\gamma \in \Gamma$ and $\ell \in L$.

Given $(z,w,\zeta) \in \mathcal{D} \times V_+ \times \mathbb{C}$ and $(g,v) \in G \times V$, we set

$$(g,v) \cdot (z,w,\zeta) = ((g,v) \cdot (z,w), j(g,z)\zeta) \qquad (7.15)$$
$$= (gz, v_{gz} + J_+w, j(g,z)\zeta),$$

where we used (6.20) and (7.8). The next lemma show that the operation (7.15) defines an action of $G \times V$ on $\mathcal{D} \times V_+ \times \mathbb{C}$.

Lemma 7.3 *The operation (7.15) satisfies*

$$(g,v) \cdot ((g',v') \cdot (z,w,\zeta)) = ((g,v)(g',v')) \cdot (z,w,\zeta)$$

for all $(g,v), (g',v') \in G \times V$ and $(z,w,\zeta) \in \mathcal{D} \times V_+ \times \mathbb{C}$.

Proof. Using (7.15), we have

$$(g,v) \cdot ((g',v') \cdot (z,w,\zeta)) = (g,v) \cdot ((g',v') \cdot (z,w), j(g',z)\zeta)$$
$$= ((g,v) \cdot ((g',v') \cdot (z,w)), j(g,g'z)j(g',z)\zeta)$$
$$= (((g,v)(g',v')) \cdot (z,w), j(gg',z)\zeta)$$
$$= ((g,v)(g',v')) \cdot (z,w,\zeta)$$

where we used the cocycle condition (7.11) for j. $\square$

We consider the action of the subgroup $\Gamma \times L \subset G \times V$ on the space $\mathcal{D} \times V_+ \times \mathbb{C}$ given by (7.15), and denote the corresponding quotient by

$$\mathfrak{B} = \Gamma \times L \backslash \mathcal{D} \times V_+ \times \mathbb{C} \tag{7.16}$$

Then, as in the case of the bundle $\mathfrak{C}$ in (7.9), the natural projection map

$$\mathcal{D} \times V_+ \times \mathbb{C} \to \mathcal{D} \times V_+$$

induces a map $\pi_{\mathfrak{B}} : \mathfrak{B} \to Y$, which has the structure of a line bundle over the Kuga fiber variety Y.

We now extend the action of $\Gamma \times L$ on $\mathcal{D} \times V_+ \times (\mathbb{R}/\mathbb{Z})$ given by (7.10) to the one on $\mathcal{D} \times V_+ \times \mathbb{C}$ by using

$$(\gamma, \ell) \cdot (z, w, \zeta) = (\gamma z, \ell_{\gamma z} + J_+ w, \mathcal{J}((\gamma, \ell, 0), (z, w))\zeta) \tag{7.17}$$

for all $(\gamma, \ell) \in \Gamma \times L$ and $(z, w, \zeta) \in \mathcal{D} \times V_+ \times (\mathbb{R}/\mathbb{Z})$. We denote the corresponding quotient by

$$\widehat{\mathfrak{C}} = \Gamma \times L \backslash \mathcal{D} \times V_+ \times \mathbb{C}. \tag{7.18}$$

Then the construction similar to the case of $\mathfrak{B}$ or $\mathfrak{C}$ provides $\widehat{\mathfrak{C}}$ with the structure of a line bundle over Y, and there is a natural embedding $\mathfrak{C} \hookrightarrow \widetilde{\mathfrak{C}}$ of the circle bundle $\mathfrak{C}$ in (7.9) into the line bundle $\widetilde{\mathfrak{C}}$ over Y.

Theorem 7.4 *Let $\mathfrak{B}$ and $\widehat{\mathfrak{C}}$ be the bundles over the Kuga fiber variety $Y = \Gamma \times L \backslash \mathcal{D} \times V_+$ given by (7.16) and (7.18), respectively. Then the space of Jacobi forms of weight j and index ν for (Γ, L, ρ, τ) is isomorphic to the space*

$$\boldsymbol{\Gamma}_0(Y, \mathfrak{B} \otimes \widehat{\mathfrak{C}}^{\otimes \nu})$$

of sections of the bundle $\mathfrak{B} \otimes \widehat{\mathfrak{C}}^{\otimes \nu}$ over Y.

Proof. From (7.15) and (7.17) it follows that the bundle $\mathfrak{B} \otimes \widehat{\mathfrak{C}}^{\otimes \nu}$ over Y can be regarded as the quotient

$$\mathfrak{B} \otimes \widehat{\mathfrak{C}}^{\otimes \nu} = \Gamma \times L \backslash \mathcal{D} \times V_+ \times \mathbb{C}$$

with respect to the action of $\Gamma \times L$ on $\mathcal{D} \times V_+ \times \mathbb{C}$ given by

$$\begin{aligned}(\gamma, \ell) \cdot &(z, w, \zeta) \\ &= (\gamma z, \ell_{\gamma z} + J_+ w, j(\gamma, z)\mathcal{J}((\gamma, \ell, 0), (z, w))^\nu \zeta).\end{aligned}$$

Let $s : Y \to \mathfrak{B} \otimes \widehat{\mathfrak{C}}^{\otimes \nu}$ be an element of $\boldsymbol{\Gamma}_0(Y, \mathfrak{B} \otimes \widehat{\mathfrak{C}}^{\otimes \nu})$. Then for $(z, w) \in \mathcal{D} \times V_+$ we have

$$s([(z, w)]) = [(z, w, \zeta_{(z,w)})]$$

for some $\zeta_{(z,w)} \in \mathbb{R}/\mathbb{Z}$, where $[(\cdot)]$ denotes the appropriate coset corresponding to the element $(\cdot)$. We define the holomorphic function $f_s : \mathcal{D} \times V_+ \to \mathbb{C}$ by

$$f_s(z,w) = \zeta_{(z,w)} \tag{7.19}$$

for all $(z,w) \in \mathcal{D} \times V_+$. For each $(\gamma, \ell) \in \Gamma \times L$, using the actions of $\Gamma \times L$ given in (7.10) and (7.15), we have

$$
\begin{aligned}
s([(z,w)]) &= s([(\gamma z, \ell_{\gamma z} + J_+ w)]) \\
&= [(\gamma z, \ell_{\gamma z} + J_+ w, \zeta_{(\gamma z, \ell_{\gamma z} + J_+ w)})] \\
&= [(\gamma, \ell)^{-1}(\gamma z, \ell_{\gamma z} + J_+ w, \zeta_{(\gamma z, \ell_{\gamma z} + J_+ w)})] \\
&= [(z, w, j(\gamma, z) \\
&\qquad \times \mathcal{J}((\gamma, \ell, 0), (z, w))^{-\nu} \zeta_{(\gamma z, \ell_{\gamma z} + J_+ w)})],
\end{aligned}
$$

where by (6.52),

$$
\begin{aligned}
\mathcal{J}((\gamma, \ell, 0), (z, w)) &= \mathbf{e}\,[J_2((\gamma, \ell, 0), (z, w))] \tag{7.20} \\
&= \mathbf{e}\,[\beta(\ell, \ell_{\gamma z})/2 + \beta(\ell, J_1 w) \tag{7.21} \\
&\qquad + \beta(\rho(\gamma)w, J_1 w)/2].
\end{aligned}
$$

Therefore we obtain

$$f_s(z, w) = j(\gamma, z)^{-1} \mathcal{J}((\gamma, \ell, 0), (z, w))^{-\nu} f_s(\gamma z, \ell_{\gamma z} + J_+ w).$$

Hence by (7.12) the function f_s satisfies the transformation formula (7.14) for a Jacobi form of weight j and index ν for (Γ, L, ρ, τ). On the other hand, suppose that $f : \mathcal{D} \times V_+ \to \mathbb{C}$ is a holomorphic function satisfying the condition (7.14). We define the map $s_f : Y \to \mathfrak{B} \otimes \widehat{\mathfrak{C}}^{\otimes \nu}$ by

$$s_f([(z, w)]) = [(z, w, f(z, w))] \tag{7.22}$$

for $(z, w) \in \mathcal{D} \times V_+$. Then this map is well-defined because, for each $(\gamma, \ell) \in \Gamma \times L$, we have

$$
\begin{aligned}
s_f([(\gamma, \ell)(z, w)]) &= s_f([(\gamma z, \ell_{\gamma z} + J_+ w)]) \\
&= [(\gamma z, \ell_{\gamma z} + J_+ w, f(\gamma z, \ell_{\gamma z} + J_+ w))] \\
&= [(\gamma z, \ell_{\gamma z} + J_+ w, j(\gamma, z) \\
&\qquad \times \mathcal{J}((\gamma, \ell, 0), (z, w))^{\nu} f(z, w)] \\
&= [(\gamma, \ell)((z, w, f(z, w))] = [(z, w, f(z, w))],
\end{aligned}
$$

which is equal to $s_f([(z, w)])$ in (7.1); hence s_f is a section of $\mathfrak{B} \otimes \widehat{\mathfrak{C}}^{\otimes \nu}$. We see easily that the holomorphic function f_{s_f} defined as in (7.19) coincides with f, and therefore the proof of the theorem is complete. □

Remark 7.5 *When the Hermitian symmetric domain $\mathcal{D}$ is the Poincaré upper half plane or a Siegel upper half space and if ρ and τ are identity maps, the interpretation of Jacobi forms as sections of a line bundle in a way similar to the result in Theorem 7.4 was investigated by Kramer and Runge. In*

[57] Kramer identified Jacobi forms on the Poincaré upper half plane with sections of a line bundle on an elliptic modular surface. He also proved the vanishing of the first cohomology of the elliptic modular surface with coefficients in that bundle and used it to derive a formula for the dimension of the space of Jacobi forms. The correspondence between Jacobi forms on a Siegel upper half space and section of a line bundle over a family of abelian varieties parametrized by a Siegel modular variety was studied by Kramer in [58] and Runge in [105]. They also considered the extension of such a line bundle over a compactification of the family of abelian varieties.

Example 7.6 *Let the automorphy factor $j_H^{-1} : G \times \mathcal{D} \to \mathbb{C}$ be defined by*

$$j_H^{-1}(g,z) = j_H(g,z)^{-1} = \det[\mathrm{ad}_{\mathfrak{p}_+}(J(g,z))]$$

for all $(g,z) \in G \times \mathcal{D}$, where j_H is as in (6.55) and J is the canonical automorphy factor for G. Then Jacobi forms of weight j_H^{-1} and index ν for (Γ, L, ρ, τ) can be used to construct a projective embedding of a Kuga fiber variety (see Theorem 6.28).

Example 7.7 *Given a nonnegative integer μ and elements $z \in \mathcal{D}$ and $g \in G$ with $\rho(g) = \left(\begin{smallmatrix} A_\rho & B_\rho \\ C_\rho & D_\rho \end{smallmatrix} \right) \in Sp(V, \beta)$, we set*

$$J_\mu(g,z) = \det(C_\rho \tau(z) + D_\rho)^\mu.$$

Then this formula determines an automorphy factor $J_\mu : G \times \mathcal{D} \to \mathbb{C}$. Jacobi forms of weight J_μ and index ν for (Γ, L, ρ, τ) were considered in [86].

Example 7.8 *Let W be a real vector space of dimension $\nu > 2$ defined over $\mathbb{Q}$, and let S be a nondegenerate symmetric bilinear form on W of signature $(2, \nu - 2)$. We consider the associated spin group $G = \mathrm{Spin}(W, S)$, which is a semisimple Lie group of Hermitian type. Then, as was described in Example 6.1, there is a homomorphism $\rho : \mathrm{Spin}(W, S) \to Sp(2^\nu, \mathbb{R})$, which induces an equivariant holomorphic map $\tau : \mathcal{D} \to \mathcal{H}_{2^\nu}$ of the symmetric domain $\mathcal{D}$ associated to G into the Siegel upper half space $\mathcal{H}_{2^\nu}$. Thus we obtain Jacobi form on the symmetric domain associated to spin groups of type $(2, n)$, and such Jacobi forms were studied recently in connection with a number of topics (see for example [12] and [26]).*

7.2 Fock Representations

Let $\widetilde{G} = G \times V \times \mathbb{R}$ be as in Section 6.2. Then the multiplication operation (6.33) restricted to the subgroup $\{1\} \times V \times \mathbb{R} \cong V \times \mathbb{R}$ of $\widetilde{G}$ is the usual multiplication operation on the Heisenberg group $V \times \mathbb{R}$. Classically a Fock representation of such a Heisenberg group is a representation in a Hilbert space of certain functions on $V_{\mathbb{C}}$ associated to a point in the corresponding

Siegel upper half space (see [107]). In this section we construct similar representations of such a Heisenberg group in Hilbert spaces associated to points in the Hermitian symmetric domain $\mathcal{D}$ by using automorphy factors associated to Jacobi forms discussed earlier in this chapter.

Throughout this section we shall adopt various notations used in Section 6.2. In particular, $\widetilde{\kappa}(\cdot, \cdot)$ the the canonical kernel function for the group $\widetilde{G} = G \times V \times \mathbb{R}$ given in (6.38). Thus we have

$$\widetilde{\kappa}((z, w), (z', w')) = \widetilde{J}((\exp \overline{(z', w')})^{-1}, (z, w))^{-1}$$

for $(z, w), (z', w') \in \widetilde{\mathcal{D}} = \mathcal{D} \times V_+$. Since $\widetilde{J} = (J_1, 0, J_2) \in \widetilde{K}_{\mathbb{C}}$, by restricting $\widetilde{\kappa}$ to $\mathcal{D} \times \mathcal{D}$ we have

$$\widetilde{\kappa}((z, w), (z', w'))$$
$$= (J_1((\exp \overline{(z', w')})^{-1}, (z, w)))^{-1}, 0, J_2((\exp \overline{(z', w')})^{-1}, (z, w)))^{-1}$$
$$= (J_1((\exp \overline{(z', w')})^{-1}, (z, w)))^{-1}, 0, -J_2((\exp \overline{(z', w')})^{-1}, (z, w))).$$

Using (6.37), we obtain

$$((\exp \overline{(z', w')})^{-1} = (\exp \overline{z}', \overline{w}', 0)^{-1} = ((\exp \overline{z}')^{-1}, -\overline{w}', 0),$$

and hence we see that

$$J_1((\exp \overline{(z', w')})^{-1}, (z, w)))^{-1} = J((\exp \overline{z}')^{-1}, z)^{-1} = \kappa(z, z'),$$

where J and κ are the canonical automorphy factor and the canonical kernel function for the group G given in (6.27) and (6.29), respectively. Thus we obtain

$$\widetilde{\kappa}((z, w), (z', w')) = (\kappa(z, z'), 0, \kappa_2((z, w), (z', w'))),$$

where

$$\kappa_2((z, w), (z', w')) = -J_2((\exp \overline{(z', w')})^{-1}, (z, w)) \tag{7.23}$$
$$= \beta(\overline{w}', \rho(\kappa_1(z, z'))^{-1}\tau(z)\overline{w}')/2$$
$$+ \beta(\overline{w}', \rho(\kappa_1(z, z'))^{-1}w)$$
$$+ \beta(\overline{\tau(z')}w, \rho(\kappa_1(z, z'))^{-1}w)/2.$$

We set

$$\mathfrak{K}((z, w), (z', w')) = \mathbf{e}\left[\kappa_2((z, w), (z', w'))\right] \tag{7.24}$$

for $(z, w), (z', w') \in \widetilde{\mathcal{D}}$.

Proposition 7.9 *Let $\mathcal{J} = \mathbf{e}\left[J_2\right]$ be as in (6.52). Then we have*

$$\mathfrak{K}((z', w'), (z, w)) = \overline{\mathfrak{K}((z, w), (z', w'))}, \tag{7.25}$$

$$\mathfrak{K}(\widetilde{g}(z, w), \widetilde{g}(z', w')) = \mathcal{J}(\widetilde{g}, (z, w))\mathfrak{K}((z, w), (z', w'))\overline{\mathcal{J}(\widetilde{g}, (z', w'))} \tag{7.26}$$

for all $(z, w), (z', w') \in \widetilde{\mathcal{D}}$ and $\widetilde{g} \in \widetilde{G}$.

Proof. Let $(z, w), (z', w') \in \widetilde{\mathcal{D}}$ and $\widetilde{g} \in \widetilde{G}$. Then by Lemma 6.9 the canonical kernel function $\widetilde{\kappa}(\cdot, \cdot)$ satisfies the relations

$$\widetilde{\kappa}((z', w'), (z, w)) = \overline{\widetilde{\kappa}((z, w), (z', w'))}^{-1}$$
$$= (\overline{\kappa(z, z')^{-1}}, 0, -\overline{\kappa_2((z, w), (z', w'))}),$$

$$\widetilde{\kappa}(\widetilde{g}(z, w), \widetilde{g}(z', w')) = \widetilde{J}(\widetilde{g}, (z, w))\widetilde{\kappa}((z, w), (z', w'))\overline{\widetilde{J}(\widetilde{g}, (z', w'))}^{-1}$$
$$= (J(g, z), 0, J_2(\widetilde{g}, (z, w)))(\kappa(z, z'), 0, \kappa_2((z, w), (z', w')))$$
$$\times (\overline{J(g, z')}^{-1}, 0, -\overline{J_2(\widetilde{g}, (z', w'))})$$
$$= (\kappa_1', 0, \kappa_2'),$$

where

$$\kappa_1' = J(g, z)\kappa(z, z')\overline{J(g, z')}^{-1},$$
$$\kappa_2' = J_2(\widetilde{g}, (z, w)) + \kappa_2((z, w), (z', w')) - \overline{J_2(\widetilde{g}, (z', w'))}.$$

In particular, we obtain

$$\kappa_2((z', w'), (z, w)) = -\overline{\kappa_2((z, w), (z', w'))},$$

$$\kappa_2(\widetilde{g}(z, w), \widetilde{g}(z', w')) = J_2(\widetilde{g}, (z, w)) + \kappa_2((z, w), (z', w')) - \overline{J_2(\widetilde{g}, (z', w'))}.$$

Thus it follows that

$$\mathfrak{K}((z', w'), (z, w)) = \mathbf{e}\left[-\overline{\kappa_2((z, w), (z', w'))}\right] = \overline{\mathfrak{K}((z, w), (z', w'))},$$

$$\mathfrak{K}(\widetilde{g}(z, w), \widetilde{g}(z', w')) = \mathbf{e}\left[J_2(\widetilde{g}, (z, w))\right]\mathbf{e}\left[\kappa_2((z, w), (z', w'))\right]\mathbf{e}\left[-\overline{J_2(\widetilde{g}, (z', w'))}\right]$$
$$= \mathcal{J}(\widetilde{g}, (z, w))\mathfrak{K}((z, w), (z', w'))\overline{\mathcal{J}(\widetilde{g}, (z', w'))};$$

hence the proof of the proposition is complete. $\qquad\square$

Lemma 7.10 *Let* $\mathfrak{L}((z, w), (z', w'))$ *be a* $\mathbb{C}$-*valued function on* $\widetilde{\mathcal{D}} \times \widetilde{\mathcal{D}}$ *that is holomorphic in* (z, w) *and satisfies* (7.25) *and* (7.26). *Then* $\mathfrak{L}$ *is a constant multiple of* $\mathfrak{K}$.

Proof. For $\widetilde{z} = (z, w), \widetilde{z}' = (z', w') \in \widetilde{\mathcal{D}}$ we set

$$\eta(\widetilde{z}, \widetilde{z}') = \mathfrak{L}(\widetilde{z}, \widetilde{z}')\mathfrak{K}(\widetilde{z}, \widetilde{z}')^{-1}.$$

Then, using (7.26), we obtain $\eta(\widetilde{g}\widetilde{z}, \widetilde{g}\widetilde{z}') = \eta(\widetilde{z}, \widetilde{z}')$ for all $\widetilde{g} \in \widetilde{G}$. Thus, if $\widetilde{z}_0 \in \widetilde{\mathcal{D}}$ is a base point, then we have

$$\eta(\widetilde{g}\widetilde{z}_0, \widetilde{z}_0) = \eta(\widetilde{z}_0, \widetilde{g}^{-1}\widetilde{z}_0)$$

for all $\widetilde{g} \in \widetilde{G}$. Since $\eta(\widetilde{z}, \widetilde{z}')$ is holomorphic in $\widetilde{z}$, by (7.25) it is antiholomorphic in $\widetilde{z}'$. Therefore, using the fact that $\widetilde{G}$ acts on $\widetilde{\mathcal{D}}$ transitively, we see that

$$\eta(\widetilde{g}\widetilde{z}_0, \widetilde{z}_0) = \eta(\widetilde{z}_0, \widetilde{g}^{-1}\widetilde{z}_0) = \eta(\widetilde{z}_0, \widetilde{z}_0)$$

for all $\widetilde{g} \in \widetilde{G}$. Thus, if $\widetilde{z}, \widetilde{z}' \in \widetilde{\mathcal{D}}$ with $\widetilde{z}' = \widetilde{g}'\widetilde{z}_0$, we obtain

$$\eta(\widetilde{z}, \widetilde{z}') = \eta((\widetilde{g}')^{-1}\widetilde{z}, \widetilde{z}_0) = \eta(\widetilde{z}_0, \widetilde{z}_0);$$

hence it follows that $\mathfrak{K}(\widetilde{z}, \widetilde{z}') = C\mathfrak{L}(\widetilde{z}, \widetilde{z}')$ with $C = \eta(\widetilde{z}_0, \widetilde{z}_0)$.   □

Given elements $z \in \mathcal{D}$ and $w, w' \in V_+$, we set

$$\mathfrak{K}_z(w, w') = \mathfrak{K}((z, w), (z, w')). \tag{7.27}$$

For each $z \in \mathcal{D}$ we denote by $\mathcal{F}_z$ the space of holomorphic functions ϕ on V_+ such that

$$\|\phi\|_z^2 = \int_{V_+} |\phi(w)|^2 \mathfrak{K}_z(w, w)^{-1} d_z w < \infty, \tag{7.28}$$

where $d_z w = \det(\operatorname{Im} \tau(z))^{-1} dw$. Thus $\mathcal{F}_z$ together with the inner product

$$\langle \phi, \psi \rangle_z = \int_{V_+} \phi(w)\overline{\psi(w)} \mathfrak{K}_z(w, w)^{-1} d_z w \tag{7.29}$$

is a Hilbert space.

For $\widetilde{g} = (g, v, t) \in \widetilde{G}$ and $\phi \in \mathcal{F}_{gz}$, we set

$$(T^{gz}(\widetilde{g}^{-1})\phi)(w) = \mathcal{J}(\widetilde{g}, (z, w))^{-1}\phi(\operatorname{pr}_2((\widetilde{g}(z, w))))$$

for all $(z, w) \in \mathcal{D} \times V_+$, where $\operatorname{pr}_2 : \mathcal{D} \times V_+ \to V_+$ is the natural projection map onto V_+; hence we have

$$\operatorname{pr}_2((\widetilde{g}(z, w))) = \operatorname{pr}_2(gz, v_{gz} + J_+^S(\rho(g), \tau(z))w)$$
$$= v_{gz} + J_+^S(\rho(g), \tau(z))w,$$

where we used (6.47).

Lemma 7.11 *For $\widetilde{g} = (g, v, t) \in \widetilde{G}$ and $\phi \in \mathcal{F}_{gz}$, we have $T^{gz}(\widetilde{g}^{-1})\phi \in \mathcal{F}_z$ and*

$$\|T^{gz}(\widetilde{g}^{-1})\phi\|_z = \|\phi\|_{gz}$$

for all $z \in \mathcal{D}$.

Proof. Let $z \in \mathcal{D}$, $\widetilde{g} = (g, v, t) \in \widetilde{G}$ and $\phi \in \mathcal{F}_{g(z)}$. Then we have

$$\|T^{gz}(\widetilde{g}^{-1})\phi\|_z^2$$
$$= \int_{V_+} |T^{gz}(\widetilde{g}^{-1})\phi(w)|^2 \mathfrak{K}_z(w, w)^{-1} d_z w$$
$$= \int_{V_+} |\mathcal{J}(\widetilde{g}, (z, w))^{-1}\phi(v_{gz} + J_+^S(\rho(g), \tau(z))w)|^2 \mathfrak{K}_z(w, w)^{-1} d_z w.$$

However, by Proposition 7.9, we have

$$\mathcal{K}_{gz}(v_{gz} + J^S_+(\rho(g), \tau(z))w, v_{gz} + J^S_+(\rho(g), \tau(z))w)$$
$$= \mathcal{J}(\tilde{g}, (z, w))\mathcal{K}_z(w, w)\overline{\mathcal{J}(\tilde{g}, (z, w))}$$
$$= |\mathcal{J}(\tilde{g}, (z, w))|^2 \mathcal{K}_z(w, w).$$

Furthermore, we have

$$d_{gz}(v_{gz} + J^S_+(\rho(g), \tau(z))w)$$
$$= \det(\operatorname{Im} \tau(gz))^{-1} d(v_{gz} + J^S_+(\rho(g), \tau(z))w)$$
$$= |J^S_+(\rho(g), \tau(z))|^2 \det(\tau(z))^{-1} d(v_{gz} + J^S_+(\rho(g), \tau(z))w),$$

noting that J^S_+ is the restriction of the canonical automorphy factor of $Sp(V, \beta)$ to V_+ as in (6.31). However, we have

$$d(v_{gz} + J^S_+(\rho(g), \tau(z))w) = |J^S_+(\rho(g), \tau(z))|^{-2},$$

which implies that $d_{gz}(v_{gz} + J^S_+(\rho(g), \tau(z))w) = d_z w$. Hence we have

$$\|T^{gz}(\tilde{g}^{-1})\phi\|^2_z = \int_{V_+} |\phi(v_{gz} + J^S_+(\rho(g), \tau(z))w)|^2$$
$$\times \mathcal{K}_{gz}(v_{gz} + J^S_+(\rho(g), \tau(z))w, v_{gz} + J^S_+(\rho(g), \tau(z))w)^{-1}$$
$$\times d_{gz}(v_{gz} + J^S_+(\rho(g), \tau(z))w)$$
$$= \int_{V_+} |\phi(w)|^2 \mathcal{K}_{gz}(w, w) d_{gz}w = \|\phi\|^2_{gz},$$

and therefore the lemma follows. □

By Lemma 7.11 we see that $T^{gz}(\tilde{g}^{-1})$ is an isometry of $\mathcal{F}_{gz}$ into $\mathcal{F}_z$, and therefore it follows that $T^z(\tilde{g})$ is an isometry of $\mathcal{F}_z$ into $\mathcal{F}_{gz}$, and for $\phi \in \mathcal{F}_z$ we have

$$(T^z(\tilde{g})\phi)(w) = (T^{g^{-1}(gz)}(\tilde{g})\phi)(w) \tag{7.30}$$
$$= \mathcal{J}(\tilde{g}^{-1}, (gz, w))^{-1}\phi(\operatorname{pr}_2(\tilde{g}^{-1}(gz, w)))$$

for all $\tilde{z} \in \mathcal{D}$ and $w \in V_+$.

Proposition 7.12 *For $\tilde{g} = (g, v, t), \tilde{g}' = (g', v', t') \in \tilde{G}$ and $\phi \in \mathcal{F}_{g(z)}$, we have*

$$T^{g'z}(\tilde{g}) \circ T^z(\tilde{g}') = T^z(\widetilde{gg}')$$

for all $z \in \mathcal{D}$.

Proof. Let $\widetilde{g} = (g, v, t), \widetilde{g}' = (g', v', t') \in \widetilde{G}$, $(z, w) \in \widetilde{\mathcal{D}} = \mathcal{D} \times V_+$ and $\phi \in \mathcal{F}_{g(z)}$. Then from (7.30) we obtain

$$(T^z(\widetilde{g}')\phi)(w) = \mathcal{J}(\widetilde{g}'^{-1}, (g'z, w))^{-1}\phi(\mathrm{pr}_2(\widetilde{g}'^{-1}(g'z, w))).$$

Applying (7.30) once again, we see that

$$
\begin{aligned}
(T^{g'z}(\widetilde{g}) \circ T^z(\widetilde{g}')(\phi))(w) = {} & \mathcal{J}(\widetilde{g}^{-1}, (gg'z, w))^{-1} \\
& \times \mathcal{J}(\widetilde{g}'^{-1}, (g'z, \mathrm{pr}_2(\widetilde{g}^{-1}(gg'z, w))))^{-1} \\
& \phi(\mathrm{pr}_2(\widetilde{g}'^{-1}(g'z, \mathrm{pr}_2(\widetilde{g}^{-1}(gg'z, w))))).
\end{aligned}
$$

On the other hand, we have

$$
\begin{aligned}
(T^z(\widetilde{g}\widetilde{g}')(\phi))(w) = {} & \mathcal{J}((\widetilde{g}\widetilde{g}')^{-1}, (gg'z, w))\phi(\mathrm{pr}_2((\widetilde{g}\widetilde{g}')^{-1}(gg'z, w))) \\
= {} & \mathcal{J}(\widetilde{g}'^{-1}, \widetilde{g}^{-1}(gg'z, w))\mathcal{J}(\widetilde{g}^{-1}, (gg'z, w)) \\
& \times \phi(\mathrm{pr}_2((\widetilde{g}\widetilde{g}')^{-1}(gg'z, w))).
\end{aligned}
$$

Since we have

$$
\begin{aligned}
\widetilde{g}^{-1}(gg'z, w) = {} & (g^{-1}gg'z, \mathrm{pr}_2(\widetilde{g}^{-1}(gg'z, w))) \\
= {} & (g'z, \mathrm{pr}_2(\widetilde{g}^{-1}(gg'z, w))),
\end{aligned}
$$

$$
\begin{aligned}
\mathrm{pr}_2(\widetilde{g}'^{-1}(g'z, \mathrm{pr}_2(\widetilde{g}^{-1}(gg'z, w)))) = {} & \mathrm{pr}_2((\widetilde{g}'^{-1}\widetilde{g}^{-1}(gg'z, w)) \\
= {} & \mathrm{pr}_2((\widetilde{g}\widetilde{g}')^{-1}(gg'z, w)),
\end{aligned}
$$

it follows that

$$(T^{g'z}(\widetilde{g}) \circ T^z(\widetilde{g}')(\phi))(w) = (T^z(\widetilde{g}\widetilde{g}')(\phi))(w),$$

and therefore the proposition follows.   □

Now we consider the subgroup $\{1\} \times V_+ \times \mathbb{R}$ of $\widetilde{G}$. We shall identify this subgroup with $\widetilde{V} = V_+ \times \mathbb{R}$. Then $\widetilde{V}$ is in fact a Heisenberg group because the restriction of the multiplication operation on $\widetilde{G}$ given by (6.33) to $\widetilde{V}$ gives us the usual multiplication operation on a Heisenberg group. For $\widetilde{u} = (u, t) \in \widetilde{V} \subset \widetilde{G}$ and $w \in V_+$ we set

$$\widetilde{u}w = \mathrm{pr}_2((\widetilde{u}(z, w)).$$

Then, using (6.47), we obtain

$$
\begin{aligned}
\widetilde{u}w = {} & \mathrm{pr}_2(((1, u, t)(z, w)) \\
= {} & u_{1z} + J_+^S(\rho(1), \tau(z))w = u_z + w.
\end{aligned}
\tag{7.31}
$$

Thus for $\widetilde{g} = \widetilde{u}$ the formula (7.30) reduces to

$$(T^z(\widetilde{u}^{-1})\phi)(w) = \mathcal{J}(\widetilde{u}, (z, w))^{-1}\phi(\widetilde{u}w) \tag{7.32}$$

for $\phi \in \mathcal{F}_z$, $z \in \mathcal{D}$ and $w \in V_+$, and $T^z(\widetilde{u}^{-1})$ is an isometry of $\mathcal{F}_z$ into itself.

Lemma 7.13 *For fixed* $z \in \mathcal{D}$ *the function* $\mathfrak{K}_z(w, w')$ *is holomorphic in* w, *and we have*

$$\mathfrak{K}_z(w', w) = \overline{\mathfrak{K}_z(w, w')}, \tag{7.33}$$

$$\mathfrak{K}_z(\widetilde{u}w, \widetilde{u}w') = J(\widetilde{u}, (z, w))\mathfrak{K}_z(w, w')\overline{J(\widetilde{u}, (z, w'))} \tag{7.34}$$

for all $w, w' \in V_+$ *and* $\widetilde{u} \in \widetilde{V}$.

Proof. Using (7.23), (7.24) and (7.27), we have

$$\mathfrak{K}_z(w, w') = \mathfrak{K}((z, w), (z, w')) = \mathbf{e}\left[\kappa_2((z, w), (z, w'))\right].$$

for $w, w' \in V_+$, where

$$\kappa_2((z, w), (z, w')) = \beta(\overline{w}', \rho(\kappa_1(z, z))^{-1}\tau(z)\overline{w}')/2 + \beta(\overline{w}', \rho(\kappa_1(z, z))^{-1}w)$$
$$+ \beta(\overline{\tau(z)}w, \rho(\kappa_1(z, z))^{-1}w)/2.$$

Thus κ_2 is holomorphic in w, and therefore $\mathfrak{K}_z(w, w')$ is holomorphic in w as well. Now (7.33) and (7.34) follows from the corresponding relations in Proposition 7.9. $\qquad\square$

Lemma 7.14 *Let* $\Psi(w, w')$ *is a function on* $V_+ \times V_+$ *that is holomorphic in* w *satisfying the conditions*

$$\Psi(w', w) = \overline{\Psi(w, w')}, \tag{7.35}$$

$$\Psi(\widetilde{u}w, \widetilde{u}w') = J(\widetilde{u}, (z, w))\Psi(w, w')\overline{J(\widetilde{u}, (z, w'))} \tag{7.36}$$

for all $w, w' \in V_+$ *and* $\widetilde{u} \in \widetilde{V}$. *Then* Ψ *is a constant multiple of* $\mathfrak{K}_z$.

Proof. This follows from Lemma 7.10. $\qquad\square$

For fixed $z \in \mathcal{D}$ the map $\phi \mapsto \phi(w)$, $\mathcal{F}_z \to \mathbb{C}$ is a continuous linear functional on $\mathcal{F}_z$, and therefore there exists an element $\xi_w^z \in \mathcal{F}_z$ such that

$$\phi(w) = \int_{V_+} \overline{\xi_w^z(w')}\phi(w')\mathfrak{K}_z(w', w')^{-1}d_z w' \tag{7.37}$$

for all $\phi \in \mathcal{F}_z$.

Lemma 7.15 *Given* $z \in \mathcal{D}$, *there is a nonzero constant* C *such that*

$$\overline{\xi_w^z(w')} = C\mathfrak{K}_z(w, w')$$

for all $w, w' \in V_+$.

Proof. For $z \in \mathcal{D}$ and $w, w' \in V_+$ we have

$$\xi_{w'}^z(w) = \int_{V_+} \overline{\xi_w^z(v)} \xi_{w'}^z(v) \mathfrak{K}_z(v,v)^{-1} d_z v,$$

which implies that

$$\overline{\xi_{w'}^z(w)} = \int_{V_+} \xi_w^z(v) \overline{\xi_{w'}^z(v)} \mathfrak{K}_z(v,v)^{-1} d_z v = \xi_w^z(w').$$

Let $\widetilde{u} = (u,t) \in \widetilde{V} \subset \widetilde{G}$ and $\phi \in \mathcal{F}_z$, so that we have

$$(T^z(\widetilde{u}^{-1})\phi)(w) = \mathcal{J}(\widetilde{u}, (z,w))^{-1}\phi(\widetilde{u}w).$$

Then we see that

$$\int_{V_+} \overline{\xi_w^z(v)} \mathcal{J}(\widetilde{u}, (z,v))^{-1}\phi(\widetilde{u}v)\mathfrak{K}_z(v,v)^{-1} d_z v$$

$$= \int_{V_+} \overline{\xi_w^z(v)}(T^z(\widetilde{u}^{-1})\phi)(v)\mathfrak{K}_z(v,v)^{-1} d_z v$$

$$= \mathcal{J}(\widetilde{u},(z,w))^{-1} \int_{V_+} \overline{\xi_{\widetilde{u}w}^z(v)}\phi(v)\mathfrak{K}_z(v,v)^{-1} d_z v$$

$$= \mathcal{J}(\widetilde{u},(z,w))^{-1} \int_{V_+} \overline{\xi_{\widetilde{u}w}^z(\widetilde{u}v)}\phi(\widetilde{u}v)\mathfrak{K}_z(\widetilde{u}v,\widetilde{u}v)^{-1} d_z v.$$

Thus we have

$$\mathcal{J}(\widetilde{u},(z,v))^{-1}\overline{\xi_w^z(v)}\mathfrak{K}_z(v,v)^{-1}$$
$$= \mathcal{J}(\widetilde{u},(z,w))^{-1}\overline{\xi_{\widetilde{u}w}^z(\widetilde{u}v)}\mathfrak{K}_z(\widetilde{u}v,\widetilde{u}v)^{-1}$$
$$= \mathcal{J}(\widetilde{u},(z,w))^{-1}\overline{\xi_{\widetilde{u}w}^z(\widetilde{u}v)}\mathcal{J}(\widetilde{u},(z,v))^{-1}\overline{\mathcal{J}(\widetilde{u},(z,v))}^{-1}\mathfrak{K}_z(v,v)^{-1}$$

for $v \in V_+$. Hence, replacing v with w', we see that

$$\overline{\xi_{\widetilde{u}w}^z(\widetilde{u}w')} = \mathcal{J}(\widetilde{u},(z,w))\overline{\xi_w^z(w')}\overline{\mathcal{J}(\widetilde{u},(z,w'))}.$$

Now the lemma follows by applying Lemma 7.14 to the function $(w,w') \mapsto \overline{\xi_w^z(w')}$. $\square$

Given an element $\widetilde{v}$ of the Heisenberg group $\widetilde{V} = V \times \mathbb{R}$, by (7.32) we obtain the isometry $T^z(\widetilde{v})$ of $\mathcal{F}_z$ into itself given by

$$(T^z(\widetilde{v})\phi)(w) = \mathcal{J}(\widetilde{v}^{-1},(z,w))^{-1}\phi(\widetilde{v}^{-1}w) \qquad (7.38)$$

for all $w \in V_+$. We now consider an operator on $\mathcal{F}_z$ associated to a function on $\widetilde{V}$. Let $\mathcal{L}(\widetilde{V})$ be the space of $\mathbb{C}$-valued continuous functions on $\widetilde{V}$ with compact support. For $F \in \mathcal{L}(\widetilde{V})$ we denote by $T^z(F)$ the operator on $\mathcal{F}_z$ defined by

$$T^z(F)\phi = \int_{\widetilde{V}} F(\widetilde{v})(T^z(\widetilde{v})\phi)d\widetilde{v}$$

for all $\phi \in \mathcal{F}_z$.

Lemma 7.16 *For $F \in \mathcal{L}(\widetilde{V})$ and $\phi \in \mathcal{F}_z$ we have*

$$(T^z(F)\phi)(w) = \int_{V_+} k_z(w, w')\phi(w')\mathfrak{K}_z(w', w')^{-1}d_z w',$$

where

$$k_z(w, w') = C \int_{\widetilde{V}} F(\widetilde{u})\overline{\mathcal{J}(\widetilde{u}, (z, w'))}^{-1}\mathfrak{K}_z(w, \widetilde{u}w')d\widetilde{u}$$

for all $w, w' \in V_+$.

Proof. Using (7.37) and Lemma 7.15, we have

$$(T^z(F)\phi)(w) = \int_{V_+} \mathfrak{K}_z(w, w')(T^z(F)\phi)(w')\mathfrak{K}_z(w', w')^{-1}d_z w'$$

for $F \in \mathcal{L}(\widetilde{V})$, $w \in V_+$ and some constant C. On the other hand, for $\widetilde{u} \in \widetilde{V}$ we have

$$(T^z(\widetilde{u})\phi)(w) = C \int_{V_+} \mathfrak{K}_z(w, v)(T^z(\widetilde{u})\phi)(v)\mathfrak{K}_z(v, v)^{-1}d_z v$$

$$= C \int_{V_+} \mathfrak{K}_z(w, v)\mathcal{J}(\widetilde{u}^{-1}, (z, v))^{-1})\phi(\widetilde{u}^{-1}v)\mathfrak{K}_z(v, v)^{-1}d_z v$$

$$= C \int_{V_+} \mathfrak{K}_z(w, \widetilde{u}w')\mathcal{J}(\widetilde{u}^{-1}, \widetilde{u}(z, w'))^{-1})$$

$$\times \phi(w')\mathfrak{K}_z(\widetilde{u}w', \widetilde{u}w')^{-1}d_z w'.$$

Thus, using the relations

$$\mathfrak{K}_z(\widetilde{u}w', \widetilde{u}w')^{-1} = \overline{\mathcal{J}(\widetilde{u}, (z, w'))}^{-1}\mathfrak{K}_z(w', \widetilde{u}w')^{-1}\mathcal{J}(\widetilde{u}, (z, w'))^{-1},$$

$$\mathcal{J}(\widetilde{u}^{-1}, \widetilde{u}(z, w'))\mathcal{J}(\widetilde{u}, (z, w')) = \mathcal{J}(\widetilde{u}^{-1}\widetilde{u}, (z, w')) = 1,$$

we obtain

$$(T^z(\widetilde{u})\phi)(w) = \int_{V_+} \mathfrak{K}_z(w, \widetilde{u}w')\overline{\mathcal{J}(\widetilde{u}, (z, w'))}^{-1}\mathfrak{K}_z(w', w')^{-1}\phi(w')d_z w'.$$

Hence we see that

$$(T^z(F)\phi)(w) = \int_{V_+} \left(\int_{\widetilde{V}} F(\widetilde{u})\overline{\mathcal{J}(\widetilde{u}, (z, w'))}^{-1}\mathfrak{K}_z(w, \widetilde{u}w')d\widetilde{u} \right)$$

$$\times \phi(w')\mathfrak{K}_z(w', w')^{-1}d_z w',$$

and therefore the lemma follows.   □

Theorem 7.17 *Let z be an element of the Hermitian symmetric domain $\mathcal{D}$ and let $\widetilde{V} \subset \widetilde{G}$ be the Heisenberg group associated to the real vector space V described above. Then the map $\widetilde{v} \mapsto T^z(\widetilde{v})$ given by (7.38) is an irreducible unitary representation of $\widetilde{V}$ on the space $\mathcal{F}_z$.*

Proof. By Proposition 7.12, for $\widetilde{v} = (v,t), \widetilde{v}' = (v',t') \in \widetilde{V}$, we have

$$T^z(\widetilde{v}) \circ T^z(\widetilde{v}') = T^z(\widetilde{v}\widetilde{v}')$$

for all $z \in D$. Furthermore, using Lemma 7.11, we see that $\|T^z(\widetilde{v})\phi\|_z = \|\phi\|_z$ for all $z \in \mathcal{D}$, $\widetilde{v} \in \widetilde{V}$ and $\phi \in \mathcal{F}_z$. Therefore the map $\widetilde{v} \mapsto T^z(\widetilde{v})$ determines a unitary representation of $\widetilde{V}$ on the space $\mathcal{F}_z$. It remains to show that T^z is irreducible. Using Lemma 7.16, we see that the image of $\mathcal{F}_z$ under T^z is dense in the ring of Hilbert-Schmidt operators on the space $\mathcal{L}^2_z(V_+)$ of square-integrable functions on V_+ with respect to the measure

$$d\mu = \mathfrak{K}_z(w,w)^{-1} d_z w = \mathfrak{K}_z(w,w)^{-1} \det(\operatorname{Im}\tau(z))^{-1} dw$$

for $w \in V_+$. This implies that the centralizer in $\operatorname{Aut}(\mathcal{L}^2_z(V_+))$ of the image group of $\widetilde{V}$ under T^z is the set $\mathbb{C}_1^\times$ of complex numbers of modulus 1. Indeed, each element λ of the centralizer commutes with every $T^z(\widetilde{u})$ for $\widetilde{u} \in \widetilde{V}$, and therefore with every $T^z(\phi)$ for $\phi \in \mathcal{F}_z$. By continuity λ commutes with every element of the Hilbert space of Hilbert-Schmidt operators on $\mathcal{L}^2_z(V_+)$. Let $\psi_1, \psi_2 \in \mathcal{L}^2_z(V_+)$, and let $\varXi$ be the Hilbert-Schmidt operator with kernel $k(w,w') = \psi_1(w)\overline{\psi_2(w')}$. Then we have $\varXi\lambda\psi = \lambda\varXi\psi$ for all $\psi \in L^2_z(V_+)$, which implies

$$\langle \psi, \psi_2 \rangle \lambda\psi_1 = \langle \lambda\psi, \psi_2 \rangle \psi_1.$$

Since ψ, ψ_1, ψ_2 are arbitrary, it follows that λ is a scalar; therefore the unitarity of λ shows the claim that $\lambda \in \mathbb{C}_1^\times$. Now let $\mathcal{F}_z^1$ be a $\widetilde{V}$ invariant subspace of $\mathcal{F}_z$ under T^z. Since T^z is unitary, there is an invariant subspace $\mathcal{F}_z^2$ such that $\mathcal{F}_z = \mathcal{F}_z^1 \oplus \mathcal{F}_z^1$. If $\varLambda$ is the scalar multiplication by $\lambda_1 \in \mathbb{C}_1^\times$ on $\mathcal{F}_z^1$ and $\lambda_2 \in \mathbb{C}_1^\times$ on $\mathcal{F}_z^2$ with $\lambda_1 \neq t_2$, then $\varLambda$ belongs to the centralizer of the image group of $\widetilde{V}$ under T^z. Hence we have $\mathcal{F}_z^2 = 0$, and therefor T^z is irreducible. $\square$

Remark 7.18 *If the Hermitian symmetric domain is the Siegel upper half space $\mathcal{H}_n$ and if ρ and τ are identity maps, the representation T^z given in Theorem 7.17 reduces to the usual Fock representation of the Heisenberg group $\widetilde{V}$ described in [107].*

7.3 Theta Functions

Let (τ, ρ) is the equivariant pair consisting of the homomorphism $\rho : G \to Sp(V, \beta)$ and the holomophic map $\tau : \mathcal{D} \to \mathcal{H}_n$ used for the construction of Kuga fiber varieties in Section 6.1. In this section we consider generalized theta functions on the Hermitian symmetric domain $\mathcal{D}$ which should reduce to usual theta functions on the Siegel upper half space $\mathcal{H}_n$ when ρ and τ are identity maps. We obtain a transformation formula for such a theta function, and show that certain types of such theta functions genenate some eigenspaces associated to the Fock representations described in Section 7.2.

We shall use the same notations as in the previous sections. Thus V is a real vector space of dimension $2n$ whose complexification is of the form $V_{\mathbb{C}} = V_+ + V_-$, and the underlying real vector space of each of V_+ and V_- is isomorphic to the real vector space V. Then there are n-dimensional subspaces V_1 and V_2 of V and an element $\alpha \in Sp(V, \beta)_{\mathbb{C}}$ such that

$$V = V_1 \oplus V_2, \quad \alpha(V_1) = V_+, \quad \alpha(V_2) = V_-.$$

Let L_0 be a lattice in V with $L = \alpha(L_0) \subset V_{\mathbb{C}}$ such that

$$\beta(L, L) \subset \mathbb{Z}, \quad L = L \cap V_+ + L \cap V_- \tag{7.39}$$

We set

$$L_+ = L \cap V_+, \quad L_- = L \cap V_-, \quad V_0 = \alpha(V) \subset V_{\mathbb{C}}.$$

Thus each element $m \in V_0$ can be written in the form $m = m_+ + m_- \in V_0$ with $m_+ = V_0 \cap V_+$ and $m_- = V_0 \cap V_-$.

Definition 7.19 *The theta function associated to $m \in V_0$ and the equivariant pair (τ, ρ) is the function $\theta_m : \mathcal{D} \times V_+ \to \mathbb{C}$ given by*

$$\theta_m(z, w) = \sum_{l_- \in L_-} \mathbf{e}\left(\beta(l_- + m_-, \tau(z)(l_- + m_-))/2 \right. \tag{7.40}$$

$$\left. + \beta(l_- + m_-, w + m_+)\right)$$

for all $(z, w) \in \mathcal{D} \times V_+$.

Example 7.20 *Let S be an $r \times r$ real symmetric positive definite matrix, and let $\tau : \mathcal{H}_k \to \mathcal{H}_{kr}$ be the Eichler embedding (see for example [25, Section II.4]) given by $\tau(Z) = S \otimes Z$ for all $Z \in \mathcal{H}_k$, where $\mathcal{H}_k$ is regarded as the set of $k \times k$ complex symmetric matrices with positive definite imaginary part. Let $\rho : Sp(k, \mathbb{R}) \to Sp(kr, \mathbb{R})$ be the homomorphism given by*

$$\rho \begin{pmatrix} A & B \\ C & D \end{pmatrix} = \begin{pmatrix} E \otimes A & S \otimes B \\ S^{-1} \otimes C & E \otimes D \end{pmatrix}, \quad \begin{pmatrix} A & B \\ C & D \end{pmatrix} \in Sp(k, \mathbb{R}),$$

where E is the $r \times r$ identity matrix. Then (τ, ρ) is an equivariant pair, and therefore (7.40) determines the associated theta function on $\mathcal{H}_k \times \mathbb{C}^{kr}$.

Example 7.21 *Let $\mathcal{H}^k$ be the product of k copies of the Poincaré upper half plane $\mathcal{H}$. We define the holomorphic map $\tau_0 : \mathcal{H}^h \to \mathcal{H}_k$ and the homomorphism $\rho_0 : Sp(1, \mathbb{R})^k \to Sp(k, \mathbb{R})$ as follows. Let $g = (g_1, \ldots, g_k)$ be an element of $Sp(1, \mathbb{R})^k$ with $g_i = \left(\begin{smallmatrix} a_i & b_i \\ c_i & d_i \end{smallmatrix}\right) \in Sp(1, \mathbb{R})$ for $1 \leq i \leq k$, and let $z = (z_1, \ldots, z_k) \in \mathcal{H}^k$. Then we set*

$$\tau_0(z) = z^*, \quad \rho_0(g) = \left(\begin{smallmatrix} a^* & b^* \\ c^* & d^* \end{smallmatrix}\right),$$

where $z^* = \mathrm{diag}\,(z_1, \ldots, z_k)$ is the $k \times k$ diagonal matrix and similarly for a^*, b^*, c^* and d^*. Let Ξ be an element of $Sp(k, \mathbb{R})^k$, and set

$$\tau(z) = \Xi\tau_0(z), \quad \rho(g) = \Xi\rho_0(g)\Xi^{-1}$$

for $z \in \mathcal{H}^k$ and $g \in Sp(k, \mathbb{R})^k$. Then (τ, ρ) is an equivariant pair, and (7.40) determines the associated theta function on $\mathcal{H}^k$. Such a function can be shown to be a Hilbert modular form under certain conditions if the results in [35] is used.

Lemma 7.22 Let $r = r_+ + r_-$ be an element of L with $r_+ \in L_+$ and $r_- \in L_-$. Then we have

$$\theta_m(z, w + r_+ + \tau(z)r_-)$$
$$= \mathbf{e}\left[-\beta(r_-, \tau(z)r_-)/2 - \beta(r_-, w + m_+)\right]\theta_{m+r}(z, w)$$

for all $(z, w) \in \mathcal{D} \times V_+$.

Proof. Given $(z, w) \in \mathcal{D} \times V_+$, we have

$$\beta(l_- + m_-, \tau(z)(l_- + m_-))/2 + \beta(l_- + m_-, w + r_+ + \tau(z)r_- + m_+)$$
$$= \beta(l_- + m_- + r_-, \tau(z)(l_- + m_- + r_-))/2$$
$$\qquad - \beta(r_-, \tau(z)(l_- + m_-))/2 - \beta(r_-, \tau(z)r_-)/2$$
$$\qquad\qquad - \beta(l_- + m_-, \tau(z)r_-)/2$$
$$\qquad + \beta(l_- + m_- + r_-, w + r_+ + m_+)$$
$$\qquad\qquad + \beta(l_- + m_-, \tau(z)r_-) - \beta(r_-, w + r_+ + m_+).$$

Since the matrix representation of $\tau(z) : V_{\mathbb{C}} \to V_{\mathbb{C}}$ is of the form $\left(\begin{smallmatrix} 0 & * \\ * & 0 \end{smallmatrix}\right)$ relative to the decomposition $V_{\mathbb{C}} = V_+ \oplus V_-$, we have $\tau(z)^{-1} = -{}^t\tau(z) = -\tau(z)$; hence we obtain

$$\beta(l_- + m_-, \tau(z)r_-) = \beta(\tau(z)^{-1}(l_- + m_-), r_-) = -\beta(\tau(z)(l_- + m_-), r_-).$$

Thus we see that

$$\beta(l_- + m_-, \tau(z)(l_- + m_-))/2 + \beta(l_- + m_-, w + r_+\tau(z)r_- + m_+)$$
$$= \beta(l_- + m_- + r_-, \tau(z)(l_- + m_- + r_-))/2$$
$$\qquad + \beta(l_- + m_- + r_-, w + r_+ + m_+)$$
$$\qquad\qquad - \beta(r_-, \tau(z)r_-)/2 - \beta(r_-, w + r_+ + m_+),$$

and therefore the lemma follows. $\qquad\qquad\qquad\qquad\qquad\qquad\square$

Given an element $l = l_+ + l_- \in L$ with $l_+ \in L_+$ and $l_- \in L_-$, we set

$$\psi_m(l) = \mathbf{e}\left[\beta(l_+, l_-)/2 + \beta(l_-, m_+) + \beta(m_-, l_+)\right]. \qquad (7.41)$$

Then ψ_m is a *quasi-character* of L in the sense that the map

$$l \mapsto \psi_m(l)\mathbf{e}\left[\beta(l_+, l_-)/2\right]$$

is a character of L. We also set $l_z = l_+ - \tau(z)l_- \in V_+$ for $z \in \mathcal{D}$ as in Section 6.1.

Theorem 7.23 *Let* $\mathcal{J} : \widetilde{G} \times \widetilde{\mathcal{D}} \to \mathbb{C}$ *be the automorphy factor given by* (6.52). *Then the theta function* θ_m *satisfies the relation*

$$\theta_m(z, w + l_z) = \psi_m(l)\mathcal{J}((1, l, 0), (z, w))\theta_m(z, w) \qquad (7.42)$$

for all $(z, w) \in \widetilde{\mathcal{D}} = \mathcal{D} \times V_+$ *and* $l \in L \subset V_0$.

Proof. Applying Lemma 7.22 for $(z, w) \in \mathcal{D} \times V_+$ and $r = l_+ - l_-$ with $r_+ = l_+ \in L_+$ and $r_- = -l_- \in L_-$, we have

$$\theta_m(z, w + l_z) = \mathbf{e}\left[-\beta(l_-, \tau(z)l_-)/2 + \beta(l_-, w + m_+)\right]\theta_{m+r}(z, w).$$

However, for $m + r = (m_+ + l_+) + (m_- - l_-)$, we have

$$\theta_{m+r}(z, w) = \sum_{k_- \in L_-} \mathbf{e}\left[\beta(k_- + m_- - l_-, \tau(z)(k_- + m_- - l_-))/2\right.$$
$$\left. + \beta(k_- + m_- - l_-, w + m_+ + l_+)\right]$$
$$= \sum_{k_- \in L_-} \mathbf{e}\left[\beta(k_- + m_-, \tau(z)(k_- + m_-))/2\right.$$
$$\left. + \beta(k_- + m_-, w + m_+ + l_+)\right]$$
$$= \sum_{k_- \in L_-} \mathbf{e}\left[\beta(k_- + m_-, \tau(z)(k_- + m_-))/2\right.$$
$$\left. + \beta(k_- + m_-, w + m_+) + \beta(k_-, l_+) + \beta(m_-, l_+)\right]$$
$$= \mathbf{e}\left[\beta(m_-, l_+)\right]\theta_m(z, w),$$

where we used the condition $\beta(L, L) \subset \mathbb{Z}$. Thus we obtain

$$\theta_m(z, w + l_z) = \mathbf{e}\left[-\beta(l_-, \tau(z)l_-)/2 + \beta(l_-, w + m_+) + \beta(m_-, l_+)\right]\theta_m(z, w).$$

Since $\beta = 0$ on $V_+ \times V_+$ and $V_- \times V_-$, we have

$$\mathcal{J}((1, l, 0), (z, w)) = \mathbf{e}\left[\beta(l, l_z)/2 + \beta(l, w)\right] = \mathbf{e}\left[\beta(l_-, l_z)/2 + \beta(l_-, w)\right]$$
$$= \mathbf{e}\left[\beta(l_-, l_+)/2 - \beta(l_-, \tau(z)l_-)/2 + \beta(l_-, w)\right].$$

Hence we see that

$$\psi(l)\mathcal{J}((1, l, 0), (z, w)) = \mathbf{e}\left[-\beta(l_-, \tau(z)l_-)/2 + \beta(l_-, w)\right.$$
$$\left. + \beta(l_-, m_+) + \beta(m_-, l_+)\right],$$

and therefore the proof of the theorem is complete. $\qquad\square$

Remark 7.24 *If we set $\widetilde{u} = (1, l, 0) \in \widetilde{G}$ and $\widetilde{z} = (z, w) \in \widetilde{\mathcal{D}}$, then (7.42) can be written in the form*

$$\theta_m(\widetilde{u}\widetilde{z}) = \mathcal{J}_\psi(\widetilde{u}, \widetilde{z})\theta_m(\widetilde{z}),$$

where $\mathcal{J}_\psi : (\Gamma \times L \times \{0\}) \times (\mathcal{D} \times V_+) \to \mathbb{C}$ is the automorphy factor given by

$$\mathcal{J}_\psi((\gamma, l, 0), (z, w)) = \psi(l)\mathcal{J}((\gamma, l, 0), (z, w))$$

for all $\gamma \in \Gamma$, $l \in L$ and $(z, w) \in \mathcal{D} \times V_+$.

Given an element $z \in \mathcal{D}$ and a quasi-character ψ of L, we denote by V_ψ^z the complex vector space consisting of all functions $f : V_+ \to \mathbb{C}$ satisfying the relation

$$T^z((1, l, 0))f = \psi(l)^{-1}f$$

for all $l \in L$, where T^z is the Fock representation in Section 7.2.

Proposition 7.25 *Let $z \in \mathcal{D}$, and let f be an element of V_ψ^z for some quasi-character ψ of L. Then $T_{(\gamma,0,0)}^z f$ is an element of $V_\psi^{\gamma z}$ for all $\gamma \in G$.*

Proof. Let $l \in L$, $\gamma \in G$ and $z \in \mathcal{D}$. Then we have

$$(1, \gamma l, 0)(\gamma, 0, 0) = (\gamma, 0, 0)(1, l, 0) = (\gamma, \gamma l, 0) \in \widetilde{G} = G \times V_0 \times \mathbb{R}.$$

Hence, using Proposition 7.12, we obtain

$$T^{\gamma z}((\gamma, 0, 0)) \circ T^z((1, l, 0)) = T^z((\gamma, 0, 0)) \circ T^z((1, l, 0)).$$

Thus for $f \in V_\psi^z$, we have

$$T^{\gamma z}((\gamma, 0, 0))(T^z((1, l, 0))f) = T^z((\gamma, 0, 0))(T^z((1, l, 0))f)$$
$$= \psi(l)^{-1}(T^z((\gamma, 0, 0))f),$$

and therefore we see that $T^z((1, l, 0))f \in V_\psi^{\gamma z}$. $\square$

Given $z \in \mathcal{D}$ and $m \in V_0$ we define the function $\theta_m^z : V_+ \to \mathbb{C}$ by

$$\theta_m^z(w) = \theta_m(z, w)$$

for all $w \in V_+$.

Proposition 7.26 *Let $z \in \mathcal{D}$, and set $\widehat{l} = (1, l, 0) \in \widetilde{G}$ for $l \in L$. Let $T^z(\widehat{l})$ be the associated operator on $\mathcal{F}_z$ given by (7.38), and let ψ_m be the quasi-character associated to $m \in V_0$ in (7.41). Then the function θ_m^z is an element of $V_{\psi_m}^z$.*

Proof. Given $z \in \mathcal{D}$, $m \in V_0$, $\widehat{l} = (1, l, 0) \in \widetilde{G}$ and $w \in V_{\mathbb{C}}$, we have

$$(T^z(\widehat{l})(\theta_m^z))(w) = \mathcal{J}(\widehat{l}^{-1}, (z, w))^{-1} \theta_m^z((\widehat{l}^{-1}(z, w))_w).$$

However, we have

$$\widehat{l}^{-1}(z, w) = (1, -l, 0)(z, w) = (z, w - l_z);$$

hence we obtain $(\widehat{l}^{-1}(z, w))_w = w - l_z$. Thus we see that

$$\begin{aligned}
(T^z(\widehat{l})(\theta_m^z))(w) &= \mathcal{J}(\widehat{l}^{-1}, (z, w))^{-1} \theta_m^z(w - l_z) \\
&= \mathcal{J}(\widehat{l}^{-1}, (z, w))^{-1} \psi_m(-l) \mathcal{J}(\widehat{l}^{-1}, (z, w)) \theta_m^z(w) \\
&= \psi_m(-l) \theta_m^z(w) = \psi_m(l)^{-1} \theta_m^z(w).
\end{aligned}$$

Thus the proposition follows. $\qquad\square$

Let L_+^* be the dual lattice of L_+ relative to β, that is,

$$L_+^* = \{v \in V_0 \mid \beta(L_+, v) \subset \mathbb{Z}\}.$$

Then by (7.39), we have $L_- \subset L_+^*$. Now we state the main theorem in this section which extends a result of Satake [107, Section 3] to the case of Hermitian symmetric domains.

Theorem 7.27 *Let Ω be the complete set of representatives L_+ modulo L_-. Then, for $z \in \mathcal{D}$ and $m \in V_0$, the set $\{\theta_{m+r}^z \mid r \in \Omega\}$ forms a basis of the complex vector space $V_{\psi_m}^z$.*

Proof. Since the set $\{\theta_{m+r}^z \mid r \in \Omega\}$ is obviously linearly independent over $\mathbb{C}$, it suffices to show that it spans the complex vector space $V_{\psi_m}^z$. Let $z \in \mathcal{D}$, $w \in V_0$ and $m = m_+ + m_-$ with $m_+ \in V_+$ and $m_- \in V_-$. Then for $l_+ \in L_+$ we have

$$\mathcal{J}((1, l_+, 0), (z, w)) = 1, \quad \psi(l_+) = \mathbf{e}\,[\beta(m_-, l_+)].$$

Thus, for $f \in V_{\psi_m}^z$, the relation $T_{(1, l_+, 0)}^z f = \psi_m(l_+)^{-1} f(w)$ reduces to

$$f(w - l_+) = \mathbf{e}\,[-\beta(m_-, l_+)] f(w).$$

Hence the function $f_e(w) = f(w) \mathbf{e}\,[-\beta(m_-, w)]$ satisfies the relation

$$\begin{aligned}
f_e(w - l_+) &= f(w - l_+) \mathbf{e}\,[-\beta(m_-, w) + \beta(m_-, l_+)] \\
&= f(w) \mathbf{e}\,[-\beta(m_-, l_+)] \mathbf{e}\,[-\beta(m_-, w)] \mathbf{e}\,[\beta(m_-, l_+)] \\
&= f_e(w).
\end{aligned}$$

Therefore we obtain a Fourier expansion of $f_e(w)$ of the form

$$f_e(w) = f(w) \mathbf{e}\,[-\beta(m_-, w)] = \sum_{r \in L_+^*} a(r) \mathbf{e}\,[\beta(r, w)],$$

which implies that

$$f(w) = \sum_{r \in L_+^*} a(r) \mathbf{e}\left[\beta(r + m_-, w)\right].$$

On the other hand, for $l_- \in L_-$, we have

$$f(w - \tau(z)l_-) = \mathbf{e}\left[-\beta(l_-, \tau(z)l_-)/2 + \beta(l_-, w) + \beta(l_-, m_+)\right] f(w).$$

By comparing the coefficients of $\mathbf{e}\left[\beta(r + m_-, w)\right]$ in the Fourier series of both sides of the above equation, we see that

$$a(r) \mathbf{e}\left[\beta(r + m_-, \tau(z)l_-)\right] = a(r - l_-) \mathbf{e}\left[-\beta(l_-, \tau(z)l_-)/2 + \beta(l_-, m_+)\right];$$

hence we have

$$a(r - l_-) = a(r) \mathbf{e}\left[\beta(r + m_-, \tau(z)l_-) + \beta(l_-, \tau(z)l_-)/2 - \beta(l_-, m_+)\right]$$

for each $r \in L_+^*$. Using this relation, we obtain

$$f(w) = \sum_{r \in \Omega} \sum_{l_- \in L_-} a(r - l_-) \mathbf{e}\left[\beta(r + m_- - l_-, w)\right]$$

$$= \sum_{r \in \Omega} \sum_{l_- \in L_-} a(r) \mathbf{e}\left[\beta(r + m_-, \tau(z)l_-) + \beta(l_-, \tau(z)l_-)/2\right.$$

$$\left. - \beta(l_-, m_+) + \beta(r + m_- - l_-, w)\right].$$

However, we have

$$\beta(r + m_-, \tau(z)l_-) + \beta(l_-, \tau(z)l_-)/2$$
$$= \beta(r + m_-, \tau(z)l_-)/2 + \beta(l_-, \tau(z)(r + m_-))/2 + \beta(l_-, \tau(z)l_-)/2$$
$$= \beta(r + m_-, \tau(z)l_-)/2 + \beta(l_-, \tau(z)(r + m_- + l_-))/2$$
$$= \beta(r + m_- + l_-, \tau(z)(r + m_- + l_-))/2 - \beta(r + m_-, \tau(z)(r + m_-))/2,$$

$$-\beta(l_-, m_+) + \beta(r + m_-, w) = \beta(r + m_- - l_-, w + m_+) + \beta(r + m_-, m_+).$$

Thus we see that

$$f(w) = \sum_{r \in \Omega} a(r) \mathbf{e}\left[\beta(r + m_- - l_-, \tau(z)(r + m_-))/2 - \beta(r + m_-, m_+)\right]$$

$$\times \sum_{l_- \in L_-} \mathbf{e}\left[\beta(r + m_- + l_-, \tau(z)(r + m_- + l_-))/2\right.$$

$$\left. + \beta(r + m_- - l_-, w + m_+)\right]$$

$$= \sum_{r \in \Omega} c(r, z) \theta_{m+r}^z(w),$$

where

$$c(r, z) = a(r) \mathbf{e}\left[\beta(r + m_-, \tau(z)(r + m_-))/2 - \beta(r + m_-, m_+)\right]$$

is a constant independent of w; hence the theorem follows.   □

7.4 Vector-Valued Jacobi Forms

In this section we extend the notion of Jacobi forms treated in Section 7.1, and discuss vector-valued Jacobi forms on symmetric domains in connection with vector-valued modular forms. Such Jacobi forms are related to torus bundles over Kuga fiber varieties.

Let $G = \mathbb{G}(\mathbb{R})$, $\mathcal{D} = G/K$, and the equivariant pair (τ, ρ) be as in Section 6.1. In particular, $\tau : \mathcal{D} \to \mathcal{H}_n$ is a holomorphic map that is equivariant with respect to the homomorphism $\rho : G \to Sp(V, \beta)$, where β is an alternating bilinear form on the real vector space V of dimension $2n$. We denote by $\mathrm{Alt}(V)$ the space of all alternating bilinear forms on V.

Definition 7.28 *A Hermitian structure on V is a pair (α, I) consisting of all elements $\alpha \in \mathrm{Alt}(V)$ and $I \in GL(V)$ with $I^2 = -1_V$ such that the bilinear map*

$$V \times V \to \mathbb{R}, \quad (v, v') \mapsto \alpha(v, Iv')$$

is symmetric and positive definite. We shall denote by $\mathrm{Herm}(V)$ the space of all Hermitian structures on V.

We $z_0 \in \mathcal{D}$ as before, and denote by $I_0 = I_{\tau(z_0)}$ the complex structure on V corresponding to the element $\tau(z_0) \in \mathcal{H}_n$. Then we see that $(\beta, I_0) \in \mathrm{Herm}(V)$, and the Siegel upper half space $\mathcal{H}_n$ of degree n can be identified with the space

$$\mathcal{H} = \mathcal{H}(V, \beta) = \{I \in GL(V) \mid (\beta, I) \in \mathrm{Herm}(V)\} \tag{7.43}$$

on which $Sp(V, \beta)$ acts by

$$(g, I) \mapsto gIg^{-1}$$

for all $g \in Sp(V, \beta)$ and $I \in \mathcal{H}(V, \beta)$. We set

$$U^* = \{\alpha \in \mathrm{Alt}(V) \mid \rho(G) \subset Sp(V, \alpha)\}.$$

Then U^* is a subspace of $\mathrm{Alt}(V)$ defined over $\mathbb{Q}$, and we have $\beta \in U^*$. Let $U = (U^*)^*$ be the dual space of U^*. Then we obtain an alternating bilinear map $A : V \times V \to U$ defined over $\mathbb{Q}$ by

$$A(v, v')(\alpha) = \alpha(v, v') \tag{7.44}$$

for all $\alpha \in U^*$ and $v, v' \in V$.

Following Satake (cf. [108, §III.5], [109]), we consider the *generalized Heisenberg group* $\mathbb{H}$ associated to A consisting of all elements of $V \times U$ together with a multiplication operation given by

$$(v, u) \cdot (v', u') = (v + v', u + u' - A(v, v')/2) \tag{7.45}$$

for all $(v, u), (v', u') \in V \times U$. Then the group G operates on $\mathbb{H}$ by

$$g \cdot (v, u) = (\rho(g)v, u) \tag{7.46}$$

for all $g \in G$ and $(v, u) \in \mathbb{H}$, and we can form the semidirect product $G \ltimes \mathbb{H}$ with respect to this operation. Thus $G \ltimes \mathbb{H}$ consists of the elements (g, v, u) of $G \times V \times U$ whose multiplication operation is given by

$$
\begin{aligned}
(g, v, u) \cdot (g', v', u') &= (gg', (v, u) \cdot (\rho(g)v', u')) \\
&= (gg', v + \rho(g)v', u + u' - A(v, \rho(g)v')/2).
\end{aligned}
\tag{7.47}
$$

If $I \in \mathcal{H} = \mathcal{H}(V, \beta)$, we extend the complex structure I on V linearly to the complexification $V_{\mathbb{C}} = V \otimes_{\mathbb{R}} \mathbb{C}$ of V, and set

$$V_+(I) = \{v \in V_{\mathbb{C}} \mid Iv = iv\}, \qquad V_-(I) = \{v \in V_{\mathbb{C}} \mid Iv = -iv\}.$$

When I is equal to I_0 considered above, we shall write $V_+ = V_+(I_0)$ and $V_- = V_-(I_0)$.

Lemma 7.29 *To each complex structure I on V there corresponds a unique complex linear map $\xi_I : V_- \to V_+$ satisfying*

$$V_-(I) = (1 + \xi_I)V_-. \tag{7.48}$$

Furthermore, the map $I \mapsto \xi_I$ determines a bijection between $\mathcal{H} = \mathcal{H}(V, \beta)$ and the set of $\mathbb{C}$-linear maps $\xi : V_- \to V_+$ such that $1 - \xi\bar{\xi}$ is positive definite and $\xi^t = \xi$, where the transpose is taken with respect to the bilinear map β.

Proof. This follows from [108, Lemma II.7.2]. $\qquad\square$

If ξ is an element of $\mathrm{Hom}_{\mathbb{C}}(V_-, V_+)$ in Lemma 7.29 corresponding to an element $I \in \mathcal{H}$, then we shall write $I = I_\xi$.

Lemma 7.30 *For each $\xi \in \mathrm{Hom}_{\mathbb{C}}(V_-, V_+)$ with $I_\xi \in \mathcal{H}$ the map*

$$\Xi_\xi : (V, I_\xi) \to V_+, \qquad v \mapsto v_+ - \xi v_-$$

determines an isomorphism of vector spaces over $\mathbb{C}$, where $v = v_+ + v_- \in V \subset V_{\mathbb{C}}$ with $v_\pm \in V_\pm$.

Proof. Given $\xi \in \mathrm{Hom}_{\mathbb{C}}(V_-, V_+)$ with $I_\xi \in \mathcal{H}$, the map Ξ_ξ is linear. Since $\dim_{\mathbb{C}} V = \dim_{\mathbb{C}} V_+$, it suffices to show that $\mathrm{Ker}\, \Xi_\xi = \{0\}$. Suppose that $v \in V$ satisfies

$$\Xi_\xi(v) = v_+ - \xi v_- = 0. \tag{7.49}$$

By (7.48) there exists $v' \in V_-(I_\xi)$ such that

$$v' = v_- + \xi v_-. \tag{7.50}$$

From (7.49) and (7.50) we see that $v' = v_+ + v_- = v \in V$. Since $V \cap V_-(I_\xi) = \{0\}$, we have $v' = v = 0$; hence it follows that $\mathrm{Ker}\, \Xi_\xi = \{0\}$ $\qquad\square$

By Lemma 7.29 we may identify the symmetric domain $\mathcal{H}$ in (7.43) with the set of elements $z \in \mathrm{Hom}_{\mathbb{C}}(V_-, V_+)$ with $z^t = z$ and $1 - z\bar{z} \gg 0$. Then the symplectic group $Sp(V, \beta)$ operates on $\mathcal{H}$ in (7.43) by

$$g(z) = (az + b)(cz + d)^{-1}$$

for all $z \in \mathcal{H}$ and

$$g = \begin{pmatrix} a & b \\ c & d \end{pmatrix} \in Sp(V, \beta);$$

here we wrote $g \in Sp(V, \beta)$ as a 2×2 block matrix with respect to the decomposition $V_{\mathbb{C}} = V_+ + V_-$. As was discussed in Section 6.2, the canonical automorphy factor J of $Sp(V, \beta)$ is the map on $Sp(V, \beta) \times \mathcal{H}$ with values in $GL(V_{\mathbb{C}})$ given by

$$J(g, z) = \begin{pmatrix} J_+(g, z) & 0 \\ 0 & J_-(g, z) \end{pmatrix} \tag{7.51}$$

for all $g \in Sp(V, \beta)$ and $z \in \mathcal{H}$, where

$$J_+(g, z) = a - g(z)c, \quad J_-(g, z) = cz + d. \tag{7.52}$$

If $\tau : \mathcal{D} \to \mathcal{H}$ is the holomorphic map equivariant with respect to the homomorphism $\rho : G \to Sp(V, \beta)$ as before and if A is as in (7.44), then we set

$$\begin{aligned}
\mathfrak{J}((g, r, s), (z, v)) = s &- A(\rho(g)r, J_+(\rho(g), \tau(z))r_z)/2 \\
&- A(\rho(g)v, J_+(\rho(g), \tau(z))v)/2 \\
&- A(\rho(g)r, J_+(\rho(g), \tau(z))v)
\end{aligned} \tag{7.53}$$

for $g \in G$, $(r, s) \in \mathbb{H}$ and $(z, v) \in \mathcal{D} \times V_+$, where

$$r_z = r_+ - zr_- \in V_+ \tag{7.54}$$

for $r = (r_+, r_-) \in V \subset V_{\mathbb{C}} = V_+ \oplus V_-$. Then it is known (see [108, §III.5]) that the group $G \ltimes \mathbb{H}$ operates on $\mathcal{D} \times V_+ \times U_{\mathbb{C}}$ by

$$(g, r, s) \cdot (z, v, u) = \big(gz, J_+(\rho(g), \tau(z))(v + r_z), u + \mathfrak{J}((g, r, s), (z, v))\big). \tag{7.55}$$

By restricting this action to $\mathcal{D} \times V_+$ we obtain the action of $G \ltimes \mathbb{H}$ on $\mathcal{D} \times V_+$ given by

$$(g, r, s) \cdot (z, v) = \big(gz, J_+(\rho(g), \tau(z))(v + r_z)\big)$$

for $g \in G$, $(r, s) \in \mathbb{H}$ and $(z, v) \in \mathcal{D} \times V_+$.

Proposition 7.31 *The map* $\mathfrak{J} : (G \ltimes \mathbb{H}) \times (\mathcal{D} \times V_+) \to U_{\mathbb{C}}$ *given by* (7.53) *satisfies*

$$\mathfrak{J}((g', r', s') \cdot (g, r, s), (z, v)) = \mathfrak{J}((g', r', s'), (g, r, s) \cdot (z, v)) + \mathfrak{J}((g, r, s), (z, v))$$

for all $(g', r', s'), (g, r, s) \in G \ltimes \mathbb{H}$ *and* $(z, v) \in \mathcal{D} \times V_+$.

Proof. Given $(g', r', s'), (g, r, s) \in G \ltimes \mathbb{H}$ and $(z, v, u) \in \mathcal{D} \times V_+ \times U_{\mathbb{C}}$, we set

$$(z_1, v_1, u_1) = (g, r, s) \cdot (z, v, u)$$
$$(z_2, v_2, u_2) = (g', r', s') \cdot (z_1, v_1, u_1)$$
$$(z_3, v_3, u_3) = ((g', r', s') \cdot (g, r, s)) \cdot (z, v, u).$$

Then by (7.55) we see that

$$u_1 = u + \mathfrak{J}((g, r, s), (z, v))$$
$$u_2 = u_1 + \mathfrak{J}((g', r', s'), (g, r, s) \cdot (z, v))$$
$$u_3 = u + \mathfrak{J}((g', r', s') \cdot (g, r, s), (z, v)).$$

Since $G \ltimes \mathbb{H}$ acts on $\mathcal{D} \times V_+ \times U_{\mathbb{C}}$, we have $u_2 = u_3$; hence the proposition follows. $\qquad\square$

Let $L_{\mathbb{H}}$ be an arithmetic subgroup of $\mathbb{H}$, and set

$$L = p_V(L_{\mathbb{H}}), \quad L_U = p_U(L_{\mathbb{H}}), \tag{7.56}$$

where $p_V : \mathbb{H} \to V$ and $p_U : \mathbb{H} \to U$ are the natural projection maps. Then L and L_U are lattices in V and U, respectively, and we have $L = L_{\mathbb{H}}/L_U$. Given elements $l, l' \in L$, we have $(l, 0), (l', 0) \in L_{\mathbb{H}}$; hence by (7.45) we see that

$$(l, 0) \cdot (l', 0) = (l + l', -A(l, l')/2) \in L_{\mathbb{H}}.$$

Since $(l + l', 0)^{-1} = (-l - l', 0) \in L_{\mathbb{H}}$, we have

$$(l + l', -A(l, l')/2) \cdot (-l - l', 0) = (0, -A(l, l')/2) \in L_{\mathbb{H}}.$$

Thus it follows that $A(L, L) \subset L_U$. Let γ be a torsion-free arithmetic subgroup of G. Using the isomorphism

$$\Gamma \ltimes L_{\mathbb{H}}/L_U \cong \Gamma \ltimes L,$$

we see that the action of $G \ltimes \mathbb{H}$ on $\mathcal{D} \times V_+ \times U_{\mathbb{C}}$ induces actions of the discrete groups $\Gamma \ltimes L_{\mathbb{H}}$, $\Gamma \ltimes L$ and Γ on the spaces $\mathcal{D} \times V_+ \times U_{\mathbb{C}}$, $\mathcal{D} \times V_+$ and $\mathcal{D}$, respectively. We denote the associated quotient spaces by

$$W = \Gamma \ltimes L_{\mathbb{H}} \backslash \mathcal{D} \times V_+ \times U_{\mathbb{C}}, \quad Y = \Gamma \ltimes L \backslash \mathcal{D} \times V_+, \quad X = \Gamma \backslash \mathcal{D}.$$

Then each of the spaces W, Y and X has a natural structure of a complex manifold, and there are natural projections

$$W \xrightarrow{\pi_1} Y \xrightarrow{\pi_2} X.$$

The complex manifold Y is the Kuga fiber variety in (6.21), and W is a *torus bundle* over Y whose fiber is isomorphic to the complex torus $U_{\mathbb{C}}/L_U$.

Let $K_{\mathbb{C}}^{\mathrm{Sp}}$ be the subgroup of $GL(V_{\mathbb{C}})$ given by

$$K_{\mathbb{C}}^{\mathrm{Sp}} = \left\{ \begin{pmatrix} p & 0 \\ 0 & q \end{pmatrix} \in GL(V_{\mathbb{C}}) \;\Big|\; p = (q^t)^{-1} \right\},$$

where the matrix is written with respect to the decomposition $V_{\mathbb{C}} = V_+ + V_-$. Then it is known that $K_{\mathbb{C}}^{\mathrm{Sp}}$ is the complexification of a maximal compact subgroup K^{Sp} of $Sp(V,\beta)$. Let J, J_+ and J_- be as in (7.51), and let $g = \begin{pmatrix} a & b \\ c & d \end{pmatrix} \in Sp(V,\beta)$ and $\zeta \in \mathcal{H}$. Since the matrix $g\zeta = (a\zeta + b)(c\zeta + d)^{-1} \in \mathcal{H}$ is symmetric, by (7.52) we obtain

$$\begin{aligned} J_+(g,\zeta) &= a - (a\zeta + b)(c\zeta + d)^{-1}c \\ &= a - (\zeta c^t + d^t)^{-1}(\zeta a^t + b^t)c \\ &= (\zeta c^t + d^t)^{-1}(\zeta(c^t a - a^t c) + d^t a - b^t c) \\ &= (\zeta c^t + d^t)^{-1} = (J_-(g,\zeta)^t)^{-1}. \end{aligned}$$

Hence it follows that $J_+(g,\zeta) \in K_{\mathbb{C}}^{\mathrm{Sp}}$.

Let $\sigma : K_{\mathbb{C}}^{\mathrm{Sp}} \to GL(\mathcal{Z})$ be a representation of $K_{\mathbb{C}}^{\mathrm{Sp}}$ in a finite-dimensional complex vector space $\mathcal{Z}$. Given a holomorphic map $f : \mathcal{D} \to \mathcal{Z}$, we set

$$(f \mid_\sigma \gamma)(z) = \sigma(J(\rho(\gamma), \tau(z)))^{-1} f(\gamma z) \tag{7.57}$$

for $\gamma \in G$ and $z \in \mathcal{D}$. Then it can be shown that

$$f \mid_\sigma \gamma \mid_\sigma \gamma' = f \mid_\sigma \gamma\gamma',$$

for $\gamma, \gamma' \in \Gamma$. Let $\Gamma \subset G$ be a torsion-free arithmetic subgroup as before.

Definition 7.32 *A holomorphic map $f : \mathcal{D} \to \mathcal{Z}$ is a modular form for Γ associated to σ if it satisfies*

$$f \mid_\sigma \gamma = f$$

for all $\gamma \in \Gamma$.

Let $\chi : U_{\mathbb{C}} \to \mathbb{C}^\times$ be a character of $U_{\mathbb{C}}$ with $\chi(s) = 1$ for all $s \in L_U$, where L_U is as in (7.56). Then by Proposition 7.31 we see that $\chi \circ \mathfrak{J} : (G \ltimes \mathbb{H}) \times (\mathcal{D} \times V_+) \to \mathbb{C}$ is an automorphy factor, that is, it satisfies

$$(\chi \circ \mathfrak{J})(\widetilde{g}\widetilde{g}', \widetilde{z}) = (\chi \circ \mathfrak{J})(\widetilde{g}, \widetilde{g}'\widetilde{z}) \cdot (\chi \circ \mathfrak{J})(\widetilde{g}', \widetilde{z})$$

for all $\widetilde{g}, \widetilde{g}' \in G \ltimes \mathbb{H}$ and $\widetilde{z} \in \mathcal{D} \times V_+$. Given a holomorphic map $F : \mathcal{D} \times V_+ \to \mathcal{Z}$, we set

$$(F \mid_{\sigma,\chi} (\gamma, r, s))(z, w) = \chi(-\mathfrak{J}((\gamma, r, s), (z, w))) \cdot \sigma(J(\rho(\gamma), \tau(z)))^{-1} \tag{7.58}$$
$$\times F(\gamma z, J_+(\rho(\gamma), \tau(z))(w + r_z))$$

for all $(z, w) \in \mathcal{D} \times V_+$ and $(\gamma, r, s) \in G \ltimes \mathbb{H}$, where r_z is as in (7.54). Using the fact that $\chi \circ \mathfrak{J}$ is an automorphy factor, we see that

$$F \mid_{\sigma,\chi} (\gamma, r, s) \mid_{\sigma,\chi} (\gamma', r', s') = F \mid_{\sigma,\chi} ((\gamma, r, s) \cdot (\gamma', r', s'))$$

for $\gamma, \gamma' \in G$ and $(r, s), (r', s') \in \mathbb{H}$.

Definition 7.33 *A holomorphic map* $F : \mathcal{D} \times V_+ \to \mathcal{Z}$ *is a Jacobi form for* $\Gamma \ltimes L_{\mathbb{H}}$ *associated to* σ *and* χ *if it satisfies*

$$F \mid_{\sigma,\chi} (\gamma, r, s) = F$$

for all $(\gamma, r, s) \in \Gamma \ltimes L_{\mathbb{H}}$.

We can obtain a family of modular forms on $\mathcal{D}$ parametrized by the rational points of $\mathbb{H}$ as is described in the next theorem.

Theorem 7.34 *Let* $F : \mathcal{D} \times V_+ \to \mathcal{Z}$ *be a Jacobi form for* $\Gamma \ltimes L_{\mathbb{H}}$ *associated to* σ *and* χ, *and let* $(r, s) \in \mathbb{H}_{\mathbb{Q}} = V_{\mathbb{Q}} \times U_{\mathbb{Q}}$ *with* $r = (r_+, r_-) \in V \subset V_{\mathbb{C}} = V_+ \oplus V_-$ *in the sense of Definition 7.33. If* $A : V \times V \to U$ *is the bilinear map in (7.44), we set*

$$f(z) = \chi(-A(r, zr_-)/2) \cdot F(z, r_z) \tag{7.59}$$

for all $z \in \mathcal{D}$. *Then* f *is a modular form for an arithmetic subgroup* $\Gamma' \subset \Gamma$ *of* G *associated to* σ *in the sense of Definition 7.32.*

Proof. Let ε be the identity element of G, and let $(r, s) \in \mathbb{H}_{\mathbb{Q}}$. Then, for each $z \in \mathcal{D}$, $J(\rho(\varepsilon), \tau(z))$ and $J_+(\rho(\varepsilon), \tau(z))$ are identity matrices, and in particular $\sigma(J(\rho(\varepsilon), \tau(z)))$ is the identity element in $GL(\mathcal{Z})$. Thus, using (7.53) and (7.58), we see that

$$\mathfrak{J}((\varepsilon, r, s), (z, 0)) = s - A(r, r_z)/2 - A(v, v)/2 - A(r, 0) = s - A(r, r_z)/2,$$

$$(F \mid_{\sigma,\chi} (\varepsilon, r, s))(z, 0) = \chi(-s + A(r, r_z)/2) F(z, r_z)$$

for all $z \in \mathcal{D}$. Hence by (7.59) we obtain

$$\begin{aligned} f(z) &= \chi(s - A(r, r_+ - zr_-)/2 - A(r, zr_-)/2)(F \mid_{\sigma,\chi} (\varepsilon, r, s))(z, 0) \\ &= \chi(s - A(r, r_+)/2)(F \mid_{\sigma,\chi} (\varepsilon, r, s))(z, 0). \end{aligned}$$

Thus it suffices to show that the function $F_{(r,s)} : \mathcal{D} \to \mathcal{Z}$ given by

$$F_{(r,s)}(z) = (F \mid_{\sigma,\chi} (\varepsilon, r, s))(z, 0)$$

is a modular form for an arithmetic subgroup $\Gamma' \subset \Gamma$ associated to σ. Given an element $\gamma \in \Gamma$, by (7.57) we have

$$\begin{aligned} (F_{(r,s)} \mid_\sigma \gamma)(z) &= \sigma(J(\rho(\gamma), \tau(z)))^{-1} F_{(r,s)}(\gamma z) \\ &= \sigma(J(\rho(\gamma), \tau(z)))^{-1}(F \mid_{\sigma,\chi} (\varepsilon, r, s))(\gamma z, 0) \\ &= ((F \mid_{\sigma,\chi} (\varepsilon, r, s)) \mid_{\sigma,\chi} (\gamma, 0, 0))(z, 0) \end{aligned}$$

for all $z \in \mathcal{D}$. However, by (7.47) we see that

$$(\varepsilon, r, s) \cdot (\gamma, 0, 0) = (\gamma, r, s) = (\gamma, 0, 0) \cdot (\varepsilon, \rho(\gamma)^{-1} r, s).$$

Using this and the fact that F is a Jacobi form for Γ, we obtain

$$(F_{(r,s)} \mid_\sigma \gamma)(z) = ((F \mid_{\sigma,\chi} (\gamma,0,0)) \mid_{\sigma,\chi} (\varepsilon, \rho(\gamma)^{-1}r, s))(z,0)$$
$$= (F \mid_{\sigma,\chi} (\varepsilon, \rho(\gamma)^{-1}r, s)(z,0)$$
$$= F_{\gamma^{-1}\cdot(r,s)}(z),$$

where $\gamma^{-1} \cdot (r,s) = (\rho(\gamma)^{-1}r, s)$ by (7.46). Let $\Gamma_{(r,s)}$ be the subgroup of Γ consisting of the elements $\gamma \in \Gamma$ satisfying

$$\gamma^{-1} \cdot (r,s) = (0, s_1) \cdot (r_2, s_2) \cdot (r, s)$$

with $s_1 \in L_U$ and $(r_2, s_2) \in L_{\mathbb{H}}$. Then $\Gamma_{(r,s)}$ is an arithmetic subgroup, and, if $\gamma \in \Gamma_{(r,s)}$, we have

$$(F_{(r,s)} \mid_\sigma \gamma)(z) = F_{\gamma^{-1}\cdot(r,s)}(z) = F_{(0,s_1)\cdot(r_2,s_2)\cdot(r,s)}(z)$$
$$= (F \mid_{\sigma,\chi} (\varepsilon,0,s_1) \mid_{\sigma,\chi} (\varepsilon,r_2,s_2) \mid_{\sigma,\chi} (\varepsilon,r,s))(z,0)$$
$$= (F \mid_{\sigma,\chi} (\varepsilon,r,s)(z,0) = F_{(r,s)}(z)$$

for all $z \in \mathcal{D}$. Hence the function $F_{(r,s)}$ is a modular form for the arithmetic subgroup $\Gamma_{(r,s)} \subset \Gamma$ associated to σ, and therefore the proof of the theorem is complete. $\square$

Twisted Torus Bundles

As was discussed in Chapter 6 equivariant holomorphic maps of Hermitian symmetric domains into Siegel upper half spaces can be used to construct Kuga fiber varieties, which can be regarded as complex torus bundles over locally symmetric spaces. In this chapter we extend the construction of such torus bundles using 2-cocycles of discrete subgroups of the semisimple Lie groups associated to the given symmetric domains (cf. [85, 87]).

Let G be a semisimple Lie group of Hermitian type, and let $\mathcal{D}$ be the associated symmetric domain, which can be identified with the quotient G/K of G by a maximal compact subgroup K. We assume that there are a homomorphism $\rho : G \to Sp(n, \mathbb{R})$ of Lie groups and a holomorphic map $\tau : \mathcal{D} \to \mathcal{H}_n$ that is equivariant with respect to ρ, where $\mathcal{H}_n$ is the Siegel upper half space of degree n. Let L be a lattice in $\mathbb{R}^{2n}$, and let Γ be a torsion-free discrete subgroup of G such that $\ell \cdot \rho(\gamma) \in L$ for all $\ell \in L$ and $\gamma \in \Gamma$, where we regarded elements of L as row vectors. If L denotes the lattice $\mathbb{Z}^{2n}$ in $\mathbb{Z}^{2n}$, as in Section 6.1, we may consider the semidirect product $\Gamma \ltimes L$ with multiplication given by

$$(\gamma_1, \ell_1) \cdot (\gamma_2, \ell_2) = (\gamma_1 \gamma_2, \ell_1 \rho(\gamma_2) + \ell_2) \tag{8.1}$$

for all $\gamma_1, \gamma_2 \in \Gamma$ and $\ell_1, \ell_2 \in L$. Then $\Gamma \ltimes L$ acts on $\mathcal{D} \times \mathbb{C}^n$ by

$$(\gamma, (\mu, \nu)) \cdot (z, w) = (\gamma z, (w + \mu \tau(z) + \nu)(C_\rho \tau(z) + D_\rho)^{-1}), \tag{8.2}$$

for all $(z, w) \in \mathcal{D} \times \mathbb{C}^n$, $(\mu, \nu) \in L \subset \mathbb{R}^n \times \mathbb{R}^n$ and $\gamma \in \Gamma$ with $\rho(\gamma) = \begin{pmatrix} A_\rho & B_\rho \\ C_\rho & D_\rho \end{pmatrix} \in Sp(n, \mathbb{R})$. If we denote the associated quotient space by

$$Y = \Gamma \ltimes L \backslash \mathcal{D} \times \mathbb{C}^n,$$

then the map $\pi_0 : Y \to X = \Gamma \backslash \mathcal{D}$ induced by the natural projection map $\mathcal{D} \times \mathbb{C}^n \to \mathcal{D}$ has the structure of a fiber bundle over the locally symmetric space X whose fibers are in fact polarized abelian varieties (see Section 6.1). The total space of such a bundle is called a Kuga fiber variety as was discussed in Chapter 6.

The torus bundle parametrized by $X = \Gamma \backslash \mathcal{D}$ described above can further be generalized if a 2-cocycle of Γ is used to modify the action of $\Gamma \ltimes L$ on $\mathcal{D} \times \mathbb{C}^n$. Indeed, given a 2-cocycle $\psi : \Gamma \times \Gamma \to L$, by replacing the multiplication operation (8.1) with

$$(\gamma_1, \ell_1) \cdot (\gamma_2, \ell_2) = (\gamma_1\gamma_2, \ell_1\rho(\gamma_2) + \ell_2 + \psi(\gamma_1, \gamma_2)), \qquad (8.3)$$

we obtain the generalized semidirect product $\Gamma \ltimes_\psi L$ of Γ and L. We denote by $\mathcal{A}(\mathcal{D}, \mathbb{C}^n)$ the space of $\mathbb{C}^n$-valued holomorphic functions on $\mathcal{D}$, and let ξ be a 1-cochain for the cohomology of Γ with coefficients in $\mathcal{A}(\mathcal{D}, \mathbb{C}^n)$ satisfying

$$\delta\xi(\gamma_1, \gamma_2)(z) = \psi(\gamma_1, \gamma_2)\begin{pmatrix} \tau(z) \\ 1 \end{pmatrix} \qquad (8.4)$$

for all $z \in \mathcal{D}$ and $\gamma_1, \gamma_2 \in \Gamma$, where δ is the coboundary operator on 1-cochains. Then an action of $\Gamma \ltimes_\psi L$ on $\mathcal{D} \times \mathbb{C}^n$ can be defined by replacing (8.2) with

$$(\gamma, (\mu, \nu)) \cdot (z, w) = (\gamma z, (w + \mu\tau(z) + \nu + \xi(\gamma)(z))(C_\rho\tau(z) + D_\rho)^{-1}).$$

If the quotient of $\mathcal{D} \times \mathbb{C}^n$ by $\Gamma \ltimes_\psi L$ with respect to this action is denoted by $Y_{\psi,\xi}$, the map $\pi : Y_{\psi,\xi} \to X = \Gamma\backslash\mathcal{D}$ induced by the natural projection $\mathcal{D} \times \mathbb{C}^n \to \mathcal{D}$ is a torus bundle over X which is called a twisted torus bundle.

In Section 8.1 we describe the multiplication operation (8.3) by introducing a 2-cocycle of Γ. The action given by (8.4) in terms of a 1-cochain is considered in Section 8.2, and this action is used in Section 8.3 to construct a complex torus bundle, called a twisted torus bundle, over a locally symmetric space. We also consider families of such torus bundles produced by different 2-cocycles and 1-cochains. In Section 8.4 we determine the cohomology $R^k\pi_*\mathcal{O}_{Y_{\psi,\xi}}$ along the fibers of $Y_{\psi,\xi}$ over X associated to the sheaf $\mathcal{O}_{Y_{\psi,\xi}}$ of holomorphic functions on $Y_{\psi,\xi}$ for $k = 0, 1$.

8.1 Two-Cocycles of Discrete Groups

As was discussed in Section 6.1, a Kuga fiber variety is obtained as a quotient by a semidirect product $\Gamma \ltimes L$ of a discrete subgroup Γ of a semisimple Lie group of Hermitian type and a lattice L. In this section we generalize the multiplication operation on $\Gamma \ltimes L$ by using a 2-cocycle of Γ.

Let G be a semisimple Lie group of Hermitian type, and let K be a maximal compact subgroup of G. Thus the quotient space $\mathcal{D} = G/K$ has the structure of a Hermitian symmetric domain. We consider an equivariant pair (τ, ρ) associated to a Kuga fiber variety as in Section 6.1. Thus $\tau : \mathcal{D} \to \mathcal{H}_n$ is a holomorphic map that is equivariant with respect to a homomorphism $\rho : G \to Sp(n, \mathbb{R})$ of Lie groups, where $Sp(n, \mathbb{R})$ and $\mathcal{H}_n$ are the symplectic group and the Siegel upper half space, respectively, of degree n. Recall that this means that the pair (τ, ρ) satisfies

$$\tau(gz) = \rho(g)\tau(z)$$

for all $z \in \mathcal{D}$ and $g \in G$.

Let L be a lattice in $\mathbb{R}^{2n}$. In this chapter we shall often consider L as a subgroup of $\mathbb{R}^n \times \mathbb{R}^n$ and write elements of L in the form (μ, ν), where $\mu, \nu \in \mathbb{R}^n$ are regarded as row vectors. Let Γ be a discrete subgroup of G such that $\ell \rho(\gamma) \in L$ for all $\ell \in L$ and $\gamma \in \Gamma$, where $\ell \rho(\gamma)$ is the matrix product of the row vector ℓ of $2n$ entries and the $2n \times 2n$ matrix $\rho(\gamma)$. Thus L has the structure of a right Γ-module, and therefore we can consider the cohomology $H^*(\Gamma, L)$ of the group Γ with coefficients in L. We denote by $\mathfrak{C}^k(\Gamma, L)$ and $\mathfrak{Z}^k(\Gamma, L)$ the spaces of the associated k-cochains and k-cocycles, respectively, and choose an element ψ of $\mathfrak{Z}^2(\Gamma, L)$. Thus ψ is a map $\psi : \Gamma \times \Gamma \to L$ satisfying

$$\psi(\gamma_1, \gamma_2)\rho(\gamma_3) + \psi(\gamma_1\gamma_2, \gamma_3) = \psi(\gamma_2, \gamma_3) + \psi(\gamma_1, \gamma_2\gamma_3) \tag{8.5}$$

$$\psi(\gamma, 1) = 0 = \psi(1, \gamma) \tag{8.6}$$

for all $\gamma_1, \gamma_2, \gamma_3, \gamma \in \Gamma$, where 1 is the identity element of Γ. We note that an element $\alpha \in \mathfrak{Z}^2(\Gamma, L)$ is a coboundary if $\alpha = \partial\beta$ for some $\beta \in \mathfrak{C}^1(\Gamma, L)$, where

$$\partial\beta(\gamma_1, \gamma_2) = \beta(\gamma_2) - \beta(\gamma_1\gamma_2) + \beta(\gamma_1)\rho(\gamma_2) \tag{8.7}$$

for all $\gamma_1, \gamma_2 \in \Gamma$.

We now consider the generalized semidirect product $\Gamma \ltimes_\psi L$ associated to ψ, which consists of the elements $(\gamma, (\mu, \nu))$ in $\Gamma \times L$ and is equipped with the multiplication operation defined by

$$(\gamma_1, (\mu_1, \nu_1)) \cdot (\gamma_2, (\mu_2, \nu_2)) = (\gamma_1\gamma_2, (\mu_1, \nu_1)\rho(\gamma_2) + (\mu_2, \nu_2) + \psi(\gamma_1, \gamma_2)) \tag{8.8}$$

for all $\gamma_1, \gamma_2 \in \Gamma$ and $(\mu_1, \nu_1), (\mu_2, \nu_2) \in L$.

Lemma 8.1 *The generalized semidirect product $\Gamma \ltimes_\psi L$ is a group with respect to the multiplication operation given by (8.8). The identity element is $(1, (0, 0))$, and the element*

$$(\gamma^{-1}, -(\mu, \nu)\rho(\gamma)^{-1} - \psi(\gamma, \gamma^{-1}))$$

is the inverse of $(\gamma, (\mu, \nu)) \in \Gamma \ltimes_\psi L$.

Proof. First, we shall show that the operation in (8.8) is associative. Let $\gamma_1, \gamma_2, \gamma_3 \in \Gamma$ and $(\mu_1, \nu_1), (\mu_2, \nu_2), (\mu_3, \nu_3) \in L$. Then by (8.8) we have

$$((\gamma_1, (\mu_1, \nu_1)) \cdot (\gamma_2, (\mu_2, \nu_2))) \cdot (\gamma_3, (\mu_3, \nu_3))$$
$$= (\gamma_1\gamma_2\gamma_3, (\mu_1, \nu_1)\rho(\gamma_2)\rho(\gamma_3) + (\mu_2, \nu_2)\rho(\gamma_3)$$
$$+ \psi(\gamma_1, \gamma_2)\rho(\gamma_3) + (\mu_3, \nu_3) + \psi(\gamma_1\gamma_2, \gamma_3)).$$

Similarly, we obtain

$$(\gamma_1, (\mu_1, \nu_1)) \cdot ((\gamma_2, (\mu_2, \nu_2)) \cdot (\gamma_3, (\mu_3, \nu_3)))$$
$$= (\gamma_1, (\mu_1, \nu_1)) \cdot (\gamma_2\gamma_3, (\mu_2, \nu_2)\rho(\gamma_3) + (\mu_3, \nu_3) + \psi(\gamma_2, \gamma_3))$$
$$= (\gamma_1\gamma_2\gamma_3, (\mu_1, \nu_1)\rho(\gamma_2)\rho(\gamma_3) + (\mu_2, \nu_2)\rho(\gamma_3) + (\mu_3, \nu_3)$$
$$+ \psi(\gamma_2, \gamma_3) + \psi(\gamma_1, \gamma_2\gamma_3)).$$

Thus the associativity follows from the cocycle condition (8.5). Given $(\gamma, (\mu, \nu)) \in \Gamma \ltimes_\psi L$, using (8.6), we have

$$(\gamma, (\mu, \nu)) \cdot (1, (0, 0)) = (\gamma, (\mu, \nu) + (0, 0) + \psi(\gamma, 1)) = (\gamma, (\mu, \nu)),$$
$$(1, (0, 0)) \cdot (\gamma, (\mu, \nu)) = (\gamma, (0, 0)\rho(\gamma) + (\mu, \nu) + \psi(1, \gamma)) = (\gamma, (\mu, \nu));$$

hence it follows that $(1, (0, 0))$ is the identity. As for the inverse, we have

$$(\gamma, (\mu, \nu)) \cdot (\gamma^{-1}, -(\mu, \nu)\rho(\gamma)^{-1} - \psi(\gamma, \gamma^{-1}))$$
$$= (1, (\mu, \nu)\rho(\gamma)^{-1} - (\mu, \nu)\rho(\gamma)^{-1} - \psi(\gamma, \gamma^{-1}) + \psi(\gamma, \gamma^{-1}))$$
$$= (1, (0, 0))$$
$$(\gamma^{-1}, -(\mu, \nu)\rho(\gamma)^{-1} - \psi(\gamma, \gamma^{-1})) \cdot (\gamma, (\mu, \nu))$$
$$= (1, -(\mu, \nu)\rho(\gamma)^{-1}\rho(\gamma) - \psi(\gamma, \gamma^{-1})\rho(\gamma) + (\mu, \nu) + \psi(\gamma^{-1}, \gamma))$$
$$= (1, \psi(\gamma^{-1}, \gamma) - \psi(\gamma, \gamma^{-1})\rho(\gamma)).$$

However, using (8.5) for $\gamma_1 = \gamma_3 = \gamma$ and $\gamma_2 = \gamma^{-1}$, we see that

$$\psi(\gamma^{-1}, \gamma) = \psi(\gamma, \gamma^{-1})\rho(\gamma),$$

and therefore it follows that $(\gamma^{-1}, -(\mu, \nu)\rho(\gamma)^{-1} - \psi(\gamma, \gamma^{-1}))$ is the inverse of $(\gamma, (\mu, \nu))$.   □

The group $\Gamma \ltimes_\psi L$ essentially depends on the cohomology class $[\psi] \in H^2(\Gamma, L)$ of ψ according to the next lemma.

Lemma 8.2 *Let $\psi, \psi' : \Gamma \times \Gamma \to L$ be 2-cocycles that are cohomologous, and let ϕ be an element of $\mathfrak{C}^1(\Gamma, L)$ such that*

$$\psi(\gamma_1, \gamma_2) = \psi'(\gamma_1, \gamma_2) + (\partial\phi)(\gamma_1, \gamma_2). \tag{8.9}$$

Then the map $\Phi : \Gamma \ltimes_\psi L \to \Gamma \ltimes_{\psi'} L$ defined by

$$\Phi(\gamma, (\mu, \nu)) = (\gamma, (\mu, \nu) + \phi(\gamma)) \tag{8.10}$$

for $\gamma \in \Gamma$ and $(\mu, \nu) \in L$ is an isomorphism.

Proof. Using (8.9), we have

$$\Phi(\gamma_1, (\mu_1, \nu_1)) \cdot \Phi(\gamma_2, (\mu_2, \nu_2))$$
$$= (\gamma_1\gamma_2, ((\mu_1, \nu_1) + \phi(\gamma_1))\rho(\gamma_2) + (\mu_2, \nu_2) + \phi(\gamma_2) + \psi'(\gamma_1, \gamma_2))$$
$$= (\gamma_1\gamma_2, (\mu_1, \nu_1)\rho(\gamma_2) + (\mu_2, \nu_2) + \psi(\gamma_1, \gamma_2) + \phi(\gamma_1\gamma_2))$$
$$= \Phi((\gamma_1, (\mu_1, \nu_1)) \cdot (\gamma_2, (\mu_2, \nu_2))).$$

for all $(\gamma_1, (\mu_1, \nu_1)), (\gamma_2, (\mu_2, \nu_2)) \in \Gamma \ltimes_\psi L$. Hence it follows that Φ is a homomorphism. If we set,

$$\Psi(\gamma', (\mu', \nu')) = (\gamma', (\mu', \nu') - \phi(\gamma)),$$

then we see easily that the map Ψ is an inverse of Φ, and therefore the lemma follows.   □

According to the following proposition, the number of cocycles ψ modulo the coboundaries is finite if the $\mathbb{R}$-rank of G is greater than two.

Proposition 8.3 *If the $\mathbb{R}$-rank of the semisimple Lie group of G is greater than two, then the cohomology group $H^2(\Gamma, L)$ is finite.*

Proof. From the short exact sequence

$$0 \to L \to V \to V/L \to 0$$

of Γ-modules, we obtain the long exact sequence

$$\cdots \to H^1(\Gamma, L) \to H^1(\Gamma, V) \to H^1(\Gamma, V/L)$$
$$\to H^2(\Gamma, L) \to H^2(\Gamma, V) \to H^2(\Gamma, V/L) \to \cdots$$

of cohomology groups of Γ. Since the $\mathbb{R}$-rank of G is greater than two, it follows from Proposition 6.4 in [16, §VII.6] that

$$H^i(\Gamma, V) = H^i(\Gamma \backslash D, \widetilde{V}) = 0$$

for $i = 1, 2$, where $\widetilde{V}$ is the local system on $\Gamma \backslash D$ defined by the representation ρ of Γ on V. Hence we have

$$H^1(\Gamma, V/L) \cong H^2(\Gamma, L).$$

However, since V/L is compact, $H^1(\Gamma, V/L)$ is also compact, while $H^2(\Gamma, L)$ is discrete. Thus it follows that $H^2(\Gamma, L)$ is finite.   □

8.2 One-Cochains Associated to 2-Cocycles

In order to introduce an action of the semidirect product associated to a 2-cocycle considered in Section 8.1, we need to introduce a certain 1-cochain of Γ. In this section we discuss some of the properties of such a cochain.

The symplectic group $Sp(n, \mathbb{R})$ acts on the Siegel upper half space $\mathcal{H}_n$ as usual by

$$g\zeta = (a\zeta + b)(c\zeta + d)^{-1} \tag{8.11}$$

for all $z \in \mathcal{H}_n$ and $g = \begin{pmatrix} a & b \\ c & d \end{pmatrix} \in Sp(n, \mathbb{R})$. For such $g \in Sp(n, \mathbb{R})$ and $\zeta \in \mathcal{H}_n$, we set

$$j(g, \zeta) = c\zeta + d.$$

Then the resulting map $j : Sp(n, \mathbb{R}) \times \mathcal{H}_n \to GL(n, \mathbb{C})$ satisfies

$$j(g'g, \zeta) = j(g', g\zeta)j(g, \zeta) \tag{8.12}$$

for all $\zeta \in \mathcal{H}_n$ and $g, g' \in Sp(n, \mathbb{R})$. Given $z \in \mathcal{D}$ and $\gamma \in \Gamma \subset G$, we set

$$j_{\rho,\tau}(\gamma, z) = j(\rho(\gamma), \tau(z)). \tag{8.13}$$

Using (8.12) and the fact that (ρ, τ) is an equivariant pair, we see that

$$j_{\rho,\tau}(\gamma'\gamma, z) = j_{\rho,\tau}(\gamma', \gamma z) j_{\rho,\tau}(\gamma, z) \tag{8.14}$$

for all $z \in \mathcal{D}$ and $\gamma, \gamma' \in \Gamma$. Thus $j_{\rho,\tau} : \Gamma \times \mathcal{D} \to GL(n, \mathbb{C})$ is an automorphy factor, and automorphic forms involving the determinant of such an automorphy factor have been studied in a number of papers (see e.g. [19] and [74]).

Let $\mathcal{A}(\mathcal{D}, \mathbb{C}^n)$ denote the space of $\mathbb{C}^n$-valued holomorphic functions on $\mathcal{D}$. Then $\mathcal{A}(\mathcal{D}, \mathbb{C}^n)$ has the structure of a double Γ-module by

$$(\gamma \cdot f)(z) = f(z), \quad (f \cdot \gamma)(z) = f(\gamma z) j_{\rho,\tau}(\gamma, z) \tag{8.15}$$

for all $\gamma \in \Gamma$ and $z \in \mathcal{D}$, where elements of $\mathbb{C}^n$ are considered as row vectors. Thus we can consider the cohomology of the group Γ with coefficients in $\mathcal{A}(\mathcal{D}, \mathbb{C}^n)$, where its group of k-cochains consists of all functions

$$\eta : \Gamma^k \to \mathcal{A}(\mathcal{D}, \mathbb{C}^n)$$

such that $\eta(\gamma_1, \dots, \gamma_k) = 0$ whenever at least one of the γ_i is 1. Then the coboundary operator

$$\delta : \mathfrak{C}^k(\Gamma, \mathcal{A}(\mathcal{D}, \mathbb{C}^n)) \to \mathfrak{C}^{k+1}(\Gamma, \mathcal{A}(\mathcal{D}, \mathbb{C}^n))$$

is given by

$$\delta\eta(\gamma_1, \dots, \gamma_{k+1}) = \gamma_1 \cdot \eta(\gamma_2, \dots, \gamma_{k+1})$$
$$+ \sum_{i=1}^{k}(-1)^i \eta(\gamma_1, \dots, \gamma_i\gamma_{i+1}, \dots, \gamma_{k+1}) + (-1)^{k+1}\eta(\gamma_1, \dots, \gamma_k) \cdot \gamma_{k+1}$$

for all $\eta \in \mathfrak{C}^k(\Gamma, \mathcal{A}(\mathcal{D}, \mathbb{C}^n))$ (see [34, Chapter 15]). In particular, for $k = 1$ we have

$$\delta\eta(\gamma_1, \gamma_2) = \gamma_1 \cdot \eta(\gamma_2) - \eta(\gamma_1\gamma_2) + \eta(\gamma_1) \cdot \gamma_2 \tag{8.16}$$

for all $\gamma_1, \gamma_2 \in \Gamma$, where the right and left actions of Γ are given by (8.15).

Given a 2-cocycle $\psi \in \mathfrak{Z}^2(\Gamma, L)$, we assume that there is an element ξ of $\mathfrak{C}^1(\Gamma, \mathcal{A}(\mathcal{D}, \mathbb{C}^n))$ satisfying

$$\delta\xi(\gamma_1, \gamma_2)(z) = \psi(\gamma_1, \gamma_2)\begin{pmatrix} \tau(z) \\ 1 \end{pmatrix} \tag{8.17}$$

for all $z \in \mathcal{D}$, where the right hand side is the matrix multiplication of the row vector $\psi(\gamma_1, \gamma_2) \in L \subset \mathbb{R}^n \times \mathbb{R}^n$ and the complex $2n \times n$ matrix $\begin{pmatrix} \tau(z) \\ 1 \end{pmatrix}$. If ψ' is another 2-cocycle that is cohomologous to ψ, the corresponding element of $\mathfrak{C}^1(\Gamma, \mathcal{A}(\mathcal{D}, \mathbb{C}^n))$ can be obtained as follows. Let $\psi' \in \mathfrak{Z}^2(\Gamma, L)$ satisfy

$$\psi = \psi' + \partial\phi \tag{8.18}$$

for some $\phi \in \mathcal{C}^1(\Gamma, L)$. Then we define the map $\xi' : \Gamma \to \mathcal{A}(\mathcal{D}, \mathbb{C}^n)$ by

$$\xi'(\gamma)(z) = \xi(\gamma)(z) - \phi(\gamma)\begin{pmatrix} \tau(z) \\ 1 \end{pmatrix} \tag{8.19}$$

for all $\gamma \in \Gamma$ and $z \in \mathcal{D}$.

Lemma 8.4 *If ψ' and ξ' are as in (8.18) and (8.19), then we have*

$$\delta\xi'(\gamma_1, \gamma_2)(z) = \psi'(\gamma_1, \gamma_2)\begin{pmatrix} \tau(z) \\ 1 \end{pmatrix}$$

for all $\gamma_1, \gamma_2 \in \Gamma$ and $z \in \mathcal{H}$.

Proof. For $\gamma_1, \gamma_2 \in \Gamma$ and $z \in \mathcal{H}$, using (8.16) and (8.19), we obtain

$$\delta\xi'(\gamma_1, \gamma_2)(z) = \xi'(\gamma_2)(z) - \xi'(\gamma_1\gamma_2)(z) + \xi'(\gamma_1)(\gamma_2 z)j_{\rho,\tau}(\gamma_2, z)$$

$$= \delta\xi(\gamma_1, \gamma_2)(z) - \phi(\gamma_2)\begin{pmatrix} \tau(z) \\ 1 \end{pmatrix}$$

$$+ \phi(\gamma_1\gamma_2)\begin{pmatrix} \tau(z) \\ 1 \end{pmatrix} - \phi(\gamma_1)\begin{pmatrix} \tau(\gamma_2 z) \\ 1 \end{pmatrix}j_{\rho,\tau}(\gamma_2, z).$$

However, we have

$$\begin{pmatrix} \tau(\gamma_2 z) \\ 1 \end{pmatrix}j_{\rho,\tau}(\gamma_2, z) = \begin{pmatrix} (a\tau(z) + b)(c\tau(z) + d)^{-1} \\ 1 \end{pmatrix}(c\tau(z) + d)$$

$$= \begin{pmatrix} a\tau(z) + b \\ c\tau(z) + d \end{pmatrix} = \rho(\gamma_2)\begin{pmatrix} \tau(z) \\ 1 \end{pmatrix}$$

if $\rho(\gamma_2) = \begin{pmatrix} a & b \\ c & d \end{pmatrix}$. Using this, (8.17) and (8.7), we see that

$$\delta\xi'(\gamma_1, \gamma_2)(z) = \psi(\gamma_1, \gamma_2)\begin{pmatrix} \tau(z) \\ 1 \end{pmatrix} - (\partial\phi)(\gamma_1, \gamma_2)\begin{pmatrix} \tau(z) \\ 1 \end{pmatrix}$$

$$= \psi'(\gamma_1, \gamma_2)\begin{pmatrix} \tau(z) \\ 1 \end{pmatrix},$$

and therefore the lemma follows.   □

Let $\psi \in 3^2(\Gamma, L)$ and $\xi \in \mathcal{C}^1(\Gamma, \mathcal{A}(\mathcal{D}, \mathbb{C}^n))$ be as in (8.17). Given elements $(\gamma, (\mu, \nu)) \in \Gamma \ltimes_\psi L$ and $(z, w) \in \mathcal{D} \times \mathbb{C}^n$, we set

$$(\gamma, (\mu, \nu)) \cdot (z, w) = (\gamma z, (w + \mu\tau(z) + \nu + \xi(\gamma)(z))j_{\rho,\tau}(\gamma, z)^{-1}), \tag{8.20}$$

where $j_{\rho,\tau} : \Gamma \times \mathcal{D} \to GL(n, \mathbb{C})$ is given by (8.13).

Lemma 8.5 *The operation given by (8.20) determines an action of the group* $\Gamma \ltimes_\psi L$ *on the space* $\mathcal{D} \times \mathbb{C}^n$.

Proof. Let $(\gamma, (\mu, \nu)), (\gamma', (\mu', \nu')) \in \Gamma \ltimes_\psi L$ and $(z, w) \in \mathcal{D} \times \mathbb{C}^n$. Then we have

$$
\begin{aligned}
(\gamma', &(\mu', \nu')) \cdot ((\gamma, (\mu, \nu)) \cdot (z, w)) \\
&= (\gamma'\gamma z, (\mu'\tau(\gamma z) + \nu' + (\mu\tau(z) + \nu + w + \xi(\gamma)(z))j_{\rho,\tau}(\gamma, z)^{-1} \\
&\qquad\qquad\qquad\qquad\qquad\qquad\qquad + \xi(\gamma')(\gamma z))j_{\rho,\tau}(\gamma', \gamma z)^{-1}) \\
&= (\gamma'\gamma z, (\mu'\tau(\gamma z) + \nu')j_{\rho,\tau}(\gamma', \gamma z)^{-1} \\
&\qquad\qquad\qquad + (\mu\tau(z) + \nu + w + \xi(\gamma)(z))j_{\rho,\tau}(\gamma'\gamma, z)^{-1} \\
&\qquad\qquad\qquad\qquad + \xi(\gamma')(\gamma z)j_{\rho,\tau}(\gamma', \gamma z)^{-1}),
\end{aligned}
$$

where we used the relation (8.14). Similarly, using (8.8) and (8.20) we have

$$
\begin{aligned}
((\gamma', &(\mu', \nu')) \cdot (\gamma, (\mu, \nu))) \cdot (z, w) \\
&= (\gamma'\gamma, (\mu', \nu')\rho(\gamma) + (\mu, \nu) + \psi(\gamma', \gamma)) \cdot (z, w) \\
&= (\gamma'\gamma z, ((\mu', \nu')\rho(\gamma)\begin{pmatrix} \tau(z) \\ 1 \end{pmatrix} + \mu\tau(z) + \nu \\
&\qquad\qquad + \psi(\gamma', \gamma)\begin{pmatrix} \tau(z) \\ 1 \end{pmatrix} + w + \xi(\gamma'\gamma)(z))j_{\rho,\tau}(\gamma'\gamma, z)^{-1}) \\
&= (\gamma'\gamma z, (\mu', \nu')\rho(\gamma)\begin{pmatrix} \tau(z) \\ 1 \end{pmatrix} j_{\rho,\tau}(\gamma'\gamma, z)^{-1} \\
&\qquad + (\mu\tau(z) + \nu + w)j_{\rho,\tau}(\gamma'\gamma, z)^{-1} + \psi(\gamma', \gamma)\begin{pmatrix} \tau(z) \\ 1 \end{pmatrix} j_{\rho,\tau}(\gamma'\gamma, z)^{-1} \\
&\qquad\qquad + (\xi(\gamma'\gamma)(z))j_{\rho,\tau}(\gamma'\gamma, z)^{-1}).
\end{aligned}
$$

However, if $\rho(\gamma) = \begin{pmatrix} a & b \\ c & d \end{pmatrix} \in Sp(n, \mathbb{R})$, we see that

$$
\begin{aligned}
\rho(\gamma)\begin{pmatrix} \tau(z) \\ 1 \end{pmatrix} &= \begin{pmatrix} a\tau(z) + b \\ c\tau(z) + d \end{pmatrix} = \begin{pmatrix} (a\tau(z) + b)(c\tau(z) + d)^{-1} \\ 1 \end{pmatrix}(c\tau(z) + d) \\
&= \begin{pmatrix} \tau(\gamma z) \\ 1 \end{pmatrix} j_{\rho,\tau}(\gamma, z).
\end{aligned}
$$

Hence, using this and (8.14), we obtain

$$
\begin{aligned}
(\mu'\tau(\gamma z) + \nu')j_{\rho,\tau}(\gamma', \gamma z)^{-1} &= (\mu', \nu')\begin{pmatrix} \tau(\gamma z) \\ 1 \end{pmatrix} j_{\rho,\tau}(\gamma', \gamma z)^{-1} \\
&= (\mu', \nu')\rho(\gamma)\begin{pmatrix} \tau(z) \\ 1 \end{pmatrix} j_{\rho,\tau}(\gamma, z)^{-1} j_{\rho,\tau}(\gamma', \gamma z)^{-1} \\
&= (\mu', \nu')\rho(\gamma)\begin{pmatrix} \tau(z) \\ 1 \end{pmatrix} j_{\rho,\tau}(\gamma'\gamma, z)^{-1}.
\end{aligned}
$$

On the other hand, using (8.15), (8.16) and (8.17), we have

$$\psi(\gamma',\gamma)\binom{\tau(z)}{1} = \delta\xi(\gamma',\gamma)(z)$$
$$= (\gamma' \cdot \xi)(\gamma)(z) - \xi(\gamma'\gamma)(z) + (\xi(\gamma') \cdot \gamma)(z)$$
$$= \xi(\gamma)(z) - \xi(\gamma'\gamma)(z) + \xi(\gamma')(\gamma z)j_{\rho,\tau}(\gamma,z).$$

Hence it follows that

$$(\gamma',(\mu',\nu')) \cdot ((\gamma,(\mu,\nu)) \cdot (z,w)) = ((\gamma',(\mu',\nu')) \cdot (\gamma,(\mu,\nu))) \cdot (z,w),$$

and therefore the proof of the lemma is complete.   □

Remark 8.6 *In [87, Lemma 3.3], there is another action of $\Gamma \ltimes_\psi L$ on $\mathcal{D} \times \mathbb{C}^n$ which apparently looks different from the one in Lemma 8.5. However, it can be shown that these two actions are equivalent. The equivalence of these actions in the case of trivial ψ and ξ can be found in Section 6.1, and similar arguments can be used to prove the equivalence in the general case.*

8.3 Families of Torus Bundles

By taking the quotient of the space $\mathcal{D} \times \mathbb{C}^n$ by the action of $\Gamma \ltimes_\psi L$ discussed in Section 8.2 we obtain a complex torus bundle over a locally symmetric space, which may be regarded as a generalized Kuga fiber variety. In this section we consider certain properties of families of such torus bundles.

We assume that the discrete subgroup $\Gamma \subset G$ does not contain elements of finite order, so that the quotient $X = \Gamma \backslash \mathcal{D}$ of $\mathcal{D}$ by the Γ-action given by (8.11) has the structure of a complex manifold, and set

$$Y_{\psi,\xi} = \Gamma \ltimes_\psi L \backslash \mathcal{D} \times \mathbb{C}^n, \tag{8.21}$$

where the quotient is taken with respect to the action in Lemma 8.5. Then the map $\pi : Y_{\psi,\xi} \to X$ induced by the natural projections $\mathcal{D} \times \mathbb{C}^n \to \mathcal{D}$ and $\Gamma \ltimes_\psi L \to \Gamma$ has the structure of a fiber bundle over X whose fiber over a point corresponding to $z \in \mathcal{D}$ is isomorphic to the complex torus

$$\mathbb{C}^n \Big/ \left(L \cdot \binom{\tau(z)}{1} \right).$$

If $\psi = 0$ and $\xi = 0$, then the the corresponding torus bundle $Y_{0,0}$ is a family of abelian varieties known as a Kuga fiber variety (cf. [61, 108]), which was discussed in Section 6.1.

Proposition 8.7 *Given $\psi \in \mathfrak{Z}^2(\Gamma, L)$ and $\xi \in \mathfrak{C}^1(\Gamma, \mathcal{A}(\mathcal{D}, \mathbb{C}^n))$, let ψ' and ξ' be as in (8.18) and (8.19). Then the map $\Phi : \Gamma \ltimes_\psi L \to \Gamma \ltimes_{\psi'} L$ given by (8.10) and the identity map on $\mathcal{D} \times \mathbb{C}^n$ induce an isomorphism $Y_{\psi,\xi} \to Y_{\psi',\xi'}$ of bundles over $X = \Gamma \backslash \mathcal{D}$.*

Proof. It suffices to show that

$$\Phi(\gamma,(\mu,\nu))\cdot(z,w) = (\gamma,(\mu,\nu))\cdot(z,w),$$

where the actions on the right and left hand sides are with respect to (ψ,ξ) and (ψ',ξ'), respectively. Indeed, we have

$$
\begin{aligned}
\Phi(\gamma,(\mu,\nu))\cdot(z,w) &= \left(\gamma z, (w+(\mu,\nu)\begin{pmatrix}\tau(z)\\1\end{pmatrix} + \phi(\gamma)\begin{pmatrix}\tau(z)\\1\end{pmatrix}\right.\\
&\qquad\qquad \left.+\,\xi'(\gamma)(z))j_{\rho,\tau}(\gamma,z)^{-1}\right)\\
&= \left(\gamma z, (w+(\mu,\nu)\begin{pmatrix}\tau(z)\\1\end{pmatrix} + \xi(\gamma)(z))j_{\rho,\tau}(\gamma,z)^{-1}\right)\\
&= (\gamma,(\mu,\nu))\cdot(z,w).
\end{aligned}
$$

and therefore the proposition follows.   □

Given a 2-cocycle $\psi : \Gamma \times \Gamma \to L$, we denote by Ξ_ψ the set of all $\xi \in \mathfrak{C}^1(\Gamma,\mathcal{A}(\mathcal{D},\mathbb{C}^n))$ satisfying (8.17). Thus, if $\psi = 0$, the set Ξ_0 coincides with the space

$$\mathfrak{Z}^1(\Gamma,\mathcal{A}(\mathcal{D},\mathbb{C}^n)) = \{\eta \in \mathfrak{C}^1(\Gamma,\mathcal{A}(\mathcal{D},\mathbb{C}^n)) \mid \delta\eta = 0\}$$

of 1-cocycles in $\mathfrak{C}^1(\Gamma,\mathcal{A}(\mathcal{D},\mathbb{C}^n))$, where $\delta\eta$ is as in (8.16). Each $\xi \in \Xi_\psi$ determines the associated torus bundle $Y_{\psi,\xi}$ over X given by (8.21). We denote by

$$\mathfrak{T}_\psi = \{Y_{\psi,\xi} \mid \xi \in \Xi_\psi\}$$

the family of torus bundles $Y_{\psi,\xi}$ parametrized by Ξ_ψ. Thus, if ψ is the zero map, the torus bundle $Y_{0,0}$ determined by $0 \in \Xi_0$ is a Kuga fiber variety. Given $\xi \in \Xi_\psi$ and $\xi' \in \Xi_{\psi'}$, if δ is the coboundary operator on $\mathfrak{C}^1(\Gamma,\mathcal{A}(\mathcal{D},\mathbb{C}^n))$, then by (8.17) we have

$$\delta(\xi+\xi')(\gamma_1,\gamma_2)(z) = (\psi+\psi')(\gamma_1,\gamma_2)\begin{pmatrix}\tau(z)\\1\end{pmatrix}$$

for all $z \in \mathcal{H}_n$ and $\gamma_1,\gamma_2 \in \Gamma$; hence we see that $\xi+\xi' \in \Xi_{\psi+\psi'}$.

Let $(\mathcal{D}\times\mathbb{C}^n)\oplus_{\mathcal{D}}(\mathcal{D}\times\mathbb{C}^n)$ be the Whitney sum of two copies of the trivial vector bundle $\mathcal{D}\times\mathbb{C}^n$ over $\mathcal{D}$, which we identify with $\mathcal{D}\times(\mathbb{C}^n\oplus\mathbb{C}^n)$. Then we can consider the map

$$s : \mathcal{D}\times(\mathbb{C}^n\oplus\mathbb{C}^n) \to \mathcal{D}\times\mathbb{C}^n$$

given by

$$s(z,v,v') = (z,v+v') \tag{8.22}$$

for all $z \in \mathcal{D}$ and $v,v' \in \mathbb{C}^n$. Let $\psi,\psi' \in \mathfrak{Z}^2(\Gamma,L)$, and let $\Gamma \ltimes_\psi L \ltimes_{\psi'} L$ be the group consisting of the elements of $\Gamma \times L \times L$ equipped with multiplication given by

$$(\gamma_1, (\mu_1, \nu_1), (\mu_1', \nu_1')) \cdot (\gamma_2, (\mu_2, \nu_2), (\mu_2', \nu_2'))$$
$$= (\gamma_1\gamma_2, (\mu_1, \nu_1)\rho(\gamma_2) + (\mu_2, \nu_2) + \psi(\gamma_1, \gamma_2),$$
$$(\mu_1', \nu_1')\rho(\gamma_2) + (\mu_2', \nu_2') + \psi'(\gamma_1, \gamma_2)).$$

Then we see that there is a group homomorphism

$$\widetilde{s} : \Gamma \ltimes_\psi L \ltimes_{\psi'} L \to \Gamma \ltimes_{\psi+\psi'} L$$

given by

$$\widetilde{s}(\gamma, \ell_1, \ell_2) = (\gamma, \ell_1 + \ell_2) \tag{8.23}$$

for all $\gamma \in \Gamma$ and $\ell_1, \ell_2 \in L$. If $\xi \in \Xi_\psi$ and $\xi' \in \Xi_{\psi'}$, we let the group $\Gamma \ltimes_\psi L \ltimes_{\psi'} L$ act on the space $\mathcal{D} \times (\mathbb{C}^n \oplus \mathbb{C}^n)$ by

$$(\gamma, (\mu, \nu), (\mu', \nu')) \cdot (z, w, w') \tag{8.24}$$
$$= (\gamma z, (\mu\tau(z) + \nu + w + \xi(\gamma)(z)) \cdot j_{\rho,\tau}(\gamma, z)^{-1},$$
$$(\mu'\tau(z) + \nu' + w' + \xi'(\gamma)(z)) \cdot j_{\rho,\tau}(\gamma, z)^{-1}).$$

Then the associated quotient space

$$Y_{\psi,\xi} \oplus_X Y_{\psi',\xi'} = \Gamma \ltimes_\psi L \ltimes_{\psi'} L \backslash \mathcal{D} \times (\mathbb{C}^n \oplus \mathbb{C}^n)$$

is the fiber product of the torus bundles $Y_{\psi,\xi} \in \mathfrak{T}_\psi$ and $Y_{\psi',\xi'} \in \mathfrak{T}_{\psi'}$ over X.

Proposition 8.8 *Let* $\xi \in \Xi_\psi$ *and* $\xi' \in \Xi_{\psi'}$ *with* $\psi, \psi' \in \mathfrak{Z}^1(\Gamma, L)$. *Then the map* s *in* (8.22) *and the morphism* $\widetilde{s}$ *in* (8.23) *induce a morphism*

$$Y_{\psi,\xi} \oplus_X Y_{\psi',\xi'} \to Y_{\psi+\psi',\xi+\xi'}$$

of torus bundles over X.

Proof. By our construction of the torus bundles involved, it suffices to show that

$$s((\gamma, (\mu, \nu), (\mu', \nu')) \cdot (z, w, w')) = \widetilde{s}(\gamma, (\mu, \nu) + (\mu', \nu')) \cdot s(z, w, w')$$

for $\gamma \in \Gamma$, $(\mu, \nu), (\mu', \nu') \in L$ and $(z, w, w') \in \mathcal{D} \times (\mathbb{C}^n \oplus \mathbb{C}^n)$. Indeed, using (8.24) and (8.23), we see that

$$s((\gamma, (\mu, \nu), (\mu', \nu')) \cdot (z, w, w'))$$
$$= (\gamma z, ((\mu + \mu')\tau(z) + (\nu + \nu')$$
$$+ (w + w') + (\xi + \eta)(\gamma)(z)) \cdot j_{\rho,\tau}(\gamma, z)^{-1})$$
$$= \widetilde{s}(\gamma, (\mu, \nu) + (\mu', \nu')) \cdot s(z, w, w'),$$

and therefore the proposition follows. $\qquad\square$

Applying Proposition 8.8 to the special case of $\psi' = 0$, we see that there is a natural morphism

$$Y_{\psi,\xi} \oplus_X Y_{0,\eta} \to Y_{\psi,\xi+\eta}$$

for $Y_{\psi,\xi} \in \mathfrak{T}_\psi$ and $\eta \in \Xi_0$.

Example 8.9 *Given an element $h \in \mathcal{A}(\mathcal{D}, \mathbb{C}^n)$, we define the 1-cochain $\eta \in \mathfrak{C}^1(\Gamma, \mathcal{A}(\mathcal{D}, \mathbb{C}^n))$ by*

$$\eta(\gamma)(z) = h(z) - h(\gamma z) j_{\rho,\tau}(\gamma, z)$$

for all $z \in \mathcal{D}$ and $\gamma \in \Gamma$. Then for $z \in \mathcal{D}$ and $\gamma', \gamma \in \Gamma$ we have

$$\begin{aligned}
\eta(\gamma)(z) &- \eta(\gamma'\gamma)(z) + \eta(\gamma')(\gamma z) j_{\rho,\tau}(\gamma, z) \\
&= (h(z) - h(\gamma z) j_{\rho,\tau}(\gamma, z)) - (h(z) - h(\gamma'\gamma z) j_{\rho,\tau}(\gamma'\gamma, z)) \\
&\quad + (h(\gamma z) - h(\gamma'\gamma z) j_{\rho,\tau}(\gamma', z)) j_{\rho,\tau}(\gamma, z) = 0;
\end{aligned}$$

hence it follows that $\eta \in \mathfrak{Z}^1(\Gamma, \mathcal{A}(\mathcal{D}, \mathbb{C}^n))$. In fact, η is a coboundary. Thus we can consider the associated torus bundle $Y_h = Y_{0,\eta}$ and a morphism $Y_{\psi,\xi} \oplus_X Y_h \to Y_{\psi,\xi+\eta}$ for each $Y_{\psi,\xi} \in \mathfrak{T}_\psi$.

We are now interested in extending the interpretation of holomorphic forms on Kuga fiber varieties as mixed automorphic forms given in Theorem 6.17 to the torus bundle $Y_{\psi,\xi}$. Let $j_H : G \times \mathcal{D} \to \mathbb{C}$ and $j_V : Sp(V, \beta) \times \mathcal{H}_n \to \mathbb{C}$ be canonical automorphy factors given by (6.55) and (6.56), respectively. Then from (6.56) and (8.13) we see that

$$\det(j_{\rho,\tau}(\gamma, z)) = j_V(\rho(\gamma), \tau(z))$$

for all $z \in \mathcal{D}$ and $\gamma \in G$.

Theorem 8.10 *Let $\pi : Y_{\psi,\xi}^m \to X$ be the m-fold fiber power of the torus bundle $Y_{\psi,\xi}$ over X in (8.21). Then the space of mixed automorphic forms on $\mathcal{D}$ of type $(j_H^{-1}, j_V^m, \rho, \tau)$ is canonically isomorphic to the space $H^0(Y_{\psi,\xi}^m, \Omega^{k+mn})$ of holomorphic $(k + mn)$-forms on $Y_{\psi,\xi}^m$, where $k = \dim_{\mathbb{C}} \mathcal{D}$.*

Proof. Using (8.21), we see that $Y_{\psi,\xi}^m$ can be regarded as the quotient

$$\Gamma \ltimes_\rho L^n \backslash \mathcal{D} \times (\mathbb{C}^n)^m$$

fibered over the locally symmetric space $X = \Gamma \backslash \mathcal{D}$ with fiber $(\mathbb{C}^n/L)^m$. Let $z = (z_1, \ldots, z_k)$ be a global coordinates for $\mathcal{D}$, and let

$$w = (w^{(1)}, \ldots, w^{(m)}) = (w_1^{(1)}, \ldots, w_n^{(1)}; \ldots; w_1^{(m)}, \ldots, w_n^{(m)})$$

be the canonical coordinates for $(\mathbb{C}^n)^m$. If Φ is a holomorphic $(k + mn)$-form on $Y_{\psi,\xi}^m$, then Φ can be considered as a holomorphic $(k + mn)$-form on $\mathcal{D} \times (\mathbb{C}^n)^m$ that is invariant under the action of $\Gamma \ltimes_\rho L^m$. Thus there is a holomorphic function $F_\Phi(z, w)$ on $\mathcal{D} \times (\mathbb{C}^n)^m$ such that

$$\Phi = F_\Phi(z, w)\, dz \wedge dw^{(1)} \wedge \cdots \wedge dw^{(m)},$$

where $z = (z_1, \ldots, z_k) \in \mathcal{D}$, $w = (w^{(1)}, \ldots, w^{(m)}) \in (\mathbb{C}^n)^m$, and $w^{(j)} = (w_1^{(j)}, \ldots, w_n^{(j)}) \in \mathbb{C}^n$ for $1 \leq j \leq m$. Given $x \in X$, the holomorphic form Φ descends to a holomorphic mn-form on the fiber $(Y_{\psi,\xi}^m)_x$ over x. The fiber $(Y_{\psi,\xi}^m)_x$ is the m-fold product of a complex torus of dimension n, and hence the dimension of the space of holomorphic mn-forms on $(Y_{\psi,\xi}^m)_x$ is one. Since any holomorphic function on a compact complex manifold is constant, the restriction of $F_\Phi(z, w)$ to the compact complex manifold $Y_{\psi,\xi}^m$ is constant. Thus $F_\Phi(z, w)$ depends only on z; and hence Φ can be written in the form

$$\Phi = f_\Phi(z)\, dz \wedge dw^{(1)} \wedge \cdots \wedge dw^{(m)},$$

where f_Φ is a holomorphic function on $\mathcal{D}$. To consider the invariance of Φ under the group $\Gamma \ltimes_\rho L^m$, we first notice that the action of $\Gamma \ltimes_\rho L^m$ on $dz = dz_1 \wedge \cdots \wedge dz_k$ is given by

$$(\gamma, (\mu, \nu)) \cdot dz = j_H(\gamma, z) dz$$

for all $(\gamma, (\mu, \nu)) \in \Gamma \ltimes_\rho L^m$, because $z \mapsto j_H(\gamma, z)$ is the Jacobian map for the transformation $z \mapsto \gamma z$ of $\mathcal{D}$. On the other hand, the action of $\Gamma \ltimes_\rho L^m$ on $dw^{(j)} = dw_1^{(j)} \wedge \cdots \wedge dw_n^{(j)}$ is given by

$$\begin{aligned}
(\gamma, (\mu, \nu)) \cdot dw^{(j)} &= d\left[(w + \mu\tau(z) + \nu + \xi(\gamma)(z)) j_{\rho,\tau}(\gamma, z)^{-1}\right] \\
&= \det(j_{\rho,\tau}(\gamma, z))^{-1} dw^{(j)} + \Psi^{(j)} \\
&= j_V(\rho(\gamma), \tau(z))^{-1} dw^{(j)} + \Psi^{(j)}
\end{aligned}$$

for $1 \leq j \leq m$, where the term $\Psi^{(j)}$ is the sum of the terms involving some dz_ℓ for $1 \leq \ell \leq k$; hence we obtain

$$(\gamma, (\mu, \nu)) \cdot \Phi = f_\Phi(\gamma z) j_H(\gamma, z) j_V(\rho(\gamma), \tau(z))^{-m} dz \wedge dw^{(1)} \wedge \cdots \wedge dw^{(m)}.$$

Thus we have

$$f_\Phi(\gamma z) = j_H(\gamma, z)^{-1} j_V(\rho(\gamma), \tau(z))^m f_\Phi(z)$$

for all $\gamma \in \Gamma$ and $z \in \mathcal{D}$. On the other hand, each mixed automorphic form on $\mathcal{D}$ of type $(j_H^{-1}, j_V^m, \rho, \tau)$ is a holomorphic function $h : \mathcal{D} \to \mathbb{C}$ satisfying

$$h(\gamma z) = j_H(\gamma, z)^{-1} j_V(\rho(\gamma), \tau(z))^n h(z)$$

for $z \in \mathcal{D}$ and $\gamma \in \Gamma$. Therefore the assignment $\Phi \mapsto f_\Phi(z)$ determines an isomorphism between the space $H^0(Y_{\psi,\xi}^m, \Omega^{k+mn})$ of holomorphic $(k + mn)$-forms on $Y_{\psi,\xi}^m$ and the space of mixed automorphic forms on $\mathcal{D}$ of type $(j_H^{-1}, j_V^m, \rho, \tau)$. $\qquad\square$

8.4 Cohomology

In this section we establish an isomorphism between the k-th cohomology along the fibers of a twisted torus bundle and the sheaf of holomorphic sections of a certain vector bundle over the base space of the torus bundle. This vector bundle is determined by an automorphy factor associated to an equivariant pair.

We fix elements $\psi \in 3^2(\Gamma, L)$ and $\xi \in \Xi_\psi$, where Ξ_ψ is as in Section 8.3, and consider the associated torus bundle $\pi : Y_{\psi,\xi} \to X$ constructed in the same section. The cohomology along the fibers of $Y_{\psi,\xi}$ over X can be provided by the direct image functors $R^i\pi_*$, which determine sheaves on X associated to sheaves on $Y_{\psi,\xi}$ (see e.g. [40, Section III.8]). We are interested in the images of the sheaf $\mathcal{O}_{Y_{\psi,\xi}}$ of holomorphic functions on $Y_{\psi,\xi}$ under such functors. Given a nonnegative integer k, $R^k\pi_*\mathcal{O}_{Y_{\psi,\xi}}$ is the sheaf on X generated by the presheaf

$$U \mapsto H^k(\pi^{-1}(U), \mathcal{O}_{Y_{\psi,\xi}})$$

for open subsets U of X. Note that by Dolbeault's theorem there is a canonical isomorphism

$$H^k(\pi^{-1}(U), \mathcal{O}_{Y_{\psi,\xi}}) \cong H^{(0,k)}(\pi^{-1}(U)).$$

Proposition 8.11 *The sheaf $R^0\pi_*\mathcal{O}_{Y_{\psi,\xi}}$ is isomorphic to the sheaf $\mathcal{O}_X$ of holomorphic functions on X.*

Proof. Let U be a sufficiently small open ball in X, and consider the section $f \in H^0(\pi^{-1}(U), \mathcal{O}_{Y_{\psi,\xi}})$ of $\mathcal{O}_{Y_{\psi,\xi}}$ on $\pi^{-1}(U)$. If $\widetilde{U} \subset \mathcal{D}$ is the inverse image of U under the natural projection map $\mathcal{D} \to X = \Gamma\backslash\mathcal{D}$, then we have

$$\pi^{-1}(U) \cong \Gamma \ltimes_\psi L\backslash\widetilde{U} \times \mathbb{C}^n. \tag{8.25}$$

Thus f may be regarded as a holomorphic function on $\widetilde{U} \times \mathbb{C}^n$ that is invariant under the action of $\Gamma \ltimes_\psi \{0\}$ and satisfies

$$f(z, w) = f(z, w + \mu\tau(z) + \nu)$$

for all $(z, w) \in \widetilde{U} \times \mathbb{C}^n$ and $(\mu, \nu) \in L$. Hence it follows that f is constant with respect to w and therefore can be identified with a Γ-invariant holomorphic function on $\widetilde{U}$ or a holomorphic function on U. □

Let $j_{\rho,\tau} : \Gamma \times \mathcal{D} \to GL(n, \mathbb{C})$ be the automorphy factor given by (8.13). Then the discrete subgroup $\Gamma \subset Sp(n, \mathbb{R})$ acts on $\mathcal{D} \times \mathbb{C}^n$ by

$$\gamma \cdot (z, w) = (\gamma z, w \cdot j_{\rho,\tau}(\gamma, z)^{-1})$$

for all $\gamma \in \Gamma$ and $(z, w) \in \mathcal{D} \times \mathbb{C}^n$. If we denote the associated quotient by

$$\mathcal{V} = \Gamma\backslash\mathcal{D} \times \mathbb{C}^n, \tag{8.26}$$

then the map $p : \mathcal{V} \to X = \Gamma \backslash \mathcal{D}$ induced by the natural projection $\mathcal{D} \times \mathbb{C}^n \to \mathcal{D}$ determines the structure of a vector bundle on $\mathcal{V}$ over X with fiber isomorphic to $\mathbb{C}^n$. By our construction we see that each holomorphic section $s : X \to \mathcal{V}$ of $\mathcal{V}$ over X can be identified with a function $\widetilde{s} : \mathcal{D} \to \mathbb{C}^n$ satisfying

$$\widetilde{s}(\gamma z) = \widetilde{s}(z) \cdot j_{\rho,\tau}(\gamma, z)^{-1} \tag{8.27}$$

for all $\gamma \in \Gamma$ and $z \in \mathcal{D}$.

Given a torus bundle $\pi : Y_{\psi,\xi} \to X$ and a sufficiently small open ball U in X, we consider a $(0,1)$-form ω on $\pi^{-1}(U)$ which determines the cohomology class $[\omega]$ in $H^{(0,1)}(\pi^{-1}(U)) = H^1(\pi^{-1}(U), \mathcal{O}_{Y_{\psi,\xi}})$. Let $\widetilde{U} \subset \mathcal{D}$ be the inverse image of U under the natural projection map $\mathcal{D} \to X$ as in (8.25), and let $z = (z_1, \ldots, z_N)$ be a local holomorphic system of coordinates on $\widetilde{U}$. Then we have

$$\omega = \sum_{\alpha=1}^{N} A_\alpha(z, w) d\bar{z}_\alpha + \sum_{\beta=1}^{n} B_\beta(z, w) d\bar{w}_\beta \tag{8.28}$$

for some $\mathbb{C}$-valued C^∞ functions $A_\alpha(z, w)$ and $B_\beta(z, w)$ on $\widetilde{U} \times \mathbb{C}^n$, where $w = (w_1, \ldots, w_n)$ is the standard coordinate system for $\mathbb{C}^n$. Let $\ell = (\mu, \nu) \in L$, and set

$$\zeta(z, \ell) = \mu \cdot \tau(z) + \nu$$

for all $z \in \mathcal{D}$. Note that $\zeta(z, \ell)$ is the same as $\xi(z, \ell) = \ell_z$ in the notation used in Chapter 6 (see Example 6.6). Then by (8.20) the action of ℓ on ω is given by

$$\ell^* \omega = \sum_{\alpha=1}^{N} A_\alpha(z, w + \zeta(z, \ell)) d\bar{z}_\alpha$$

$$+ \sum_{\beta=1}^{n} B_\beta(z, w + \zeta(z, \ell)) \left(d\bar{w}_\beta + \sum_{\alpha=1}^{N} \frac{\partial \overline{\zeta(z, \ell)}_\beta}{\partial \bar{z}_\alpha} d\bar{z}_\alpha \right).$$

Since $\ell^* \omega = \omega$, we obtain

$$A_\alpha(z, w) = A_\alpha(z, w + \zeta(z, \ell)) + \sum_{\beta=1}^{n} B_\beta(z, w + \zeta(z, \ell)) \frac{\partial \overline{\zeta(z, \ell)}_\beta}{\partial \bar{z}_\alpha},$$

$$B_\beta(z, w) = B_\beta(z, w + \zeta(z, \ell)) \tag{8.29}$$

for all $\ell \in L$.

Lemma 8.12 *Let $\omega^{(1)}$ be the $(0,1)$-form on $\widetilde{U} \times \mathbb{C}^n$ given by (8.28). Then there exists a $(0,1)$-form on $\widetilde{U} \times \mathbb{C}^n$ of the form*

$$\omega^{(1)} = \sum_{\alpha=1}^{N} \left(A_\alpha(z, w) - \frac{\partial f(z, w)}{\partial \bar{z}_\alpha} \right) d\bar{z}_\alpha + \sum_{\beta=1}^{n} C_\beta(z) d\bar{w}_\beta \tag{8.30}$$

such that $[\omega^{(1)}] = [\omega]$ in $H^{(0,1)}(\pi^{-1}(U))$, where $f(z,w)$ and $C_\beta(z)$ for $1 \leq \beta \leq n$ are C^∞ functions on $\widetilde{U} \times \mathbb{C}^n$ and $\widetilde{U}$, respectively.

Proof. We first rewrite the $(0,1)$-form ω in (8.28) as

$$\omega = \sum_{\alpha=1}^{N} A_\alpha(z,w)d\bar{z}_\alpha + \widetilde{\Phi}(z,w) \tag{8.31}$$

by setting

$$\widetilde{\Phi}(z,w) = \sum_{\beta=1}^{n} B_\beta(z,w)d\bar{w}_\beta.$$

Then, for fixed $z \in \widetilde{U} \subset \mathcal{D}$, by (8.29) we see that the $(0,1)$-form $\widetilde{\Phi}(z,w)$ on $\mathbb{C}^n$ is L-invariant and satisfies $\bar{\partial}_w \widetilde{\Phi}(z,w) = 0$. Thus for each $z \in \widetilde{U}$ we obtain a $\bar{\partial}_w$-closed $(0,1)$-form $\Phi(z)$ that is cohomologous to $\widetilde{\Phi}(z,w)$ on the complex torus

$$\mathbb{C}^n \Big/ \left(L \cdot \begin{pmatrix} \tau(z) \\ 1 \end{pmatrix} \right),$$

which is the fiber of the bundle $Y_{\psi,\xi}$ over the image of z in X. From harmonic theory we see that there are C^∞ functions $C_\beta(z)$ on $\widetilde{U}$ with $1 \leq \beta \leq n$ such that

$$\Phi^0(z) = \sum_{\beta=1}^{n} C_\beta(z)d\bar{w}_\beta \tag{8.32}$$

is a harmonic form in w that is cohomologous, for each fixed z, to $\Phi(z)$ in $H^{(0,1)}(\pi^{-1}(z))$. Hence there is a C^∞ function $f(z,w)$ on $\widetilde{U} \times \mathbb{C}^n$ such that $f(z, w + \zeta(z,\ell)) = f(z,w)$ for all $\ell \in L$ and

$$\Phi(z) - \Phi^0(z) = \bar{\partial}_w f(z,w) = \sum_{\beta=1}^{n} \frac{\partial f(z,w)}{\partial \bar{w}_\beta} d\bar{w}_\beta \tag{8.33}$$

$$= \bar{\partial}f(z,w) - \sum_{\alpha=1}^{N} \frac{\partial f(z,w)}{\partial \bar{z}_\alpha} d\bar{z}_\alpha.$$

We now define $\omega^{(1)}$ by (8.30), so that

$$\omega^{(1)} = \sum_{\alpha=1}^{N} \left(A_\alpha(z,w) - \frac{\partial f(z,w)}{\partial \bar{z}_\alpha} \right) d\bar{z}_\alpha + \Phi^0(z) \tag{8.34}$$

by (8.32). Then from (8.31), (8.33) and (8.34), we obtain

$$\omega - \omega^{(1)} = \bar{\partial}f;$$

hence it follows that $[\omega] = [\omega^{(1)}]$ in $H^{(0,1)}(\pi^{-1}(U))$.   □

Lemma 8.13 *There are C^∞ functions $F_{j,\alpha}(z)$ with $1 \le j \le n$ and $1 \le \alpha \le N$ such that the $(0,1)$-form on $\tilde{U} \times \mathbb{C}^n$ given by*

$$\omega^{(2)} = \sum_{\alpha=1}^{N} F_\alpha(z)(^t\overline{w} - {}^t w)d\overline{z}_\alpha + C_\beta(z)d\overline{w}_\beta \qquad (8.35)$$

satisfies $[\omega^{(2)}] = [\omega^{(1)}] = [\omega]$ in $H^{(0,1)}(\pi^{-1}(U))$, where the functions $C_\beta(z)$ are as in (8.30), $\omega = (\omega_1, \ldots, \omega_n)$, and $F_\alpha(z) = (F_{\alpha,1}(z), \ldots, F_{\alpha,n}(z))$ for each $\alpha \in \{1, \ldots, n\}$.

Proof. From (8.30) we may write

$$\omega^{(1)} = \sum_{\alpha=1}^{N} D_\alpha(z, w)d\overline{z}_\alpha + \sum_{\beta=1}^{n} C_\beta(z)d\overline{w}_\beta \qquad (8.36)$$

with

$$D_\alpha(z, w) = A_\alpha(z, w) - \frac{\partial f(z, w)}{\partial \overline{z}_\alpha}.$$

Since $\omega^{(1)}$ is a $\overline{\partial}$-closed form, we have

$$0 = \overline{\partial}\omega^{(1)} = \sum_{\alpha=1}^{N}\sum_{\lambda=1}^{N} \frac{\partial D_\alpha(z, w)}{\partial \overline{z}_\lambda}d\overline{z}_\lambda \wedge d\overline{z}_\alpha + \sum_{\alpha=1}^{N}\sum_{\epsilon=1}^{n} \frac{\partial D_\alpha(z, w)}{\partial \overline{w}_\epsilon}d\overline{w}_\epsilon \wedge d\overline{z}_\alpha$$
$$+ \sum_{\beta=1}^{n}\sum_{\lambda=1}^{N} \frac{\partial C_\beta(z)}{\partial \overline{z}_\lambda}d\overline{z}_\lambda \wedge d\overline{w}_\beta;$$

hence we obtain

$$\frac{\partial C_\beta(z)}{\partial \overline{z}_\lambda} = \frac{\partial D_\lambda(z, w)}{\partial \overline{w}_\beta}, \qquad \frac{\partial D_\alpha(z, w)}{\partial \overline{z}_\lambda} = \frac{\partial D_\lambda(z, w)}{\partial \overline{z}_\alpha} \qquad (8.37)$$

for $1 \le \alpha, \lambda \le N$ and $1 \le \beta \le n$. Thus we have

$$D_\lambda(z, w) = \sum_{\beta=1}^{n} F_{\lambda,\beta}(z)\overline{w}_\beta + P_\lambda(z, w), \qquad F_{\lambda,\beta}(z) = \frac{\partial C_\beta(z)}{\partial \overline{z}_\lambda}, \qquad (8.38)$$

where $P_\lambda(z, w)$ is a holomorphic function in w. Since $\ell^*\omega^{(1)} = \omega^{(1)}$ for each $\ell \in L$, by (8.36) we obtain

$$D_\lambda(z, w) = D_\lambda(z, w + \zeta(z, \ell)) + \sum_{\beta=1}^{n} C_\beta(z)\frac{\partial \overline{\zeta(z, \ell)}_\beta}{\partial \overline{z}_\lambda} \qquad (8.39)$$

for all $\ell \in L$. Hence, if we set

$$\tilde{P}_\lambda^0(z) = \sum_{\beta=1}^{n}\left(C_\beta(z)\frac{\partial\overline{\zeta(z,\ell)}_\beta}{\partial\bar{z}_\lambda} + F_{\lambda,\beta}(z)\overline{\zeta(z,\ell)}_\beta\right)$$

for $1 \le \lambda \le N$, by (8.38) and (8.39) we have

$$P_\lambda(z,w) - P_\lambda(z,w+\zeta(z,\ell)) = \tilde{P}_\lambda^0(z), \qquad (8.40)$$

which is a function of z only. Thus $P_\lambda(z,w)$ must be of the form

$$P_\lambda(z,w) = P_\lambda^0(z) + \sum_{\beta=1}^{n} P_{\lambda,\beta}^1(z)w_\beta \qquad (8.41)$$

for each λ. Using (8.40) and (8.41) for $w = \zeta(z,\ell)$, we see that the functions $P_{\lambda,\beta}^1(z)$ satisfy

$$\sum_{\beta=1}^{n} P_{\lambda,\beta}^1(z)\zeta(z,\ell)_\beta = -\sum_{\beta=1}^{n}\left(F_{\lambda,\beta}(z)\overline{\zeta(z,\ell)}_\beta + C_\beta(z)\frac{\partial\overline{\zeta(z,\ell)}_\beta}{\partial\bar{z}_\lambda}\right). \qquad (8.42)$$

Since $\zeta(z,\ell) = \mu\cdot\tau(z) + \nu$ for $\ell = (\mu,\nu)$, using $\mu = 0$ and $\nu = (\nu_1,\ldots,\nu_n)$ with $\nu_j \ne 0$ and $\nu_k = 0$ for $k \ne j$, from (8.42) we obtain

$$P_{\lambda,j}^1(z) = -F_{\lambda,j}(z) \qquad (8.43)$$

for each $j \in \{1,\ldots,n\}$. Thus (8.42) reduces to

$$\sum_{\beta=1}^{n} F_{\lambda,\beta}(z)(\mu\cdot\tau(z))_\beta = \sum_{\beta=1}^{n}\left(F_{\lambda,\beta}(z)(\mu\cdot\overline{\tau(z)})_\beta + C_\beta(z)\frac{\partial(\mu\cdot\overline{\tau(z)})_\beta}{\partial\bar{z}_\lambda}\right)$$

for $\ell = (\mu,0)$. By considering μ with only one nonzero entry μ_j for each j we see that

$$F_\lambda(z)\tau(z) = F_\lambda(z)\overline{\tau(z)} + C(z)\frac{\partial\overline{\tau(z)}}{\partial\bar{z}_\lambda},$$

where the $F_\lambda(z)$ and $C(z)$ are row vectors given by

$$F_\lambda(z) = (F_{\lambda,1}(z),\ldots,F_{\lambda,n}(z)), \qquad C(z) = (C_1(z),\ldots,C_n(z))$$

and the products are matrix products. Thus we have

$$2iF_\lambda(z)(\operatorname{Im}\tau(z)) = C(z)\frac{\partial\overline{\tau(z)}}{\partial\bar{z}_\lambda}. \qquad (8.44)$$

By (8.37) and (8.38) we have

$$\frac{\partial F_\alpha(z)}{\partial\bar{z}_\lambda}{}^t\overline{w} + \frac{\partial P_\alpha(z,w)}{\partial\bar{z}_\lambda} = \frac{\partial D_\alpha(z,w)}{\partial\bar{z}_\lambda} = \frac{\partial D_\lambda(z,w)}{\partial\bar{z}_\alpha}$$
$$= \frac{\partial F_\lambda(z)}{\partial\bar{z}_\alpha}{}^t\overline{w} + \frac{\partial P_\lambda(z,w)}{\partial\bar{z}_\alpha},$$

where $\overline{w}$ is regarded as a row vector and ${}^t\overline{w}$ is its transpose. Hence it follows that

$$\frac{\partial F_\alpha(z)}{\partial \overline{z}_\lambda} = \frac{\partial F_\lambda(z)}{\partial \overline{z}_\alpha}, \qquad \frac{\partial P_\alpha(z,w)}{\partial \overline{z}_\lambda} = \frac{\partial P_\lambda(z,w)}{\partial \overline{z}_\alpha}.$$

Thus, using this and (8.41), we obtain

$$\frac{\partial P_\alpha^0(z)}{\partial \overline{z}_\lambda} + \frac{\partial P_\alpha^1(z)}{\partial \overline{z}_\lambda}{}^t w = \frac{\partial P_\lambda^0(z)}{\partial \overline{z}_\alpha} + \frac{\partial P_\lambda^1(z)}{\partial \overline{z}_\alpha}{}^t w$$

with $P_\epsilon^1 = (P_{\epsilon,1}^1, \dots, P_{\epsilon,n}^1)$ for $\epsilon = \alpha, \lambda$, which implies that

$$\frac{\partial P_\alpha^0(z)}{\partial \overline{z}_\lambda} = \frac{\partial P_\lambda^0(z)}{\partial \overline{z}_\alpha}.$$

By (8.36), (8.38), (8.41) and (8.43) we see that

$$\omega^{(1)} = \sum_{\alpha=1}^N (F_\alpha(z){}^t\overline{w} + P_\alpha^0(z) + P_\alpha^1(z){}^t w)d\overline{z}_\alpha + \sum_{\beta=1}^n C_\beta(z)d\overline{w}_\beta$$

$$= \sum_{\alpha=1}^N (F_\alpha(z)({}^t\overline{w} - {}^t w) + P_\alpha^0(z))d\overline{z}_\alpha + C(z)d{}^t\overline{w}.$$

Hence, if we define the $(0,1)$-form $\omega^{(2)}$ on $\widetilde{U} \times \mathbb{C}^n$ by (8.35), we obtain

$$\omega^{(1)} = \omega^{(2)} + \sum_{\alpha=1}^N P_\alpha^0(z)d\overline{z}_\alpha.$$

Since $\sum_{\alpha=1}^N P_\alpha^0(z)d\overline{z}_\alpha$ is a closed 1-form on $\pi^{-1}(U)$, it is exact by Poincaré's lemma; hence it follows that $[\omega] = [\omega^{(1)}] = [\omega^{(2)}]$ in $H^{(0,1)}(\pi^{-1}(U))$.   □

Lemma 8.14 *The $(0,1)$-form $\omega^{(2)}$ on $\widetilde{U} \times \mathbb{C}^n$ given by (8.35) can be written in the form*

$$\omega^{(2)} = \phi(z)\overline{\partial}((\operatorname{Im}\tau(z))^{-1}\operatorname{Im}{}^t w)$$

for some $\mathbb{C}^n$-valued holomorphic function ϕ on $\widetilde{U}$.

Proof. By (8.38) and (8.44) we have

$$\frac{\partial C(z)}{\partial \overline{z}_\lambda}(\operatorname{Im}\tau(z)) = F_\lambda(z)(\operatorname{Im}\tau(z))$$

$$= \frac{1}{2i}C(z)\frac{\partial\overline{\tau(z)}}{\partial\overline{z}_\lambda} = \frac{1}{2i}C(z)\frac{\partial}{\partial\overline{z}_\lambda}(\overline{\tau(z)} - \tau(z))$$

$$= -C(z)\frac{\partial}{\partial\overline{z}_\lambda}(\operatorname{Im}\tau(z)).$$

Thus we obtain

$$\frac{\partial}{\partial \bar{z}_\lambda}(C(z)\operatorname{Im}\tau(z)) = 0,$$

and therefore we see that the function $C(z)\operatorname{Im}\tau(z)$ is holomorphic. Now we define the vector-valued holomorphic function ϕ on $\widetilde{U} \subset \mathcal{D}$ by

$$\phi(z) = (\phi_1(z),\ldots,\phi_n(z)) = -2iC(z)\operatorname{Im}\tau(z). \tag{8.45}$$

Using this, (8.38) and (8.35), we obtain

$$\omega^{(2)} = \sum_{\alpha=1}^{N} \frac{\partial C(z)}{\partial \bar{z}_\alpha}(^t\overline{w} - {}^tw)d\bar{z}_\alpha + C(z)d^t\overline{w} \tag{8.46}$$

$$= -\frac{1}{2i}\sum_{\alpha=1}^{N}\phi(z)\frac{\partial(\operatorname{Im}\tau(z))^{-1}}{\partial\bar{z}_\alpha}(^t\overline{w} - {}^tw)d\bar{z}_\alpha - \frac{1}{2i}\phi(z)(\operatorname{Im}\tau(z))^{-1}d^t\overline{w}$$

$$= -\frac{1}{2i}\phi(z)((\bar{\partial}(\operatorname{Im}\tau(z))^{-1})(^t\overline{w} - {}^tw) + (\operatorname{Im}\tau(z))^{-1}\bar{\partial}(^t\overline{w} - {}^tw))$$

$$= \phi(z)\bar{\partial}((\operatorname{Im}\tau(z))^{-1}\operatorname{Im}{}^tw),$$

which prove the lemma. □

Theorem 8.15 *Let $\mathcal{V}$ be the vector bundle over X given by (8.26). Then the sheaf $R^1\pi_*\mathcal{O}_{Y_{\psi,\xi}}$ is isomorphic to the sheaf $\widetilde{\mathcal{V}}$ of holomorphic sections of $\mathcal{V}$ over X.*

Proof. We shall first show that ϕ in (8.45) corresponds to a holomorphic section of the bundle $\mathcal{V}$ over U. By (8.27) it suffices to show that the function $\phi : \widetilde{U} \to \mathbb{C}^n$ in (8.46) satisfies

$$\phi(\gamma z) = \phi(z)j_{\rho,\tau}(\gamma, z)^{-1} \tag{8.47}$$

for all $\gamma \in \Gamma$ and $z \in \widetilde{U}$. If $\gamma \in \Gamma$, using (8.20), we see that the action of $(\gamma, 0) \in \Gamma \ltimes_\psi L$ on $d\overline{w}$ is given by

$$(\gamma, 0)^*d\overline{w} = d(\overline{(w + \xi(\gamma)(z))j_{\rho,\tau}(\gamma,z)^{-1}})$$

$$= d\overline{w} \cdot \overline{j_{\rho,\tau}(\gamma,z)}^{-1} + \overline{w} \cdot d(\overline{j_{\rho,\tau}(\gamma,z)}^{-1}) + d(\overline{\xi(\gamma)(z)} \cdot \overline{j_{\rho,\tau}(\gamma,z)}^{-1})$$

$$= d\overline{w} \cdot \overline{j_{\rho,\tau}(\gamma,z)}^{-1} + \text{(terms in } d\bar{z}_\alpha),$$

where we used the fact that the functions $\xi(\gamma)(z)$ and $j_{\rho,\tau}(\gamma,z)$ are holomorphic in z. Since $\omega^{(2)}$ in (8.35) can be written in the form

$$\omega^{(2)} = C(z)d^t\overline{w} + \text{(terms in } d\bar{z}_\alpha)$$

and since $(\gamma,0)^*$ takes terms in $d\bar{z}_\alpha$ to themselves, we see that

$$(\gamma,0)^*\omega^{(2)} = C(\gamma z)^t\overline{j_{\rho,\tau}(\gamma,z)}^{-1}d^t\overline{w} + \text{(terms in } d\bar{z}_\alpha).$$

We now compare terms in $d\overline{w}_\beta$ in the relation $(\gamma, 0)^* \omega^{(2)} = \omega^{(2)}$ to obtain

$$C(\gamma z)^t \overline{j_{\rho,\tau}(\gamma, z)}^{-1} = C(z).$$

Using this, (8.45), and the fact that

$$\operatorname{Im} \tau(\gamma z) = {}^t \overline{j_{\rho,\tau}(\gamma, z)}^{-1} \cdot \operatorname{Im} \tau(z) \cdot j_{\rho,\tau}(\gamma, z)^{-1},$$

we obtain

$$\begin{aligned}
\phi(\gamma z) &= -2iC(\gamma z) \operatorname{Im} \tau(\gamma z) \\
&= -2iC(z) \operatorname{Im} \tau(z) j_{\rho,\tau}(\gamma, z)^{-1} \\
&= \phi(z) j_{\rho,\tau}(\gamma, z)^{-1}.
\end{aligned}$$

Hence it follows that ϕ can be regarded as a holomorphic section of $\mathcal{V}$ over U. On the other hand, if $\widehat{\varphi}$ is a holomorphic section of $\mathcal{V}$ over U represented by a vector-valued holomorphic function $\varphi : \widetilde{U} \to \mathbb{C}^n$, we denote by $\omega_{\widehat{\varphi}}$ the $(0, 1)$-form on $\widetilde{U} \times \mathbb{C}^n$ given by

$$\omega_{\widehat{\varphi}} = \varphi(z) \overline{\partial}((\operatorname{Im} \tau(z))^{-1} \operatorname{Im} {}^t w).$$

Denoting by $\boldsymbol{\Gamma}(U, \mathcal{V})$ the space of holomorphic sections of $\mathcal{V}$ over U and using (8.46), we see that the map

$$\boldsymbol{\Gamma}(U, \mathcal{V}) \to H^{(0,1)}(\pi^{-1}(U))$$

sending $\widehat{\varphi}$ to the cohomology class $[\omega_{\widehat{\varphi}}]$ of $\omega_{\widehat{\varphi}}$ is surjective. Thus we obtain the corresponding surjective map

$$\mathfrak{F} : \widetilde{\mathcal{V}} \to R^1 \pi_* \mathcal{O}_{Y_{\psi,\xi}}$$

of sheaves on X. In order to show that $\mathfrak{F}$ is injective, given $x \in X$, we denote by T_x and $\mathcal{V}_x$ the fibers of the bundles $Y_{\psi,\xi}$ and $\mathcal{V}$, respectively, over x. Then $\mathcal{V}_x$ and $H^1(T_x, \mathcal{O}) = H^{(0,1)}(T_x)$ are the fibers of the sheaves $\widetilde{\mathcal{V}}$ and $R^1 \pi_* \mathcal{O}_{Y_{\psi,\xi}}$, respectively. Thus, using the fact that both $\mathcal{V}_x$ and $H^1(T_x, \mathcal{O})$ are isomorphic to the n-dimensional space $\mathbb{C}^n$, we see that the surjectivity of $\mathfrak{F}$ implies its injectivity. Hence it follows that $\mathfrak{F}$ is an isomorphism of sheaves on X, and the proof of the theorem is complete. $\qquad\square$

Corollary 8.16 *Let $\widetilde{\mathcal{V}}$ be as in Theorem 8.15, and let k be a positive integer. Then there is an isomorphism*

$$R^k \pi_* \mathcal{O}_{Y_{\psi,\xi}} \cong \wedge^k(\widetilde{\mathcal{V}})$$

of sheaves on X.

Proof. Since $\pi^{-1}(x)$ is a complex torus for each $x \in X$, from a well-known result (see e.g. Corollary 1 in [97, p. 8]) we see that

$$H^k(\pi^{-1}(x), \mathcal{O}_{Y_{\psi,\xi}}) \cong \wedge^k(H^1(\pi^{-1}(x), \mathcal{O}_{Y_{\psi,\xi}}))$$

for each $x \in X$. Hence we obtain

$$R^k \pi_* \mathcal{O}_{Y_{\psi,\xi}} \cong \wedge^k(R^k \pi_* \mathcal{O}_{Y_{\psi,\xi}}),$$

and therefore the corollary follows by combining this with the isomorphism in Theorem 8.15. $\square$

References

1. S. Abdulali, *Conjugates of strongly equivariant maps*, Pacific J. Math. **165** (1994), 207–216.
2. S. Addington, *Equivariant holomorphic maps of symmetric domains*, Duke Math. J. **55** (1987), 65–88.
3. A. Ash, D. Mumford, M. Rapoport, and Y. Tai, *Smooth compactification of locally symmetric varieties*, Math. Sci. Press, Brookline, 1975.
4. W. Baily, *Introductory lectures on automorphic forms*, Princeton Univ. Press, Princeton, 1973.
5. W. Baily and A. Borel, *Compactification of arithmetic quotients of bounded symmetric domains*, Ann. of Math. **84** (1966), 442–528.
6. P. Bayer and J. Neukirch, *On automorphic forms and Hodge theory*, Math. Ann. **257** (1981), 135–155.
7. R. Berndt, *On automorphic forms for the Jacobi group*, Jb. d. Dt. Math.-Verein. **97** (1995), 1–18.
8. R. Berndt and S. Böcherer, *Jacobi forms and discrete series representations of the Jacobi group*, Math. Z. **204** (1990), 13–44.
9. R. Berndt and R. Schmidt, *Elements of the representation theory of the Jacobi group*, Birkhäuser, Boston, 1998.
10. B. Birch, *Elliptic curves over $\mathbb{Q}$: a progress report*, Proc. Sympos. Pure Math., vol. 20, Amer. Math. Soc., Providence, 1971, pp. 396–400.
11. R. Borcherds, *Automorphic forms on $O_{s+2,2}(\mathbb{R})$ and generalized Kac-Moody algebras*, Proc. Internat. Congress of Mathematicians (Zürich, 1994), Birkhäuser, Boston, 1995, pp. 744–752.
12. _____, *Automorphic forms on $O_{s+2,2}(\mathbb{R})$ and infinite products*, Invent. Math. **120** (1995), 161–213.
13. A. Borel, *Introduction to automorphic forms*, Proc. Sympos. Pure Math., vol. 9, Amer. Math. Soc., Providence, 1966, pp. 199–210.
14. _____, *Automorphic forms on $SL_2(\mathbb{R})$*, Cambridge Univ. Press, Cambridge, 1997.
15. A. Borel and H. Jacquet, *Automorphic forms and automorphic representations*, Proc. Sympos. Pure Math., vol. 33, Part I, Amer. Math. Soc., Providence, 1979, pp. 189–202.
16. A. Borel and N. Wallach, *Continuous cohomology, discrete subgroups and representations of reductive groups*, Princeton Univ. Press, Princeton, 1980.
17. Y. Choie, H. Kim, and M. Knopp, *Construction of Jacobi forms*, Math. Z. **219** (1995), 71–76.
18. Y. Choie and M. H. Lee, *Mellin transforms of mixed cusp forms*, Canad. Math. Bull. **42** (1999), 263–273.

19. Y. Choie and M. H. Lee, *Mixed Siegel modular forms and special values of certain Dirichlet series*, Monatsh. Math. **131** (2000), 109–122.

20. D. Cox and S. Zucker, *Intersection numbers of sections of elliptic surfaces*, Invent. Math. **53** (1979), 1–44.

21. P. Deligne, *Travaux de Shimura, Sém. Bourbaki, Exposé 389*, Lecture Notes in Math., vol. 244, Springer-Verlag, Heidelberg, 1971.

22. M. Eichler, *Eine Verallgemeinerung der Abelschen Integrals*, Math. Z. **67** (1957), 267–298.

23. M. Eichler and D. Zagier, *The theory of Jacobi forms*, Progress in Math., vol. 55, Birkhäuser, Boston, 1985.

24. G. Faltings and C. L. Chai, *Degeneration of abelian varieties*, Springer-Verlag, Berlin, 1990.

25. E. Freitag, *Singular modular forms and theta relations*, Lecture Notes in Math., vol. 1487, Springer-Verlag, Heidelberg, 1991.

26. E. Freitag and C. Hermann, *Some modular varieties of low dimension*, Adv. Math. **152** (2000), 203–287.

27. S. Gelbart, *Automorphic forms on adele groups*, Princeton Univ. Press, Princeton, 1975.

28. S. Gelbart and F. Shahidi, *Analytic properties of automorphic L-functions*, Academic Press, New York, 1988.

29. R. Godement, *Introduction à la théorie de Langlands*, Sém. Bourbaki **19** (1966/67), Exposé 321.

30. R. Goodman and N. Wallach, *Whittaker vectors and conical vectors*, J. Funct. Anal. **39** (1980), 199–279.

31. B. B. Gordon, *Algebraic cycles in families of abelian varieties over Hilbert-Blumenthal surfaces*, J. Reine Angew. Math. **449** (1994), 149–171.

32. P. Griffiths and J. Harris, *Principles of algebraic geometry*, John Wiley & Sons, New York, 1978.

33. R. Gunning, *Lectures on modular forms*, Princeton Univ. Press, Princeton, 1962.

34. M. Hall, Jr., *The theory of groups*, Macmillan, New York, 1959.

35. W. Hammond, *The modular groups of Hilbert and Siegel*, Proc. Sympos. Pure Math., vol. 9, Amer. Math. Soc., Providence, 1966, pp. 358–360.

36. Harish-Chandra, *Representations of semi-simple Lie groups: VI*, Amer. J. Math. **78** (1956), 564–628.

37. _____, *Automorphic forms on semisimple Lie groups*, Springer-Verlag, Heidelberg, 1968.

38. M. Harris, *Functorial properties of toroidal compactification of locally symmetric varieties*, Proc. London Math. Soc. **59** (1989), 1–22.

39. _____, *Automorphic forms of $\bar{\partial}$-cohomology type as coherent cohomology classes*, J. Differential Geom. **32** (1990), 1–63.

40. R. Hartshorne, *Algebraic geometry*, Springer-Verlag, Heidelberg, 1977.

41. H. Hida, *Elementary theory of L-functions and Eisenstein series*, Cambridge Univ. Press, Cambridge, 1993.

42. K. Hulek, C. Kahn, and S. Weintraub, *Moduli spaces of abelian surfaces: compactification, degenerations and theta functions*, Walter de Gruyter, Berlin, 1993.

43. B. Hunt and W. Meyer, *Mixed automorphic forms and invariants of elliptic surfaces*, Math. Ann. **271** (1985), 53–80.

44. J.-I. Igusa, *Theta functions*, Springer-Verlag, Heidelberg, 1972.
45. H. Jacquet, *Fonctions de Whittaker associées aux groupes de Chevalley*, Bull. Soc. Math. France **95** (1967), 243–309.
46. D. Johnson and J. Millson, *Deformation spaces associated to hyperbolic manifolds*, Progress in Math., vol. 67, Birkhäuser, Boston, 1987, pp. 48–106.
47. S. Katok, *Closed geodesics, periods and arithmetic of modular forms*, Invent. Math. **80** (1985), 469–480.
48. S. Katok and J. Millson, *Eichler-Shimura homology, intersection numbers and rational structures on spaces of modular forms*, Trans. Amer. Math. Soc. **300** (1987), 737–757.
49. H. Klingen, *Metrisierungstheorie und Jacobiformen*, Abh. Math. Sem. Univ. Hamburg **57** (1987), 165–178.
50. _____, *Über Kernfunktionen für Jacobiformen und Siegelsche Modulformen*, Math. Ann. **285** (1989), 405–416.
51. N. Koblitz, *Congruences for periods of modular forms*, Duke Math. J. **54** (1987), 361–373.
52. K. Kodaira, *On compact analytic surfaces II*, Ann. of Math. **77** (1963), 563–626.
53. W. Kohnen, *Cusp forms and special values of certain Dirichlet series*, Math. Z. **207** (1991), 657–660.
54. _____, *Jacobi forms and Siegel modular forms: Recent results and problems*, Enseign. Math. **39** (1993), 121–136.
55. M. Kontsevich, *Product formulas for modular forms on SO(2, n) (after R. Borcherds)*, Sem. Bourbaki 1996/97, Asterisque, No. 245, 1997, pp. 41–56.
56. B. Kostant, *On Whittaker vectors and representation theory*, Invent. Math. **48** (1978), 101–184.
57. J. Kramer, *A geometric approach to the theory of Jacobi forms*, Compositio Math. **79** (1991), 1–19.
58. _____, *An arithmetic theory of Jacobi forms in higher dimensions*, J. Reine Angew. Math. **458** (1995), 157–182.
59. A. Krieg, *Jacobi forms of several variables and the Maaß space*, J. Number Theory **56** (1996), 242–255.
60. T. Kubota, *Elementary theory of Eisenstein series*, John Wiley, New York, 1973.
61. M. Kuga, *Fiber varieties over a symmetric space whose fibers are abelian varieties I, II*, Univ. of Chicago, Chicago, 1963/64.
62. M. Kuga and G. Shimura, *On the zeta-function of a fiber variety whose fibers are abelian varieties*, Ann. of Math. **82** (1965), 478–539.
63. H. Lange and Ch. Birkenhake, *Complex abelian varieties*, Springer-Verlag, Berlin, 1992.
64. R. Langlands, *Eisenstein series*, Proc. Sympos. Pure Math., vol. 9, Amer. Math. Soc., Providence, 1966, pp. 235–252.
65. _____, *Euler products*, Yale Univ. Press, New Haven, 1967.
66. _____, *On the functional equations satisfied by Eisenstein series*, vol. 544, Springer-Verlag, Heidelberg, 1976.
67. H. Laufer, *On rational singularities*, Amer. J. Math. **94** (1972), 597–608.
68. M. H. Lee, *Mixed cusp forms and holomorphic forms on elliptic varieties*, Pacific J. Math. **132** (1988), 363–370.
69. _____, *Conjugates of equivariant holomorphic maps of symmetric domains*, Pacific J. Math. **149** (1991), 127–144.

70. _____, *Mixed cusp forms and Poincaré series*, Rocky Mountain J. Math. **23** (1993), 1009–1022.

71. _____, *Mixed Siegel modular forms and Kuga fiber varieties*, Illinois J. Math. **38** (1994), 692–700.

72. _____, *Twisted torus bundles over arithmetic varieties*, Proc. Amer. Math. Soc. **123** (1995), 2251–2259.

73. _____, *Mixed automorphic forms on semisimple Lie groups*, Illinois J. Math. **40** (1996), 464–478.

74. _____, *Mixed automorphic vector bundles on Shimura varieties*, Pacific J. Math. **173** (1996), 105–126.

75. _____, *Siegel cusp forms and special values of Dirichlet series of Rankin type*, Complex Variables Theory Appl. **31** (1996), 97–103.

76. _____, *Mixed Hilbert modular forms and families of abelian varieties*, Glasgow Math. J. **39** (1997), 131–140.

77. _____, *Generalized Jacobi forms and abelian schemes over arithmetic varieties*, Collect. Math. **49** (1998), 121–131.

78. _____, *Hilbert cusp forms and special values of Dirichlet series of Rankin type*, Glasgow Math. J. **40** (1998), 71–78.

79. _____, *Periods of mixed automorphic forms and differential equations*, Appl. Math. Lett. **11** (1998), 15–19.

80. _____, *Eisenstein series and Poincaré series for mixed automorphic forms*, Collect. Math. **51** (2000), 225–236.

81. _____, *Jacobi forms on symmetric domains and torus bundles over abelian schemes*, J. Lie Theory **11** (2001), 545–557.

82. _____, *Mixed automorphic forms and Whittaker vectors*, Quart. J. Math. Oxford Ser. (2) **52** (2001), 471–484.

83. _____, *Pseudodifferential operators associated to linear ordinary differential equations*, Illinois J. Math. **45** (2001), 1377–1388.

84. _____, *Rational equivariant holomorphic maps of symmetric domains*, Arch. Math. **78** (2002), 275–282.

85. _____, *Cohomology of complex torus bundles associated to cocycles*, Canad. J. Math. **55** (2003), 839–855.

86. _____, *Theta functions on Hermitian symmetric domains and Fock representations*, J. Austral. Math. Soc. Ser. A **74** (2003), 201–234.

87. M. H. Lee and D. Y. Suh, *Torus bundles over locally symmetric varieties associated to cocycles of discrete groups*, Monatsh. Math. **59** (2000), 127–141.

88. H. Maass, *Über eine neue Art von nichtanalytischen automorphen Funktionen und did Bestimmung Dirichletscher Reihen durch Funktionalgleichungen*, Math. Ann. **121** (1949), 141–183.

89. Y. Manin, *Cyclotomic fields and modular curves*, Russian Math. Surveys **26** (1971), 7–78.

90. _____, *Parabolic points and zeta functions of modular curves*, Math. USSR Izv. **6** (1972), 19–64.

91. _____, *Periods of parabolic forms and p-adic Hecke series*, Math. USSR Sb. **21** (1973), 371–393.

92. _____, *Non-archimedian integration and Jacquet-Langlands p-adic L-functions*, Russian Math. Surveys **31** (1976), 5–54.

93. R. Miatello and N. Wallach, *Automorphic forms constructed from Whittaker vectors*, J. Funct. Anal. **86** (1989), 411–487.

94. J. S. Milne, *Canonical models of (mixed) Shimura varieties and automorphic vector bundles*, Automorphic forms, Shimura Varieties and *L*-functions, vol. 1, Academic Press, Boston, 1990, pp. 283–414.

95. T. Miyake, *Modular forms*, Springer-Verlag, Heidelberg, 1989.

96. D. Mumford, *Families of abelian varieties*, Proc. Sympos. Pure Math., vol. 9, Amer. Math. Soc., Providence, 1966, pp. 347–351.

97. _____, *Abelian varieties*, Oxford Univ. Press, Oxford, 1970.

98. _____, *Tata lectures on theta III*, Birkhäuser, Boston, 1991.

99. S. Murakami, *Cohomology of vector-valued forms on symmetric spaces*, Univ. of Chicago, Chicago, 1966.

100. M. Osborne and G. Warner, *The theory of Eisenstein systems*, Academic Press, New York, 1990.

101. A. Panchishkin, *Non-archimedian L-functions of Siegel and Hilbert modular forms*, Springer-Verlag, Heidelberg, 1991.

102. I. Piatetskii-Shapiro, *Geometry of classical domains*, Gordon and Breach, New York, 1969.

103. R. Pink, *Arithmetical compactification of mixed Shimura varieties*, Bonner Mathematische Schriften, vol. 209, Universität Bonn, Bonn, 1990.

104. M. Razar, *Values of Dirichlet series at integers in the critical strip*, Lecture Notes in Math., vol. 627, Springer-Verlag, Heidelberg, 1977, pp. 1–10.

105. B. Runge, *Theta functions and Siegel-Jacobi forms*, Acta Math. **175** (1995), 165–196.

106. I. Satake, *Clifford algebras and families of Abelian varieties*, Nagoya Math. J. **27** (1966), 435–446.

107. _____, *Fock representations and theta-functions*, Ann. of Math. Studies, vol. 66, Princeton Univ. Press, 1971, pp. 393–405.

108. _____, *Algebraic structures of symmetric domains*, Princeton Univ. Press, Princeton, 1980.

109. _____, *Toric bundles over abelian schemes*, Algebraic cycles and related topics (Kitasakado, 1994), World Scientific, River Edge, 1995, pp. 43–49.

110. G. Schiffmann, *Intégrales d'entrelacement et fonctions de Whittaker*, Bull. Soc. Math. France **99** (1971), 3–72.

111. J. Shalika, *The multiplicity one theorem for GL(n)*, Ann. of Math. **100** (1974), 171–193.

112. G. Shimura, *Sur les intégrales attachés aux formes automorphes*, J. Math. Soc. Japan **11** (1959), 291–311.

113. _____, *Moduli of abelian varieties and number theory*, Proc. Sympos. Pure Math., vol. 9, Amer. Math. Soc., Providence, 1966, pp. 312–332.

114. _____, *Introduction to the arithmetic theory automorphic functions*, Princeton Univ. Press, Princeton, 1971.

115. T. Shioda, *On elliptic modular surfaces*, J. Math. Soc. Japan **24** (1972), 20–59.

116. N. Skoruppa, *Developments in the theory of Jacobi forms*, Conference on automorphic forms and their applications. Khabarovsk 1988 (Kuznetsov, ed.), 1990, pp. 165–185.

117. V. Šokurov, *Holomorphic differential forms of higher degree on Kuga's modular varieties*, Math. USSR Sb. **30** (1976), 119–142.

118. _____, *Shimura integrals of cusp forms*, Math. USSR Izv. **16** (1981), 603–646.

119. _____, *The study of the homology of Kuga varieties*, Math. USSR Izv. **16** (1981), 399–418.

120. P. Stiller, *Special values of Dirichlet series, monodromy, and the periods of automorphic forms*, Mem. Amer. Math. Soc., vol. 299, 1984.
121. M. Taylor, *Noncommutative harmonic analysis*, Amer. Math. Soc., Providence, 1986.
122. G. Warner, *Harmonic analysis on semi-simple Lie groups I*, Springer-Verlag, Heidelberg, 1972.
123. T. Yamazaki, *Jacobi forms and a Maass relation for Eisenstein series*, J. Fac. Sci. Univ. Tokyo Sect. IA Math. **33** (1986), 295–310.
124. C. Ziegler, *Jacobi forms of higher degree*, Abh. Math. Sem. Univ. Hamburg **59** (1989), 191–224.
125. S. Zucker, *Hodge theory with degenerating coefficients: L^2-cohomology in the Poincaré metric*, Ann. of Math. **109** (1979), 415–476.

Index

Printing and Binding: Strauss GmbH, Mörlenbach

Lecture Notes in Mathematics

For information about Vols. 1–1665
please contact your bookseller or Springer-Verlag

Vol. 1764: A. Cannas da Silva, Lectures on Symplectic Geometry. XII, 217 pages. 2001.

Vol. 1765: T. Kerler, V. V. Lyubashenko, Non-Semisimple Topological Quantum Field Theories for 3-Manifolds with Corners. VI, 379 pages. 2001.

Vol. 1766: H. Hennion, L. Hervé, Limit Theorems for Markov Chains and Stochastic Properties of Dynamical Systems by Quasi-Compactness. VIII, 145 pages. 2001.

Vol. 1767: J. Xiao, Holomorphic Q Classes. VIII, 112 pages. 2001.

Vol. 1768: M.J. Pflaum, Analytic and Geometric Study of Stratified Spaces. VIII, 230 pages. 2001.

Vol. 1769: M. Alberich-Carramiñana, Geometry of the Plane Cremona Maps. XVI, 257 pages. 2002.

Vol. 1770: H. Gluesing-Luerssen, Linear Delay-Differential Systems with Commensurate Delays: An Algebraic Approach. VIII, 176 pages. 2002.

Vol. 1771: M. Émery, M. Yor (Eds.), Séminaire de Probabilités 1967-1980. A Selection in Martingale Theory. IX, 553 pages. 2002.

Vol. 1772: F. Burstall, D. Ferus, K. Leschke, F. Pedit, U. Pinkall, Conformal Geometry of Surfaces in S^4. VII, 89 pages. 2002.

Vol. 1773: Z. Arad, M. Muzychuk, Standard Integral Table Algebras Generated by a Non-real Element of Small Degree. X, 126 pages. 2002.

Vol. 1774: V. Runde, Lectures on Amenability. XIV, 296 pages. 2002.

Vol. 1775: W. H. Meeks, A. Ros, H. Rosenberg, The Global Theory of Minimal Surfaces in Flat Spaces. Martina Franca 1999. Editor: G. P. Pirola. X, 117 pages. 2002.

Vol. 1776: K. Behrend, C. Gomez, V. Tarasov, G. Tian, Quantum Comohology. Cetraro 1997. Editors: P. de Bartolomeis, B. Dubrovin, C. Reina. VIII, 319 pages. 2002.

Vol. 1777: E. García-Río, D. N. Kupeli, R. Vázquez-Lorenzo, Osserman Manifolds in Semi-Riemannian Geometry. XII, 166 pages. 2002.

Vol. 1778: H. Kiechle, Theory of K-Loops. X, 186 pages. 2002.

Vol. 1779: I. Chueshov, Monotone Random Systems. VIII, 234 pages. 2002.

Vol. 1780: J. H. Bruinier, Borcherds Products on O(2,1) and Chern Classes of Heegner Divisors. VIII, 152 pages. 2002.

Vol. 1781: E. Bolthausen, E. Perkins, A. van der Vaart, Lectures on Probability Theory and Statistics. Ecole d' Eté de Probabilités de Saint-Flour XXIX-1999. Editor: P. Bernard. VIII, 466 pages. 2002.

Vol. 1782: C.-H. Chu, A. T.-M. Lau, Harmonic Functions on Groups and Fourier Algebras. VII, 100 pages. 2002.

Vol. 1783: L. Grüne, Asymptotic Behavior of Dynamical and Control Systems under Perturbation and Discretization. IX, 231 pages. 2002.

Vol. 1784: L.H. Eliasson, S. B. Kuksin, S. Marmi, J.-C. Yoccoz, Dynamical Systems and Small Divisors. Cetraro, Italy 1998. Editors: S. Marmi, J.-C. Yoccoz. VIII, 199 pages. 2002.

Vol. 1785: J. Arias de Reyna, Pointwise Convergence of Fourier Series. XVIII, 175 pages. 2002.

Vol. 1786: S. D. Cutkosky, Monomialization of Morphisms from 3-Folds to Surfaces. V, 235 pages. 2002.

Vol. 1787: S. Caenepeel, G. Militaru, S. Zhu, Frobenius and Separable Functors for Generalized Module Categories and Nonlinear Equations. XIV, 354 pages. 2002.

Vol. 1788: A. Vasil'ev, Moduli of Families of Curves for Conformal and Quasiconformal Mappings.IX, 211 pages. 2002.

Vol. 1789: Y. Sommerhäuser, Yetter-Drinfel'd Hopf algebras over groups of prime order. V, 157 pages. 2002.

Vol. 1790: X. Zhan, Matrix Inequalities. VII, 116 pages. 2002.

Vol. 1791: M. Knebusch, D. Zhang, Manis Valuations and Prüfer Extensions I: A new Chapter in Commutative Algebra. VI, 267 pages. 2002.

Vol. 1792: D. D. Ang, R. Gorenflo, V. K. Le, D. D. Trong, Moment Theory and Some Inverse Problems in Potential Theory and Heat Conduction. VIII, 183 pages. 2002.

Vol. 1793: J. Cortés Monforte, Geometric, Control and Numerical Aspects of Nonholonomic Systems. XV, 219 pages. 2002.

Vol. 1794: N. Pytheas Fogg, Substitution in Dynamics, Arithmetics and Combinatorics. Editors: V. Berthé, S. Ferenczi, C. Mauduit, A. Siegel. XVII, 402 pages. 2002.

Vol. 1795: H. Li, Filtered-Graded Transfer in Using Noncommutative Gröbner Bases. IX, 197 pages. 2002.

Vol. 1796: J.M. Melenk, hp-Finite Element Methods for Singular Perturbations. XIV, 318 pages. 2002.

Vol. 1797: B. Schmidt, Characters and Cyclotomic Fields in Finite Geometry. VIII, 100 pages. 2002.

Vol. 1798: W.M. Oliva, Geometric Mechanics. XI, 270 pages. 2002.

Vol. 1799: H. Pajot, Analytic Capacity, Rectifiability, Menger Curvature and the Cauchy Integral. XII,119 pages. 2002.

Vol. 1800: O. Gabber, L. Ramero, Almost Ring Theory. VI, 307 pages. 2003.

Vol. 1801: J. Azéma, M. Émery, M. Ledoux, M. Yor (Eds.), Séminaire de Probabilités XXXVI. VIII, 499 pages. 2003.

Vol. 1802: V. Capasso, E. Merzbach, B.G. Ivanoff, M. Dozzi, R. Dalang, T. Mountford, Topics in Spatial Stochastic Processes. Martina Franca, Italy 2001. Editor: E. Merzbach. VIII, 253 pages. 2003.

Vol. 1803: G. Dolzmann, Variational Methods for Crystalline Microstructure - Analysis and Computation. VIII, 212 pages. 2003.

Vol. 1804: I. Cherednik, Ya. Markov, R. Howe, G. Lusztig, Iwahori-Hecke Algebras and their Representation Theory. Martina Franca, Italy 1999. Editors: V. Baldoni, D. Barbasch. X, 103 pages. 2003.

Vol. 1805: F. Cao, Geometric Curve Evolution and Image Processing. X, 187 pages. 2003.

Vol. 1806: H. Broer, I. Hoveijn. G. Lunther, G. Vegter, Bifurcations in Hamiltonian Systems. Computing Singularities by Gröbner Bases. XIV, 169 pages. 2003.

Vol. 1807: V. D. Milman, G. Schechtman, Geometric Aspects of Functional Analysis. Israel Seminar 2000-2002. VIII, 429 pages. 2003.

Vol. 1808: W. Schindler, Measures with Symmetry Properties.IX, 167 pages. 2003.

Vol. 1809: O. Steinbach, Stability Estimates for Hybrid Coupled Domain Decomposition Methods. VI, 120 pages. 2003.

Vol. 1810: J. Wengenroth, Derived Functors in Functional Analysis. VIII, 134 pages. 2003.

Vol. 1811: J. Stevens, Deformations of Singularities. VII, 157 pages. 2003.

Recent Reprints and New Editions